Zbigniew H. Stachurski

Fundamentals of Amorphous Solids

Related Titles

Che, M., Védrine, J.C. (eds.)

Characterization of Solid Materials and Heterogeneous Catalysts

From Structure to Surface Reactivity

2012
ISBN: 978-3-527-32687-7
also available in electronic formats

Mittemeijer, E.J., Welzel, U. (eds.)

Modern Diffraction Methods

2013
ISBN: 978-3-527-32279-4
also available in electronic formats

Dubois, J., Belin-Ferré, E. (eds.)

Complex Metallic Alloys

Fundamentals and Applications

2011
ISBN: 978-3-527-32523-8
also available in electronic formats

Zolotoyabko, E.

Basic Concepts of Crystallography

2011
ISBN: 978-3-527-33009-6

Raveau, B., Seikh, M.

Cobalt Oxides

From Crystal Chemistry to Physics

2012
ISBN: 978-3-527-33147-5
also available in electronic formats

Leng, Y.

Materials Characterization

Introduction to Microscopic and Spectroscopic Methods

Second Edition
2013
ISBN: 978-3-527-33463-6
also available in electronic formats

Zbigniew H. Stachurski

Fundamentals of Amorphous Solids

Structure and Properties

Verlag GmbH & Co. KGaA

The Author

Prof. Dr. Zbigniew H. Stachurski
Australian National University
College of Engingeering and Computer Science
Research School of Engineering
0200 Canberra
Australia

All books published by **Wiley-VCH** are carefully produced. Nevertheless, authors, editors, and publisher do not warrant the information contained in these books, including this book, to be free of errors. Readers are advised to keep in mind that statements, data, illustrations, procedural details or other items may inadvertently be inaccurate.

Library of Congress Card No.: applied for

British Library Cataloguing-in-Publication Data
A catalogue record for this book is available from the British Library.

Bibliographic information published by the Deutsche Nationalbibliothek
The Deutsche Nationalbibliothek lists this publication in the Deutsche Nationalbibliografie; detailed bibliographic data are available on the Internet at http://dnb.d-nb.de.

© 2015 by Higher Education Press.
All rights reserved.

Published by Wiley-VCH Verlag GmbH & Co. KGaA, Boschstr. 12, 69469 Weinheim, Germany, under exclusive license granted by HEP **for all media and languages excluding Chinese** and throughout the world excluding Mainland China, and with non-exclusive license for electronic versions in Mainland China.

Print ISBN: 978-3-527-33707-1
ePDF ISBN: 978-3-527-68217-1
ePub ISBN: 978-3-527-68219-5
Mobi ISBN: 978-3-527-68218-8

Cover Design Formgeber, Mannheim, Germany
Typesetting Laserwords Private Limited, Chennai, India
Printing and Binding Markono Print Media Pte Ltd., Singapore

Printed on acid-free paper

Contents

Preface *XI*

1	**Spheres, Clusters and Packing of Spheres** *1*	
1.1	Introduction *1*	
1.2	Geometry of Spheres *9*	
1.2.1	A Sphere and Its Neighbours *9*	
1.2.2	Neighbours by Touching *10*	
1.2.3	Hard and Soft Spheres *12*	
1.3	Geometry of Clusters *15*	
1.3.1	Regular Clusters *15*	
1.3.2	Irregular Clusters *17*	
1.3.3	Coordination of $(1+k)$ Clusters *18*	
1.3.3.1	Blocking Model for Cluster Formation *19*	
1.3.3.2	Fürth Model for Cluster Formation *22*	
1.3.4	Configuration of $(1+k)$ Clusters *25*	
1.3.4.1	Regular Clusters *25*	
1.3.4.2	Irregular Clusters *26*	
1.3.4.3	Closing Vector Based on Radial Vector Polygon *27*	
1.3.4.4	Physical Meaning of the Closing Vector, ζ *32*	
1.3.4.5	Spherical Harmonics *33*	
1.4	Geometry of Sphere Packings *34*	
1.4.1	Fixed and Loose Packings *36*	
1.4.2	Ordered Packing *36*	
1.4.3	Disordered Packing *37*	
1.4.4	Random Packing *38*	
1.4.5	Random Sequential Addition of Hard Spheres *38*	
1.4.6	Random Closed Packing of Spheres *40*	
1.4.7	Neighbours by Voronoi Tessellation *43*	
1.4.8	Neighbours by Coordination Shell *47*	
1.4.8.1	Pair Distribution Function *49*	
1.4.8.2	The Probability of Contacts *50*	
1.4.8.3	Contact Configuration Function *50*	

1.5	Ideal Amorphous Aolid (IAS) *51*
1.6	Construction of an Ideal Amorphous Solid Class I *53*
1.7	Elementary Theory of Amorphousness *59*
1.7.1	Background *59*
1.7.2	The Axioms *60*
1.7.3	Conjecture *61*
1.7.4	The Rules *62*
1.7.5	Statistical Correspondence *63*
1.8	Classes of Ideal Amorphous Solids *64*
1.8.1	IAS Class I: Random Close Packing of Individual Atoms *64*
1.8.2	IAS Class II: Random Close Packing of Linear Model Chains *64*
1.8.3	IAS Class III: Random Close Packing of Three-Dimensionally cross-Linked Chains *65*
1.9	Imperfections in IAS *66*
1.9.1	Geometrical (local) Flaws *67*
1.9.2	Statistical (global) Flaws *68*
1.9.3	The Effect of Flaws on the Density of IAS *69*
1.9.4	Short and Medium Range Order *72*
	References *72*
	Books on Crystallography *74*
	Books on Glasses *74*
	Books on Random Walks *74*
	Books on Sphere Packings *74*
	Books on Crystal Imperfections *74*
2	**Characteristics of Sphere Packings** *75*
2.1	Geometrical Properties *75*
2.1.1	The Coordination Distribution Function, $\Psi(k)$ *75*
2.1.2	Tetrahedricity *76*
2.1.3	Voronoi Polyhedra Notation *78*
2.1.4	Topology of Clusters *79*
2.1.4.1	Ordered Clusters *80*
2.1.4.2	Irregular Clusters *81*
2.1.5	The Configuration Distribution Function, $\Phi_k(\zeta)$ *82*
2.1.6	The Volume Fraction *83*
2.1.6.1	Regular Polyhedra *84*
2.1.6.2	Irregular Polyhedra *84*
2.1.7	The Packing Fraction *85*
2.1.7.1	The Average Packing Fraction for the Round Cell *86*
2.1.7.2	The Local Packing Fraction *87*
2.1.7.3	The Limits of Packing Fraction *87*
2.1.8	Representative Volume Element *91*
2.1.9	Density of Single Phase *92*
2.1.9.1	Density of Crystalline Solid *92*
2.1.9.2	Density of Amorphous Solid *92*

2.1.10	Density of a Composite	*93*
2.1.11	Solidity of Packing	*94*
2.2	X-ray Scattering	*96*
2.2.1	Introduction	*96*
2.2.2	Geometry of Diffraction and Scattering	*96*
2.2.3	Intensity of a Scattered Wave	*97*
2.2.3.1	Amorphous Solid	*100*
2.2.3.2	Ehrenfest Formula	*100*
2.2.3.3	Polyatomic Solid	*101*
2.2.4	Factors Affecting Integrated Scattered Intensity	*102*
2.2.4.1	Integrated Intensity of Powder Pattern Lines from Crystalline Body	*102*
2.2.4.2	Integrated Scattered Intensity from Monoatomic Body	*102*
2.3	Glass Transition Measured by Calorimetry	*103*
	References	*104*
3	**Glassy Materials and Ideal Amorphous Solids**	*107*
3.1	Introduction	*107*
3.1.1	Solidification	*108*
3.1.1.1	Solidification by Means of Crystallization	*110*
3.1.1.2	Solidification through Vitrification	*111*
3.1.2	Cognate Groups of Amorphous Materials (Glasses)	*114*
3.1.2.1	Metallic Glasses	*114*
3.1.2.2	Inorganic Glasses	*116*
3.1.2.3	Organic Glasses	*117*
3.1.2.4	Amorphous Thin Films	*119*
3.2	Summary of Models of Amorphous Solids	*119*
3.2.1	Lattice with Atomic Disorder	*120*
3.2.2	Disordered Clusters on Lattice	*121*
3.2.3	Geometric Models for Amorphous Networks	*121*
3.2.4	Packing of Regular but Incongruent Clusters	*121*
3.2.5	Irregular Clusters – Random Packing	*123*
3.2.6	Molecular Dynamics	*123*
3.2.7	Monte Carlo Method	*124*
3.3	IAS Model of a-Argon	*125*
3.3.1	IAS Parameters	*126*
3.3.2	Round Cell Simulation and Analysis	*127*
3.3.2.1	Coordination Distribution Function	*129*
3.3.2.2	Voronoi Volume and Configuration Distribution Functions	*129*
3.3.2.3	Radial Distribution Function	*129*
3.3.2.4	X-ray Scattering from the IAS Model	*131*
3.3.2.5	Crystalline and Amorphous Cluster	*132*
3.3.3	Summary of a-Ar IAS Structure	*132*
3.4	IAS Model of a-NiNb Alloy	*133*
3.4.1	Introduction	*133*

3.4.2	IAS Model of a-NiNb Alloy *133*	
3.4.2.1	Coordination Distribution Functions *133*	
3.4.2.2	Voronoi Volume Distribution *135*	
3.4.2.3	Pair Distribution Function *135*	
3.4.2.4	Probability of Contacts *135*	
3.4.3	X-ray Scattering from a-NiNb Alloy *136*	
3.4.3.1	Experimental Results *136*	
3.4.3.2	Theoretical Results *138*	
3.4.4	Density of a-Ni62-Nb38 Alloy *138*	
3.4.4.1	Crystalline Alloy *138*	
3.4.4.2	Amorphous Alloy *139*	
3.4.5	Summary of a-NiNb IAS Structure *140*	
3.5	IAS Model of a-MgCuGd Alloy *140*	
3.5.1	Physical Properties of the Elements *140*	
3.5.2	IAS Simulation of a-MgCuGd Alloy *140*	
3.5.2.1	Coordination Distribution Functions *143*	
3.5.2.2	Configuration Distribution Function *143*	
3.5.2.3	Radial Distribution Function *143*	
3.5.2.4	Probability of Contacts *147*	
3.5.2.5	Cluster Composition According to IAS *147*	
3.5.2.6	Cluster Composition According to MD *148*	
3.5.3	X-ray Scattering from a-Mg65-Cu25-Gd10 Alloy *149*	
3.5.3.1	Flat Plate X-ray Scattering Pattern *149*	
3.5.3.2	Calibration based on Si Powder Pattern *149*	
3.5.3.3	Uncertainties and Corrections *153*	
3.5.4	Density of Mg65-Cu25-Gd10 Alloy *155*	
3.5.4.1	Crystalline Alloy *155*	
3.5.4.2	Amorphous Alloy *156*	
3.5.5	Summary of a-MgCuGd IAS Structure *156*	
3.6	IAS Model of a-ZrTiCuNiBe Alloy *157*	
3.6.1	Transmission Electron Microscopy *157*	
3.6.2	IAS Simulation of Amorphous a-ZrTiCuNiBe Alloy *159*	
3.6.2.1	Coordination Distribution Function *159*	
3.6.2.2	Voronoi Volume Distribution *161*	
3.6.2.3	Radial Distribution Function *161*	
3.6.3	Atomic Probe of the a-ZrTiCuNiBe Alloy *162*	
3.6.3.1	Probability of Contacts *162*	
3.6.4	Selected Clusters from the a-ZrTiCuNiBe Alloy *165*	
3.6.5	X-ray Scattering from the a-ZrTiCuNiBe Alloy *165*	
3.6.6	Density of ZrTiCuNiBe Alloy *167*	
3.6.6.1	Crystalline Alloy *167*	
3.6.6.2	Amorphous Alloy *168*	
3.6.6.3	Vitreloy Alloys *169*	
3.6.7	Summary of a-ZrTiCuNiBe IAS Structure *169*	
3.7	IAS Model of a-Polyethylene (a-PE) *169*	

3.7.1	Radial Distribution Function	*171*
3.7.2	X-ray Scattering	*173*
3.7.2.1	Short-Range Order	*174*
3.7.3	Summary of a-PE IAS Structure	*174*
3.8	IAS Model of a-Silica (a-SiO$_2$)	*176*
3.8.1	Molecular Parameters for SiO$_2$	*176*
3.8.2	IAS and United Atom Models for SiO$_2$	*176*
3.8.3	Summary of a-SiO$_2$ IAS Structure	*179*
3.9	Chalcogenide Glasses	*179*
3.9.1	As12–Ge33–Se55 Chalcogenide Glass	*180*
3.9.2	Measured Coordination Distribution	*181*
3.9.3	Measured X-ray Scattering	*182*
3.9.4	Glass-Transition Temperature of AsGeSe Glasses	*184*
3.9.5	Models of Atomic Arrangements in AsGeSe Glass	*184*
3.9.5.1	IAS Model of AsGeSe Glass	*184*
3.9.5.2	Other Models of AsGeSe Glass	*185*
3.9.6	Summary of a-AsGeSe IAS Structure	*186*
3.10	Concluding Remarks	*186*
3.10.1	Chapter 3	*186*
3.10.2	Chapter 2	*187*
	References	*187*
4	**Mechanical Behaviour**	*191*
4.1	Introduction	*191*
4.2	Elasticity	*193*
4.2.1	Phenomenology	*193*
4.2.2	Continuum Mechanics	*195*
4.2.2.1	Calculation of Average Elastic Constants – Aggregate Theory	*198*
4.2.2.2	Green's Elastic Strain Energy	*199*
4.2.3	Atomistic Elasticity	*200*
4.2.3.1	Calculation of an Elastic Constant for Single Crystal of Argon	*200*
4.3	Elastic Properties of Amorphous Solids	*202*
4.3.1	Elastic Modulus of Amorphous Argon	*202*
4.4	Fracture	*203*
4.4.1	Phenomenology	*203*
4.4.2	Continuum Mechanics	*204*
4.4.2.1	Definition of Fracture Mechanics: Fracture Toughness	*204*
4.4.2.2	Elastic Strain Energy Release	*207*
4.4.2.3	Solid Surface Energy	*208*
4.4.2.4	Griffith's Fracture Stress	*209*
4.4.2.5	The Role of Defects	*210*
4.4.3	Atomistic Fracture Mechanics of Solids	*212*
4.4.3.1	Theoretical Cleavage Strength	*212*
4.4.3.2	Theoretical Shear Strength	*213*
4.5	Plasticity	*215*

4.5.1	Phenomenology	215
4.5.2	Continumm Mechanics	217
4.5.2.1	Tresca Yield Criterion	217
4.5.2.2	Huber–von Mises Criterion	217
4.5.3	Atomistic Mechanics of Crystalline Solids	218
4.5.3.1	Strain Hardening	218
4.5.3.2	Grain Boundary Strengthening	220
4.5.3.3	Solid Solution Hardening	221
4.5.3.4	Precipitation Hardening	223
4.5.3.5	Mechanisms of Plastic Flow in Crystalline Materials	224
4.5.3.6	Displacement of Atoms Around Dislocations	224
4.5.3.7	Critical Shear Stress to Move Dislocation	226
4.6	Plasticity in Plasticity: Amorphous Solids	227
4.6.1	Plastic Deformation by Shear Band Propagation	228
4.7	Superplasticity	231
4.7.1	Phenomenology	231
4.7.2	Continuum Mechanics	233
4.7.3	Superplasticity in Bulk Metallic Glasses	234
4.7.3.1	Calculation of Strain Rate for Superplasticity	234
4.7.4	Concordant Deformation Mechanism	236
4.7.4.1	Density Variation in Amorphous Solids	238
4.7.4.2	The 'Inclusion' Problem	240
4.7.4.3	The System without Transformation	240
4.7.4.4	The System with Transformation	241
4.7.4.5	Conclusions	242
4.8	Viscoelasticity	243
4.8.1	Phenomenology	244
4.8.2	Time- and Temperature-Dependent Behaviour	245
4.8.2.1	Definitions of Viscosity	245
4.8.2.2	Order of Magnitude Calculations	246
4.8.3	Temperature Effect on Viscoelastic Behaviour	247
4.8.3.1	Arrhenius Behaviour	247
4.8.3.2	Vogel-Fulcher–Tammann Behaviour	247
	References	248

Index 253

Preface

This book is intended primarily for students of materials science and related fields who want to acquire a fundamental understanding of the atomic arrangements in amorphous solids. A concomitant aim of the book is to provide an appropriate and consistent methodology and vocabulary for describing the atomic structure of amorphous solids.

The book may also be of interest to theoreticians, for this is a relatively new field of science, open to further evolution and requiring formal proofs of some of the concepts contained herein.

The first three chapters of this book focus mainly on the atomic arrangements in amorphous solids based on the ideal amorphous solid model used as geometrical foundation. Amorphous polymers and inorganic glasses have the most complex atomic arrangements whereas metallic glasses can be considered as being the simplest model of atomic arrangements in amorphous solids.

The twentieth-century may be called by science historians as the Renaissance of Materials Science because many new analytical methods were invented to discover and to ascertain the atomic arrangements in solids (for example, ranging from X-ray diffraction to high resolution transmission electron microscopy), thereby opening up the new science of structure–property relationships. Some of these relationships are described in this book.

The fourth chapter places the mechanical behaviour of amorphous materials in the general context of the mechanics of solids. In many instances, solutions from continuum mechanics of isotropic materials can be applied directly to homogeneous amorphous solids, as they are theoretically isotropic. The chapter presents these solutions in the light of mechanical behaviour of polycrystalline solids.

A good book should contain all the information that the reader is looking for, or at least it should point accurately to other sources where the information can be found. I believe this book has a coherent structure that conveys an important message about the distinction between ideal amorphous solid, ideal crystalline solids and that of real amorphous materials with the inevitable characteristic defects and imperfections in their atomic arrangements.

Many of my colleagues and friends have influenced my decision to write this book. In the first place, I wish to thank Professors Qinghua Qin, Richard Welberry and Witold Brostow for their encouragement. I am grateful to Professors

Jun Shen and Gang Wang who introduced me to the field of glassy metals, and Professors Kevin Kendall and Christian Kloc for their constant support. I wish to extend special thanks to Dr. Xiao Hua Tan for comments on the manuscript.

September 2014 *Canberra*

1
Spheres, Clusters and Packing of Spheres

1.1
Introduction

Imagine, Design, Create, Explore

Theory of amorphousness is a science about the structural arrangement of atoms in amorphous solids. It is part of Materials Science, which includes the closely related theory of crystallography. Whilst theory of crystallography is well established, theory of amorphousness is beginning to emerge as a body of science in its own right.

Arrangement of objects leads to creation of patterns. The invention of a repeating pattern as a thoughtful and creative process has ancient beginnings, at first in art as discovered by archaeology and evidenced in mosaics existing in ancient buildings, and later in science as known from old manuscripts; for example the five ideal Greek solids. There are three types of patterns that can fill in Euclidean space contiguously, and to infinity. These are:

- patterns with translational symmetry that possess an underlying lattice
- patterns with fivefold rotational symmetry but without translational symmetry
- random patterns with no lattice and no rotational symmetry.

Two-dimensional examples of such patterns are shown in Figure 1.1. All three are used as conceptual models for atomic arrangements in solids.

Crystalline solids have been known and appreciated since antiquity. In modern times the intrinsic elements of symmetry in single crystals of minerals were given attention in 1822 by R. J. Haüy (pronounced $\bar{a}$-wee, $\bar{a}$ as in 'aside') in 'Traité de Cristallographie'. Shortly after, the theoretical treatments of W. H. Miller in 1839 on *hkl* notation, A. Bravais in 1845 on 14 lattices, A. Schönflies in 1892 and W. Barlow in 1898 on 230 space groups (with many contributions from others) resulted in a complete theory of geometrical crystallography. Perfectly regular and ordered structures of infinite extent are described by geometrical crystallography as perfect (*ideal*) solids, with positions and arrangements of all atoms defined precisely along specific lattices. Theory of crystallography provides a datum from which the ideal atomic arrangements (and defects) in real materials can be determined. By comparison, no such universal laws or rules are well known for the atomic structure in amorphous solids.

Fundamentals of Amorphous Solids: Structure and Properties, First Edition. Zbigniew H. Stachurski.
© 2015 Wiley-VCH Verlag GmbH & Co. KGaA. Published by Wiley-VCH Verlag GmbH & Co. KGaA.
Boschstr. 12, 69469 Weinheim, Germany, under exclusive license granted by HEP for all media and languages excluding Chinese and throughout the world excluding Mainland China, and with non-exclusive license for electronic versions in Mainland China

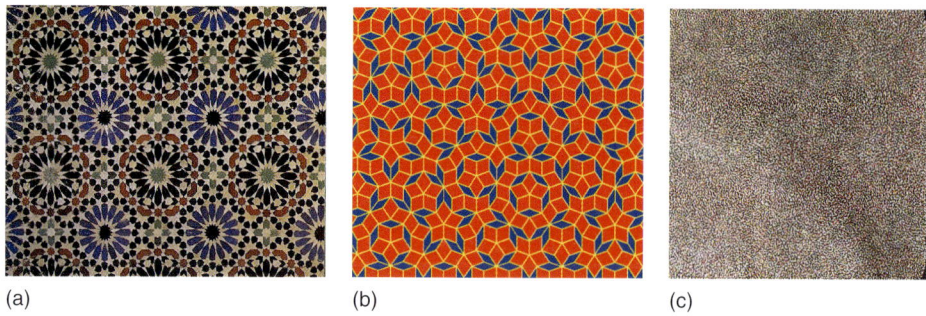

Figure 1.1 (a–c) Fragments of two-dimensional patterns representing the three formats of atomic arrangement in solids: crystalline (tiles from Morocco), quasi-crystalline (computer pattern generated by T.R. Welberry of ANU) and amorphous (Aboriginal painting by Ada Ross, Australia).

In a historical perspective, it would be interesting to contemplate the following question: if Pythagoras were a statistician rather than deducing perfect harmony from ratios of pure numbers on strings, would we have had a theory of amorphousness in solid-state developed ages ago? Looking back in time, one can draw a direct line from the modern theory of geometric crystallography to the philosophy of pure numbers and rational ratios of antiquity. The René-Just Haüy description of packing of elementary blocks to form a single crystal with a simple relationship between its crystal faces and packing arrangement derives directly from the deductive Pythagorean notion of perfect harmony based on the relationship between the length of the string and perfect harmonic notes as $1:2, 2:3, 3:4$, and son on. The relationship between atomic planes and crystal lattice is also expressed by simple ratios of whole numbers, the reciprocals of which are known as *Miller indices*. By comparison, relatively little is known about the structure of amorphous solids. Our knowledge of amorphous structures seems incomplete when compared with that of crystalline solids. In particular, the concept of an ideal amorphous solid as a datum and the corresponding theory have been lacking hitherto.

Until quite recently, amorphous solids were described as disordered crystalline solids, with some degree of order intermediate between a liquid and a solid. This was based on the understanding that glasses are free from the constraints that govern the arrangements of atomic clusters in crystalline materials, so there is a degree of ambiguity in the way that neighbouring clusters can be positioned and oriented.

A possible implication deriving from this view is that amorphous solids originate from the corresponding ordered crystalline state. In the field of geometric crystallography, a disordered crystalline structure implies the presence of defects which are defined relative to the perfectly ordered structure. Therefore, disordered materials are crystalline materials that, in principle, can be restored to the perfect crystalline state by the reversal of defects. It is conjectured that this cannot be done in amorphous materials and that a different approach and terminology

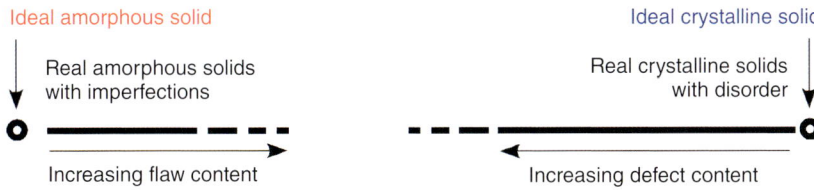

Figure 1.2 A view of the structure of solids along an undefined, somewhat arbitrary variable. The circles indicate the positions of the ideal (perfect) structures; the lines indicate the spectrum of structures in real solids.

should be used to describe their structure, namely, *random* atomic arrangement. To emphasize this point, we note that in the field of statistics, when describing a set of *random* data, it would be unfitting to refer to that set as *disordered* data. Hence, it is proposed that amorphous structures, based on irregular packing of spheres, should be referred to as having a random arrangement of atoms rather than a disordered atomic arrangement. To promote this view, a drawing is shown in Figure 1.2 with the contemplated relative positions of the two types of solids and the envisaged discontinuity between the random and ordered types of atomic arrangements that must exist. The very small gap between the circle and the line on the crystalline side indicates that almost perfect single crystals can be grown. The larger gap on the amorphous side indicates that the structure of glasses may not be as close to the ideal amorphous solid as described later in this book. A discontinuity in the line near the middle is meant to indicate that even highly disordered crystalline solids are not the same as highly flawed amorphous glasses, and vice versa. This view is very close to that expressed by Kazunobu Tanaka *et al.* in the introduction to their book on 'Amorphous Silicon'.

A scientifically satisfactory explanation of the amorphous state continues to be a challenge, and for this reason, we advance and promote the theoretical concept of an *ideal amorphous solid* as a partial solution to this enigma. The right approach to a definition of amorphousness is through an appropriate geometric and topological model of the ideal amorphous solid, as described in this chapter.

However, the usage of the word 'disordered' appears in dictionaries to mean unpredictable, opposite to law and order. So, this seems to be also a matter of habit and semantics, rather than a question of pure logic. Nevertheless, an appropriate and consistent vocabulary conjures up a clear vision of the atomic arrangements and helps to define the field of science of amorphous solids, separate and distinct from the field of crystallography.

The study of atomic arrangements in amorphous solids was stimulated in the 1960s by theoretical work of J. D. Bernall on the structure of liquids, concurrent with experimental random packing of spheres by G. D. Scott. In the last few decades, research into atomic arrangements in amorphous solids has separated into two main streams: (i) more refined and detailed studies of packing of spheres and molecules and (ii) atomistic simulations by molecular dynamics (MD), including *ab-initio* methods. The understandings we gain from the two approaches are of different nature. In the latter approach, a unique definition of an amorphous atomic structure cannot be achieved because in a simulated thermodynamic system with suppressed self-assembly tendencies every simulation, even repeated on the same system, must result in a different atomic arrangement. Modelling amorphous materials by these methods is equivalent to random packing with extreme cooling rates of the order of 10^{15} K s^{-1}. Nevertheless, these methods are successful and appropriate to simulate the structure of real amorphous solids with atomic arrangements containing imperfections. In the former case, simulations and geometrical modelling follow the methodology of representing atoms by hard spheres and creating representations of random atomic arrangement, naturally quite different from crystalline structures.

The earliest concept of atoms appears in a written record from Leucippus of Miletus (once an ancient Greek city on the western coast of Anatolia) and Democritus of Abdera (city–state on the coast of Thrace, its foundation attributed to Heracles), Greek philosophers of 5-4th century BC (Taylor, 1999). They conjectured that as matter is divided into smaller and smaller parts, there must be a limit to this division; namely atoms, the smallest indivisible objects. Otherwise, if there were no limit to the division, then the parts could be divided into "nothingness", and therefore, matter would not exist (*reductio ad absurdum* method of logic). Their theory envisaged atoms as invisible and indivisible particles, not as perfectly shaped as spheres but in the form of odd shapes with hooks and protrusions to render various properties of matter, such as taste, colour, fluidity and friction, as described by the Roman poet, Lucretius, in his *De Rerum Natura* (first century BC didactic poem on Epicurean philosophy). Coincidentally, modern view of atoms also shows electronic orbitals as having various shapes and protrusions, although the complete atom, encompassing all the orbitals, is imagined as having a spherical shape (Figure 1.3).

The concept of representing atoms by spheres has evolved gradually and over a long period of time. In 1611, Joannis Kepler drew hexagonal close packing of spheres to illustrate a compact solid (Kepler, 1611) and suggested that the hexagonal symmetry of snowflakes is due to the regular packing of the constituent particles. Some 50 years later, Robert Hooke wrote that crystals are composed of close packed 'spheroids'. At that time it was thought that spherical atomic particles must be close packed to form a rigid solid. Layers of round spherical objects,

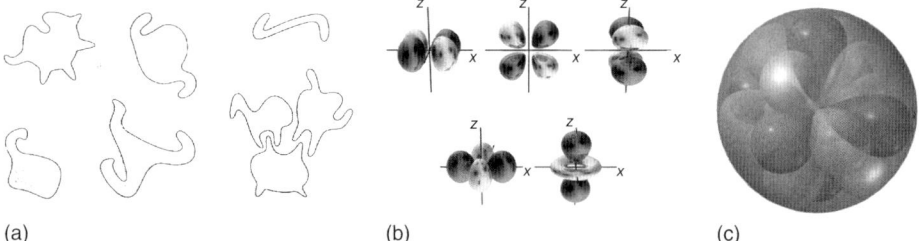

Figure 1.3 (a) Atoms of Leccipus and Democritus as depicted by Lucretius (Adapted from *Scientific American*). (b) Quantum mechanics view of electron clouds around atoms. (c) View of an atom as a sphere encompassing all electronic orbitals.

close packed in a hexagonal arrangement in repeating stacks of ABABAB... pattern, formed such an arrangement with fundamental sixfold symmetry, known in crystallography as hexagonal close packed (hcp). Soon, a variation of this layered packing was discovered, with a stacking sequence of ABCABC.., which gives a so-called face centred cubic (fcc) arrangement with characteristic fourfold and threefold symmetry.

The representation of atoms as spheres is not only intuitive, it is also in part justified by the Born–Oppenheimer approximation, which states that

$$\Psi_{total} = \psi_{electronic} \times \psi_{nuclear} \tag{1.1}$$

In simple terms, the approximation allows the wave function of a molecule, Ψ_{total}, to be separated into its electronic and nuclear components. The success of the BO approximation relies on the fact that spinning and oscillations of the electrons are several orders of magnitude higher than the frequency of oscillations of the nuclei. Consequently, the electrons surround the nuclei like clouds, on average spherically distributed around the central core. This is supported by many instances, for example, by the behaviour of colliding atoms which rebound in a way similar to that of billiard balls on a snooker table. In metals the electrons in the inner shells are strongly bound, and the electron density in the ionic core, which holds most of the electrons, satisfies the spherical distribution of electronic density, $\rho(\mathbf{r})$.

Early writing on mineralogy, especially on gemstones, comes from ancient Babylonia, the ancient Greco-Roman world, ancient and medieval China and Sanskrit texts from ancient India and the ancient Islamic World. Books on the subject include the *Naturalis Historia* of Pliny the Elder, and *Kitab al Jawahir* (Book of Precious Stones) by Muslim scientist Al Biruni. The German Renaissance specialist Georgius Agricola wrote works such as *De re metallica* (On Metals, 1556) and *De Natura Fossilium* (On the Nature of Rocks, 1546) which begin the scientific approach to the subject. Systematic scientific studies of minerals and rocks developed in post-Renaissance Europe. Figure 1.4 shows portraits of people who contributed to the early stages of crystallography.

Figure 1.4 Portraits by various artists: Joannis Kepler, Dutch astronomer (1571–1630), Robert Hooke, English natural philosopher (1635–1703), Niels Stensen, Danish geologist (1638–1686) and René Just Haüy, French mineralogist (1743–1822) reproduced from Wikipedia.

Steno gave the first accurate observations on a type of crystal in his 1669 book *De solido intra solidum naturaliter contento*. The principle in crystallography, known as *Steno's law of constant angles* or simply as *Steno's law*, states that the angles between corresponding faces on crystals are the same for all specimens of the same mineral. Steno's seminal work paved the way for the law of the rationality of the crystallographic indices of French mineralogist René-Just Haüy in 1801.

Mineralogy played an important role in the eighteenth century in establishing the principles of crystallography, in which crystals are represented by ordered packing of spheres in a unit cell. At the start of nineteenth century René-Just Haüy published a book on crystallography and substantiated the law of rational indices (Haüy, 1821), which later gave rise to Miller indices. In the book, single crystals were envisaged as ordered packing of spheres in regular polyhedra with angles between their faces corresponding precisely to the angles found in natural crystals. Hence, different arrangements resulted in different angles for different crystals – a glimpse into the nature of crystals.

Mineralogists and crystallographers focused on models of ideal solids with ordered, symmetrical arrangements, whereas mathematicians continued to ponder, amongst other things, about random, irregular packings, and physicists used the results to describe fluidity of liquids (Bernal, 1959), and now the atomic arrangements in amorphous materials.

> Excerpt from:
> "RENE-JUST HAÜY AND HIS INFLUENCE"
> by HERBERT P. WHITLOCK
> New York State Museum
>
> Essai d'une théorie sur la structure des crystaux (1784)
> Trait de cristallographie (1822, 2 vols.)
> Volume 3, pages 92–98, 1918

We know that in the house of his friend, M. Defrance, Haüy dropped the now historic group of prismatic crystals of calcite and gathered from the ruin of a fine specimen the cleavage pieces to him recognizable as of the same form as other crystals of calcite; it thus appears that he had inevitably thrust upon him the key to the mystery of the mathematical inter-relation of these forms. But without a mind prepared to interpret this chance occurrence, without the imagination reaching out to its interpretation, the incident would have meant no more to him than to his friend who stood beside him. Bergmann, although unknown to Haüy, had an almost identical incident called to his attention by his pupil Gahn but had failed to fully realize its significance. Bergmann did not voice the cry, which on the lips of his illustrious successor has become historic, 'Tout est trouvé'.

Returning to his cabinet, Haüy lost no time in verifying the principle which was thus revealed to him. Under his hammer were sacrificed successively a scalenohedral crystal of calcite of the form known as *dog tooth spar* and another of a low rhombohedral habit; in each case, the primitive cleavage rhombohedron appeared amongst the fragments, as he expected that it would. With the idea of developing the 'primitive form' from other species, he ruthlessly attacked the other treasured specimens of his little collection and his sacrifice was fully justified by the results, for the cleavage fragments in many instances furnished him with the basis, significantly termed by him *le noyau*, upon which the complicatedly modified crystal combinations were, as it were, built up. He conceived the theory of modified forms, built up from the primitive by diminishing layers of crystal particles (décroissements), each successive layer having a definite relation to the preceding one and primitive nucleus.

By the early nineteenth century, the size of atoms was estimated by chemists to be very small, deduced from the knowledge of one gramme-mole (Amedeo Avogadro, Italian savant (1776–1856)) and density (Archimedes (287–c.212 BC)). However, precise atomic dimensions were only determined with the application of X-rays to the diffraction from crystalline solids in the first decade of the twentieth century. In Bragg's law, the distance between interatomic planes relates simply to the inverse of the angle of the diffracted beam, which allows for great accuracy of measurement of atomic dimensions. It turned out that atomic diameters are of the order of 10^{-10} m, and at that time, it was proposed to define a special unit of measure for that purpose, called *angström* Å, named after the Swedish physicist Anders Jonas Angström (1814–1874). For example, the atomic radius of phosphorus (P) is almost exactly 1 Å.

In the 1950s, Robert Corey and Linus Pauling at the California Institute of Technology created accurate scale models (1 inch = 1 Å) of molecules using hardwood spheres. Since then, textbooks on crystallography and materials science are ubiquitous in the use of spheres to represent atoms arranged in unit cells and on crystalline lattices. In chemistry, a space-filling model is a three-dimensional molecular model where the atoms are represented by spheres with sizes proportional to the radii of the atoms, as shown in Figure 1.5.

In mathematical geometry, the packing of hard (non-intersecting) spheres, both ordered and disordered, is a challenging subject with a long history. It

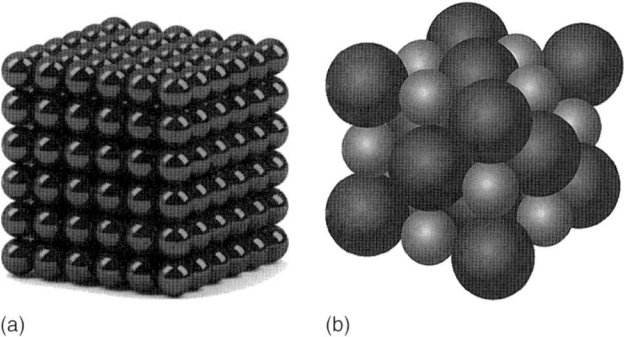

(a) (b)

Figure 1.5 (a and b) Models of packing of spherical atoms as may appear in classical textbooks on crystallography and/or materials science.

has been studied over the whole range of packing fractions from near 0 to the maximum of 0.74 in three-dimensional space and also in higher dimensions. For packing fractions close to zero (simulating rarefied gases or stars in the Universe), the spheres can be considered as points distributed in space according to some law, for example the Poisson's linear point process. For small packing fractions (gases at normal or increased pressure), binomial distribution seems more appropriate. For packing fractions between approximately 0.40 and 0.74, the assemblies of spheres are models of liquids, amorphous (glasses), polycrystalline, and single-crystal solids. In this book, we shall be concerned with dense packings only (0.60–0.74). Examples of such packings from simulations are shown in Figure 1.6a and b.

This chapter contains the elementary geometrical and mathematical aspects required to define spheres, clusters of spheres and touching and non-touching neighbours, leading to assemblage of spheres usually referred to as *packing of spheres*. Methods for packing are described with a special emphasis on random arrangement of spheres. For this purpose, we approach the method of packing without any reference to a *lattice* (which is an essential crystalline concept). However, when appropriate, examples from regular and crystallographic arrangements are included for comparison and emphasis by contrast. To further remove any suggestion or inclination to lattice-like concepts, the packed aggregate of spheres will be presented as spherical in shape, and called *round cell*, rather than a cubic simulation cell used ubiquitously in computer simulations. The use of round cell should dissuade from any involuntary thought to perceive a lattice or to think of the edges of the cubic cell as indications of a lattice. To understand amorphous solids, one must disengage oneself from any crystallographic predilection.

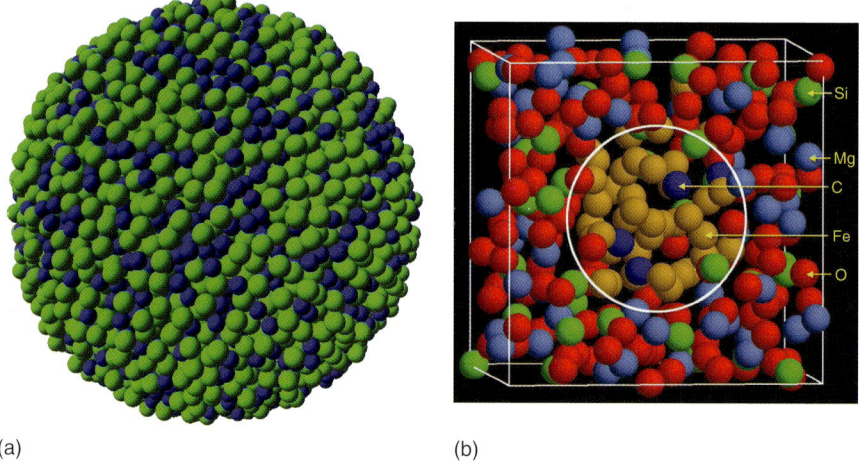

(a)   (b)

Figure 1.6 (a and b) Example of a round cell with random packing of spheres of two sizes, in comparison to a cubic cell with seemingly random packing of spheres of five sizes, from computer simulations.

1.2 Geometry of Spheres

1.2.1 A Sphere and Its Neighbours

In a general way, we call S_n a centrally symmetric convex body. Then a sphere of diameter, $D = 2R$, is defined in Euclidean space, E^n, by

$$S_n = \left\{ \mathbf{x} = (x_1, x_2, ..., x_n) : \sum_{i=1}^{n} (x_i)^2 \leq \left(\frac{D}{2}\right)^2 \right\} \quad (1.2)$$

A sphere is a geometrical solid such that every point on its surface is at an equal distance from its centre. In three dimensions, points for which $(x_i)^2 = (x^2 + y^2 + z^2)_i < (D/2)^2$ is true, constitute the interior of the sphere, and all points for which $(x_i)^2 = (D/2)^2$ is true form the surface of the sphere.

In Cartesian space (E^3), if X is a set of points representing the centres of spheres, then we call,

$$S_3 + X \quad (1.3)$$

a packing of spheres in three dimensions.

Information about the character of the packing is contained entirely in the set X. The content of the set, which forms a list or a matrix, can be combined with additional information, such as *id* numbers of the spheres and the sphere's radii. Equation 1.3 is presented at this point to introduce the concept of neighbouring spheres. It will be considered in greater detail in the context of sphere packings later in this chapter.

Packing of spheres implies neighbours. We look for neighbours of spheres within primary clusters formed by the spheres. Let m be a positive integer, then an h-neighbour of a sphere (S_3) exists if,

$$X_i = \{x \in X : |x, x_i| \leq h\} \quad (1.4)$$

$$\text{where} \quad X_i = \{x_{ij} : j = 1, ..., m_i\} \quad (1.5)$$

In Equation 1.4, $|x, x_i| = |(x_i^2 - x_j^2)^{1/2}|$ denotes two-norm Euclidean distance between two points (centres of two spheres).

For hard, impenetrable (non-overlapping) spheres, $(x_i^2 - x_j^2)^{1/2} < D$, is disallowed for any i, j.

The case, $(x_i^2 - x_j^2)^{1/2} = D$, indicates touching neighbours i, j.

The number of neighbours and the distances between them define the geometrical character of the cluster and hence the packing.

The number of nearest neighbours is involved in considerations in condensed matter physics where the interactions between neighbours govern the physical properties. In dense packings, knowing about the number of nearest neighbours is important, and therefore, neighbouring spheres are categorized according to their distance from the chosen central sphere. In particular, there are three categories that are used in physics of solids, and these are defined and described in detail:

- touching neighbours (by contact)
- neighbours by Voronoi tessellation (within a limited distance)
- neighbours by coordination shells (first, second, etc.) from radial distribution function

1.2.2
Neighbours by Touching

The first category is decided simply by the condition, $h = D$. Then, the h-neighbour is a touching neighbour. In this relationship, it is assumed that one sphere is at the origin ($\mathbf{x} = 0$) and the other sphere is the neighbour.

The sphere at the origin will be referred to as the *inner sphere* and the touching sphere(s) will be called the *outer sphere(s)*. Within a packing, each outer sphere is in turn an inner sphere of its own cluster.

Two touching spheres meet at one point, the *contact point*, which belongs to the surfaces of both spheres, and lies on a straight line joining the centres of the two spheres. Conical projection of the outer sphere onto the surface of the inner sphere delineates a 'shadow' cup extended over the conical angle, $\pi/3$, associated with that contact point, as can be seen in Figure 1.7(a).

For two (or more) outer spheres touching the inner sphere, the contact points must have a minimum angular separation, that is their shadow cups must not overlap. We note that the contact points can be identified by radial vectors, $\vec{R}_j$, $j = 1 \ldots k$, drawn from the centre of the inner sphere to the points on its surface, as shown in Figure 1.7(b).

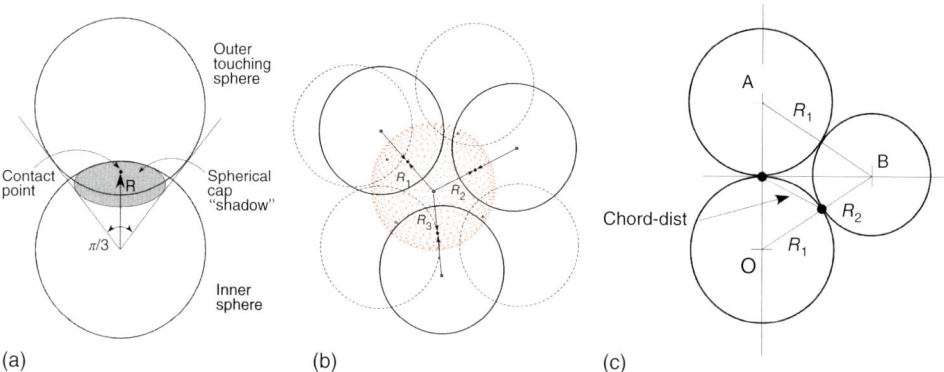

Figure 1.7 (a) Two spheres touching at a point, showing shadow of the outer sphere as a spherical cup on the inner sphere. (b) The exclusion angle shown for three outer spheres. (c) Geometrical relationship for the chord distance between the contact points of unequal spheres. $OA = 2R_1$, whereas $OB = AB = R_1 + R_2$.

- Then the *angular separation*, expressed as

$$\text{angle}, \phi(R_i, R_j) = \arccos\left(\langle R_i, R_j\rangle / \frac{1}{4}D^2\right) \qquad (1.6)$$

between any two vectors emanating from the centre of the sphere, cannot be less than $\frac{1}{3}\pi$.
- In general, the condition

$$\frac{1}{3}\pi \leq \phi(\vec{R}_i, \vec{R}_j) \leq \pi, (1 \leq i < j \leq k); \qquad (1.7)$$

must hold for any set of k outer spheres.

Another way to express the same condition is by means of the chord distance between the contact points on the surface of the inner sphere. In relation to Figure 1.7c, if the inner sphere (at O) has one of the touching spheres (at A) of radius, R_1, and the other sphere (at B) of radius, R_2, then the condition is expressed as follows:

$$\text{chord distance} \geq 2R_1 \sin\left[\frac{1}{2}\arcsin\sqrt{1 - \left(\frac{R_1}{R_1+R_2}\right)^2}\right] \qquad (1.8)$$

If $R_2 = R_1$, then from Equation 1.8 we calculate, $\frac{1}{2}\arcsin\sqrt{3/4} = 30°$, so the minimum allowed chord distance $= R_1$, which corresponds to a minimum angle of separation of 60°, in agreement with condition 1.7.

If $R_2 < R_1$, then the minimum distance will be correspondingly less. Finally, if $R_2 = 0$, then from Equation 1.8 we calculate, $\arcsin\sqrt{0} = 0°$, so the minimum allowed chord distance is zero, as it should be.

Naturally, a sphere can have more than one touching neighbour. The number of touching neighbours, denoted by k, can vary in the range: $k_{min} \leq k \leq k_{max}$. In general,

- $k_{min} = 1$, but $k_{min} = 4$ when the packing is to represent a solid; see section on 'fixed and loose spheres',
- $k_{max} = 12$ for spheres of the same size according to the so-called Kepler conjecture, and by common experience, but with formal proof achieved only recently (Hales, 1998).

Spheres of unequal sizes

Calculations have been carried out for k_{max} in clusters in which the spheres are of unequal size. As the size of the inner sphere changes relative to that of the outer spheres, then the maximum possible number of outer spheres changes as is given in Table 1.1.

For clusters comprising multi-sized spheres, it is possible to estimate the maximum number of touching neighbours from Table 1.1 by interpolation.

Table 1.1 Maximum possible number of outer spheres of radius, R_2, that can be in contact with inner sphere of radius R_1 over a selected range (Clare and Kepert, 1986).

Ratio, R_2/R_1	1.56	1.35	1.20	1.10	1.00	0.92	0.87	0.82	0.79
k_{max}	8	9	10	11	12	13	14	15	16

1.2.3
Hard and Soft Spheres

Hard spheres are characterized by perfect rigidity, and consequently, a precisely defined radius, R. There is no attraction between hard spheres, but when brought in contact, the repulsive force diverges to infinity. In physical terms, this is described by a potential which varies with distance as shown in the following equation:

$$V_{hs} = \begin{cases} \infty & \text{for } r < R, \\ 0 & \text{for } r \geq R \end{cases} \quad (1.9)$$

The force between the spheres is derived from the potential: $f_{hs} = -\partial V_{hs}/\partial r$,

$$f_{hs} = \begin{cases} -\infty & \text{for } r < R, \\ 0 & \text{for } r \geq R \end{cases} \quad (1.10)$$

The variations of the potential and the corresponding force as a function of distance are shown in Figure 1.8.

Soft spheres are characterized by both repulsive and attractive potentials, of which the Lennard–Jones potential is a good example:

$$V_{ss} = V_{LJ} = 4\epsilon_0 \left[\left(\frac{\sigma}{r}\right)^m - \left(\frac{\sigma}{r}\right)^n \right] \quad (1.11)$$

In Equation 1.11, ϵ_0 is the depth of the potential energy well and σ is a parameter such that $V_{LJ} = 0$ when $r = \sigma$. The minimum, $V_{LJ} = -\epsilon_0$, occurs when $r = r_0 = (m/n)^{(1/(m-n))} \sigma$. The derivative of V_{LJ} with respect to r gives the interatomic force as

$$f_{LJ} = \frac{4\epsilon_0}{\sigma} \left[-m\left(\frac{r}{\sigma}\right)^{-m-1} + n\left(\frac{r}{\sigma}\right)^{-n-1} \right] \quad (1.12)$$

Both r_0 and ϵ_0 are specific to the given pair of interacting atoms. The exponents, m and n, define the strength of the repulsive and attractive forces, respectively. The most commonly used values that represent physical interactions are $m = 12$ and $n = 6$. Then, $r_0 = 2^{1/6} \sigma = 1.1225 \sigma$.

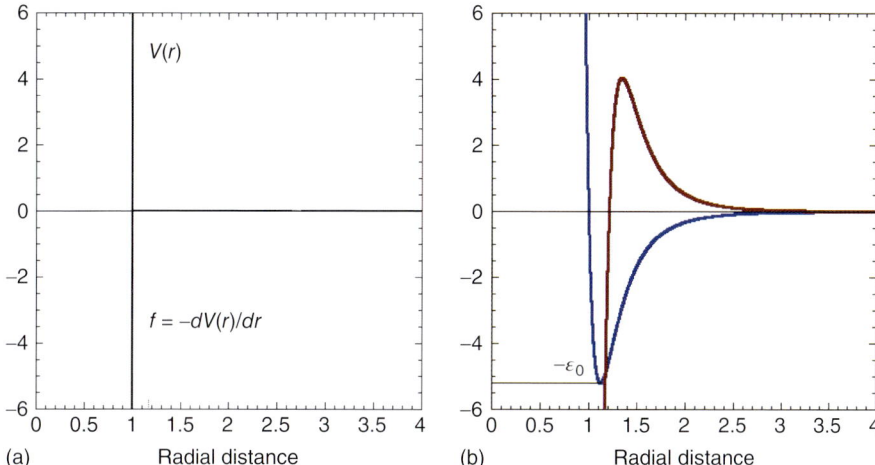

Figure 1.8 (a) Variation with distance of the potential and interatomic force for hard spheres. (b) Variation of L–J potential energy (with a minimum) and interatomic force (with a maximum) between two atoms.

The equilibrium separation of the two atoms defines the atomic radius. This assumption is reasonably accurate for metallic and van der Waals solids but may not be so for substances with covalent or ionic bonds in which the electronic structure of the atoms overlaps.

As a rule, hard spheres are used for modelling atomic arrangements by geometric packing, whereas soft spheres are used in molecular dynamics (MD) simulation of atomic arrangements and their properties.

Forces between particles

Force is the action of one body on another body. This may be direct as in body (surface) contact or through a force field (gravitation, magnetic, electrostatic, atomic, etc.).

Work, W, done by an applied force, $\mathbf{F}$, is evaluated at the start and end of the trajectory of the point of application. This means that there is a function $V(\mathbf{x})$, called a *potential*, that can be evaluated at the two points $\mathbf{x}(t_1)$ and $\mathbf{x}(t_2)$ to obtain the work over any trajectory between these two points. By convention, we define this function with a negative sign so that positive work is a reduction in the potential, that is,

$$W = \int_C \mathbf{F} \cdot d\mathbf{x} = \int_{\mathbf{x}(t_1)}^{\mathbf{x}(t_2)} \mathbf{F} \cdot d\mathbf{x} = V(\mathbf{x}(t_1)) - V(\mathbf{x}(t_2)) \tag{1.13}$$

The function $V(\mathbf{x})$ is called the *potential energy* associated with the applied force.

The application of the Nabla operator to the work function gives

$$\nabla W = -\nabla V = -\left(\frac{\partial V}{\partial x}, \frac{\partial V}{\partial y}, \frac{\partial V}{\partial z}\right) = \mathbf{F} \tag{1.14}$$

Because the potential V defines a force $\mathbf{F}$ at every point x in space, the set of forces is called a *force field*. A particle in a forcefield may experience a force acting on it, depending on the character of the field and the properties of the particle.

Types of forces

- Long range – gravitational forces: a particle possessing mass, m will experience a force when in a gravitational field.

$$F = \frac{G(m_1\, m_2)}{r_{12}^2}$$

- Electrical and magnetic forces: a particle possessing charge, q will experience a force when in an electrical field.

$$F = \frac{Q(q_1\, q_2)}{r_{12}^2}$$

- Short range – interatomic forces as described in the following.

Interatomic forces

All intermolecular forces are directional as a rule, except those between two noble gas atoms. The induction and dispersion interactions are always attractive, irrespective of orientation, but the electrostatic interaction changes sign upon rotation of the molecules, that is depending on the mutual orientation of the molecules.

Thermal motion averages out electrostatic forces to a large extent because of rotation of the molecules. The thermal averaging effect is much less pronounced for the attractive induction and dispersion forces.

London dispersion forces, named after the German–American physicist Fritz London, are forces that arise between instantaneous multipoles in molecules without permanent multipole moments. These forces dominate the interaction of nonpolar molecules and also play a less significant role in van der Waals forces.

The van der Waals force between two spheres of constant radii is a function of separation as the force on an object is the negative of the derivative of the potential energy function, $F_{\text{VW}}(r) = -(dV(r))/dr$.

The Lennard–Jones potential, expressed in Equation 1.11, is a mathematically simple model that approximates the interaction between a pair of neutral atoms or molecules. The repulsive r^{-12} term describes Pauli repulsion at short ranges because of overlapping electron orbitals. It is used because it approximates the Pauli repulsion well and is convenient for computational efficiency of calculating r^{12} as the square of r^6. The attractive r^{-6} term describes attraction at long ranges.

The Buckingham potential, expressed in Equation 1.15, is a formula that describes the Pauli repulsion energy and van der Waals energy, for the interaction of two atoms.

$$\Phi_{12}(r) = A \exp(-Br) - \frac{C}{r^6} \tag{1.15}$$

A, B and C are constants of the particular atomic pair. The two terms on the right-hand side constitute a repulsion and an attraction because their first derivatives with respect to r are negative and positive, respectively. R. A. Buckingham proposed this as a simplification of the Lennard–Jones potential in a theoretical study of gaseous helium, neon and argon.

With the L–J potential, the number of atoms bonded to an atom does not affect the bond strength. The bond energy per atom therefore increases linearly with the number of bonds per atom. Experiments show instead that the bond energy per atom increases quadratically with the number of bonds.

$$V(r) = 4\epsilon[12\sigma^{12}/r^{13} - 6\sigma^6/r^7] \qquad (1.16)$$

Many-atom potentials developed in the 1980s allow to model dense solids where bonds become weaker as a consequence of Pauli principle, due to local environment becoming crowded. Density functional theory (DFT) is a quantum mechanical modelling method used to model the structure of atoms, molecules and condensed phases.

1.3 Geometry of Clusters

The inner sphere (1) and its (k) touching neighbours form a *primary* or local ($1 + k$) cluster. Thus, a cluster composed of identical spheres is specified uniquely by four parameters.

1) the radius of the sphere (or spheres)
2) the number of the touching spheres, k, also called the *coordination number*
3) the disposition (or positioning) of the outer spheres, defined by the contact points
4) the orientation of the cluster in space

1.3.1 Regular Clusters

In Mathematics, a regular shape refers to a polygon that has all its sides equal and all its angles also equal. Perfect examples are squares and an equilateral triangle, where all the angles and sides are equal. A regular solid object has equal sides and equal angles, so that all its faces are regular polygons.

The smallest regular primary cluster is composed of $(1 + 4)$ spheres and has tetrahedral geometry, as shown in Figure 1.9.

The geometry of this cluster is specified by

1) the radius of each sphere = 1
2) the coordination number of the inner sphere, $k = 4$
3) the angular separation between each pair of adjacent outer spheres, $\phi = 2\arcsin\sqrt{2/3} = 109.47°$
4) the orientation of the cluster can be specified with respect to any of its four axes of twofold and four axes of threefold symmetry.

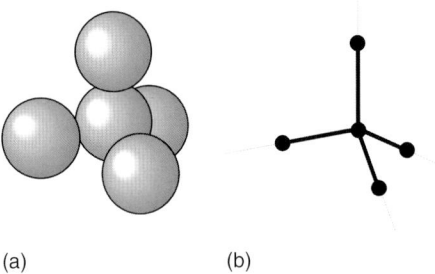

(a) (b)

Figure 1.9 (a and b) Regular cluster of tetrahedral geometry ($k = 4$) and its minimum outline representation.

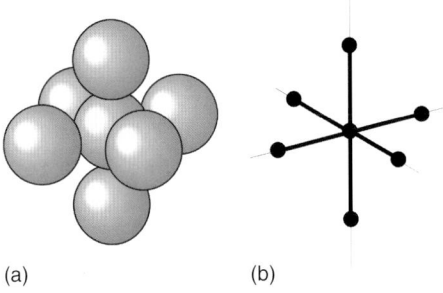

(a) (b)

Figure 1.10 Regular cluster of cubic geometry ($k = 6$) and its minimum outline representation.

Rotation of the cluster around any of the rotational axes makes the new cluster indistinguishable from its original image, which is the property of rotational symmetry.

Another example of a regular cluster is of cubic geometry, as shown in Figure 1.10

The geometry of this cluster is specified by:

1) the radius of each sphere = 1
2) the coordination number of the inner sphere, $k = 8$
3) the angular separation between adjacent spheres, $\phi = \pi/2 = 90°$
4) the orientation of the cluster can be specified with respect to either its four × threefold, three × fourfold or six × twofold axes of symmetry.

We can generalize the property of rotational symmetry by stating that the rotation of any regular cluster by the angular separation makes the rotated cluster indistinguishable from its original image.

Clusters derived from crystalline structures have a well-defined number of outer spheres. The very special and unique arrangements of spheres in these clusters arise because the number of neighbours and separation angles assume precise values, as given in Table 1.2.

Table 1.2 Separation angles between nearest outer spheres in selected regular clusters.

Cluster name	Neighbours, k	Separation angle, ϕ
hcp, fcc and icosahedral	12	$\pi/3 = 60°$
Body-centred cubic (bcc)	8	$\pi/3 = 60°$
Simple cubic (sc)	6	$\pi/2 = 90°$
Tetrahedral (th)	4	$2\arcsin\sqrt{2/3} = 109.47°$

It can be accepted, either from prior knowledge or even only intuitively at this stage, that regular clusters cannot be arrived at by random addition of spheres to the inner sphere; there must be some special forces at play to influence the existence of these special, unique arrangements. In the field of crystallization, these forces derive from minimization of Gibbs free energy; in physical flow of balls, it is the gravitational force that contributes to formation of ordered clusters, and so on.

That such forces are required can be understood with the help of the following example. The probability of throwing the number, say 3, with a standard dice is 1/6. One can make a dice with 12 faces, then the probability of throwing 3 is 1/12; for a dice of 24 faces, the probability is 1/24. In the special case when the dice has an infinite number of faces (i.e. a perfect sphere), the probability becomes $1/\infty = 0$. Therefore, a perfect cluster of fcc structure, or any other ordered cluster of spheres which requires achieving singular values of the separation angles, cannot be created by a random processes.

In mathematical statistics, if X is a continuous random variable, then it has a probability density function $f(x)$, and therefore, its probability of falling into a given interval, say $[a, b]$ is given by the integral,

$$\Pr[a \leq X \leq b] = \int_a^b f(x)\, dx \qquad (1.17)$$

In particular, the probability for X to take any single value a, such that $a \leq X \leq a$, is zero, because an integral with coinciding upper and lower limits is always equal to zero.

1.3.2
Irregular Clusters

If external force fields do not exist, formation of a cluster around a central sphere by addition of spheres at random can only result in a random configuration of the cluster. The number of ways that packing arrangements can occur depends on the space that is free to be occupied and the number of spheres available for packing.

An example of an irregular cluster is shown in Figure 1.11. The radius of each sphere can be specified (say, 1), and the coordination number of the inner sphere can be determined (say, $k = 7$). However, an irregular cluster has no axis of symmetry; therefore, the separation angles of the outer spheres will be different for

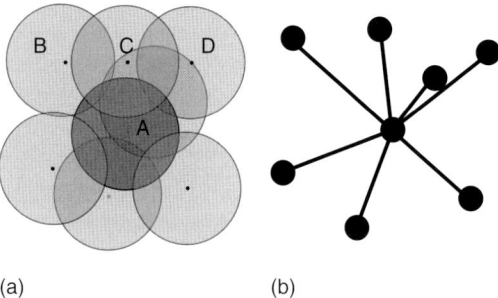

Figure 1.11 (a and b) A cluster with irregular geometry.

each pair of outer spheres, and the orientation of the cluster in space can only be specified by giving the positions (contact points) of all outer spheres.

Furthermore, as a random assemblage of spheres is composed of random (irregular) clusters where each primary cluster is different from every other primary cluster, it follows that in random packing the clusters must have:

- a distribution of the number of touching neighbours; and
- a distribution of their positions (or equivalently, a distribution of separation angles).

Next we present statistical models for the formation of random clusters.

1.3.3
Coordination of $(1 + k)$ Clusters

The term *coordination number* was defined originally in 1893 by Alfred Werner (Swiss chemist) as the total number of neighbours of a central atom in a molecule. The emphasis is on bonding structure in molecules or ions and the coordination number of an atom is determined by simply counting the other atoms to which it is bonded (by either single or multiple bonds). For example, $[Cr(NH_3)_2Cl_2Br_2]^-$ has Cr^{3+} as its central cation, which has a coordination number of 6.

Coordination number was adopted in crystallography and materials science to be the number of atoms touching a given atom in the interior of a crystal. For example, iron at room temperature has a body-centred cubic (bcc) crystal structure. Each iron atom in the interior occupies the centre of a cube formed by eight neighbouring iron atoms. The bulk coordination number for this structure is therefore 8. However, for an atom at a surface of a crystal, it is defined as the surface coordination number. For example, an atom lying on a [110] plane of iron exposed to the surface will have only six touching atoms, and correspondingly, its coordination number will be 6.

In random packing of spheres, two mathematical models are presented for predicting the coordination number, or rather for the distribution of coordination numbers in a large sample of random clusters:

- the so-called blocking model formed by sequential addition of spheres to the central inner sphere until no more spheres can be added
- the so-called distributed loose sphere model based on the fact that, as a rule, random clusters have lower density of packing than the possible maximum for close packed ordered clusters.

1.3.3.1 Blocking Model for Cluster Formation

Consider the construction of a $(1 + k)$ random cluster by sequential addition of spheres to the surface of the inner sphere. The problem can be approached with the following question: What is the probability, $P(k)$, that k spheres can be added in this way onto random positions of the inner sphere, subject to the exclusion condition.

For two spheres touching the inner sphere, the contact points must have a minimum angular separation, as specified by Equation 1.7. Bearing in mind this limitation, we consider the addition of a sphere as equivalent to occupying a spherical cap on the inner sphere, centred on the contact point and resting on a cone with a vertex angle of $\phi = \pi/3$. The solid angle of a cone with apex angle $\phi/2$ is equal to the area of a spherical cap on a unit sphere,

$$\Omega = 2\pi(1 - \cos\phi/2) \qquad (1.18)$$

Then, the probability that another sphere can be added is equal to 1 for as long as there is a spherical cap with a minimum area Ω left free on the surface of the inner sphere. This is always true for $1 \leq k \leq 5$, regardless of the placement of the k spheres.

For $k \geq 6$, the probability will vary, depending on the disposition of the previously added spheres. As each spherical cap has an associated concentric exclusion zone extended to an angle $2\pi/3$, the minimum probability, $P_{min}(k)$, will occur when the zones are distributed (positioned) as far from each other as possible (i.e. equi-spaced), and the maximum probability, $P_{max}(k)$, will occur when their respective exclusion zones overlap to their maximum extent. Therefore, the actual probability will depend on k and be limited to

$$P_{min}(k) \leq P(k) \leq P_{max}(k) \qquad (1.19)$$

Naturally, for $k = 1, P(k) = 1$ and for $k = 12, P(k) = 0$. One must find the values of $P(k)$ for all other values of k.

Apparently, there is no general analytical solution to this problem. Instead, it is relatively simple to set up a computer program to carry out such sequential addition of spheres and to sum the frequencies of the successful events. Such computations have been carried out and the results, based on 10^5 cluster samples, are shown in Figure 1.12(a). By this method, we find that the probability of random clusters with coordination of eight to nine is over 85% and the average coordination number is $\overline{k}_B = 8.29$.

The computational process involves two stages. In stage one, the surface of the inner sphere is divided into n equally spaced points, as shown in Figure 1.13. This is a well-established procedure in mathematical geometry (Conway and Sloane,

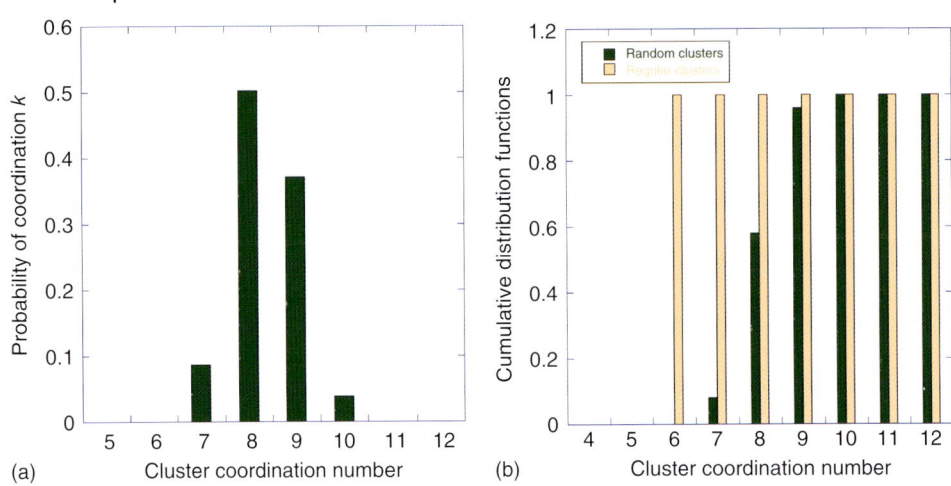

Figure 1.12 Statistics for 10^5 random clusters. (a) Probability of coordination in random clusters. (b) The corresponding blocking function.

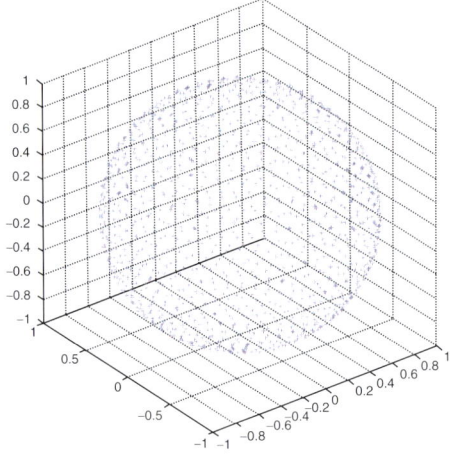

Figure 1.13 A sphere whose surface has 1000 equi-spaced points on its surface.

1998), and a ready made code has been published by Leopardi (2006). Then, from the created list of n points, a particular point is selected at random, and condition 1.7 is tested. If it is satisfied, then an outer sphere is added to (made to touch) the inner sphere at that point. If the condition is not satisfied, then the point is rejected and the program continues to loop with further spheres added until no more can be added. The data of such randomly created clusters are collected, and the summary statistics are shown in Figure 1.12(a).

We call this discrete function the *coordination number distribution function* derived from the blocking model, denoted $\Psi_B(k)$.

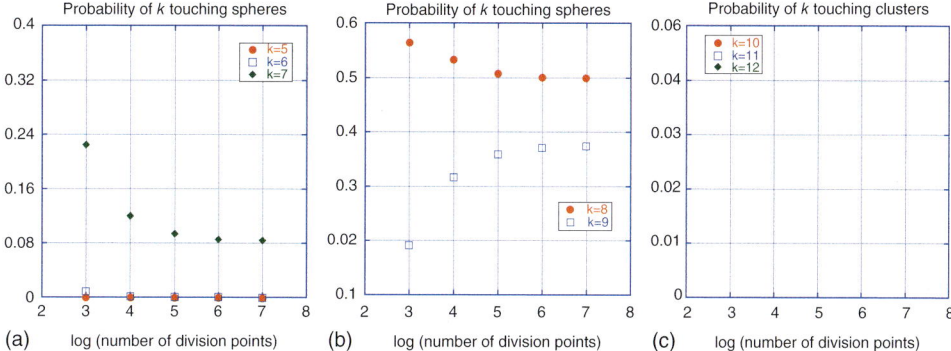

Figure 1.14 (a–c) Coordination dependence on a number of division points.

The coordination probability of clusters created by this method depends on the number n of the sphere division points. This dependence is shown in Figure 1.14. If the number is small (for example 8), then all created clusters will be regular with coordination 8. If the number is $n = 20$, then the created clusters will be random but the points for addition of spheres are quickly exhausted and the distribution will be limited. With increasing n, the randomness of positioning of the spheres increases. In principle, a perfectly random distribution would be obtained for $n = \infty$. The plot in Figure 1.12(a) gives the asymptotic values of probability for $n \geq 10^7$.

We can redefine the cluster creation problem with reference to the so-called *blocking number* (Barlow, 1883, Zong, 1999). For a regular cluster of equal size spheres, the blocking number is 6. It is known that a regular simple cubic pattern is sufficient to block the inner sphere from the addition of any more outer spheres. However, a random arrangement of six outer spheres is unlikely to completely block the inner sphere, and the blocking will vary from cluster to cluster. By definition, complete blocking occurs when there is not a single spherical cap area left on the surface of the inner sphere.

The concept can be generalized to both random and regular clusters by defining a *blocking function*,

$$\hat{b}(k) = \sum_k P(k) \tag{1.20}$$

For regular clusters, the blocking function, $\hat{b}$, has a sharp transition: $\hat{b} = 0$ for $k \leq 5$ and $\hat{b} = 1$ for $k \geq 6$, as indicated in Figure 1.12b by the gold bars.

For irregular clusters, the corresponding blocking function, also included in the graph, is shown by the dark green bars. This blocking function assumes the values $\hat{b}(k \leq 5) = 0$ and then increases gradually in the interval $6 \leq k \leq 9$, for $k = 11$ $\hat{b}(k) \cong 1$, finally achieving the value $\hat{b}(k = 12) = 1$.

Suppose we consider the blocking function (Equation 1.20) to be the CDF of the probability of occurrence of clusters with k outer spheres,

$$\text{CDF}(k) = \hat{b}(k) = \sum_k P(k) \tag{1.21}$$

Table 1.3 Probability of coordination and the corresponding blocking function for clusters created by random sequential addition of spheres.

k	$P(X \leq k)$	$CDF = \Psi_B(k)$
1	0	0
2	0	0
3	0	0
4	0	0
5	0	0
6	0.009	0.009
7	0.08	0.089
8	0.50	0.589
9	0.37	0.959
10	0.038	0.997
11	0.003	0.9999
12	0	1.0

Then, the following Table 1.3 can be constructed from the results displayed in Figure 1.12.

The cumulative distribution function, CDF, of a real-valued random variable X is the function given by

$$\text{CDF}_X(x) = P(X \leq x) \tag{1.22}$$

where the right-hand side represents the probability that the random variable X takes on a value less than or equal to x. The probability that X lies in the semi-closed interval $(a, b]$, where $a < b$, is therefore:

$$P(a < X \leq b) = \text{CDF}_X(b) - \text{CDF}_X(a) \tag{1.23}$$

If X is a purely discrete random variable, then it attains values $x_1, x_2, \ldots$ with probability $p_i = P(x_i)$, and the CDF of X will be discontinuous at the points x_i and constant in between:

$$\text{CDF}(x) = \sum_{x_i \leq x} P(X = x_i) = \sum_{x_i \leq x} p(x_i) \tag{1.24}$$

1.3.3.2 Fürth Model for Cluster Formation

Another possible distribution comes from an argument initially proposed by R. Furth (1964) in connection with studies of the structure of liquids by J. D. Bernal. We describe this process via the following three variables:

1) k, the number of neighbour-touching spheres relative to each inner sphere;
2) x, the maximum number of *sites* for spheres in dense packing that can be in contact with the inner sphere; and
3) s, the number of *virtual sites* available to spheres in a less dense packing.

In general, $s \geq x \geq k$. From elementary combinatorics, k spheres can be placed in x sites in $\binom{x}{k}$ ways, and $x - k$ (empty) sites can be chosen from $s - x$ in $\binom{s-x}{x-k}$

ways, whereas x sites can be distributed over s sites in $\binom{s}{x}$ ways. The probability of k spheres touching the inner sphere amongst the s sites subject to the maximum nearest-neighbour constraint, is thus

$$P(k) = \Psi_F(k, s) = \binom{x}{k}\binom{s-x}{x-k}\binom{s}{x}$$
$$= \frac{[x!(s-x)!]^2}{k!s![(x-k)!]^2(s+k-2x)!} \quad (1.25)$$

Now, assume that the number of possible sites is related to the available space around the inner sphere. Specifically, suppose that

$$s = x(v_h''/v_h') \quad (1.26)$$

where $v_h = V_V - v_s = v_s(1 - p_f)/p_f$, V_V if the volume of the Voronoi cell, v_s is the volume of an inscribed sphere, and p_f is the packing fraction. In relation 1.26, the prime corresponds to close packing with $s = x = k = 12$ and a corresponding maximum packing density, and the double prime corresponds to a less dense random packing for which $s > x > k$, for which the function $\Psi_F(k)$ will have a non-degenerate distribution. Combining these relations with Equation 1.26 gives a formula for s:

$$s = x\left(\frac{p_f'(1-p_f'')}{p_f''(1-p_f')}\right) \quad (1.27)$$

For close packing, $x = 12$ and $p_f' = \pi/\sqrt{18} \approx 0.74$. For lower density random packing, we assume $p_f'' = 0.62$. Substitution of these values in formula 1.27 gives $s \approx 20.9$. Taking the closest maximum whole number, $s = 21$, formula 1.25 can be now evaluated. The result is a multi-valued discrete function showing a prevalence of coordination numbers 6, 7 and 8 as shown in Figure 1.15. Other cases for $s = 17$ to 20 have been evaluated and are given in Table 1.4.

We call this discrete function the Fürth coordination number distribution function derived from the loose sphere model, denoted $\Psi_F(k)$.

Consequent on random packing is variation of coordinations in clusters, with admissible values in the range, $4 \leq k \leq 12$. If $c(k)$ is the fraction of clusters with a given contact number, k, then for the whole packed body of spheres, $\sum c(k) = 1$, summed for $k = 4,...,12$. Then, the average number of contacts per sphere is $\overline{k} = \sum k\, c(k)$, typically a non-integer, and possibly an irrational number. The average coordination number for this distribution is $\overline{k}_F = 6.91$.

In contrast to the blocking model, this model allows coordination numbers of 4 and 5, with 7 being the most frequent. In support of this observation, we can say that clusters with such low values of touching spheres are possible if they are stabilized by the next nearest-neighbour spheres, that is spheres existing within the s virtual sites around the inner sphere. It will be shown later that these are so-called Voronoi nearest neighbours.

In view of the distribution of coordination numbers, one can assume that in a packing, clusters with k values higher than $\overline{k}$ will be surrounded by clusters with

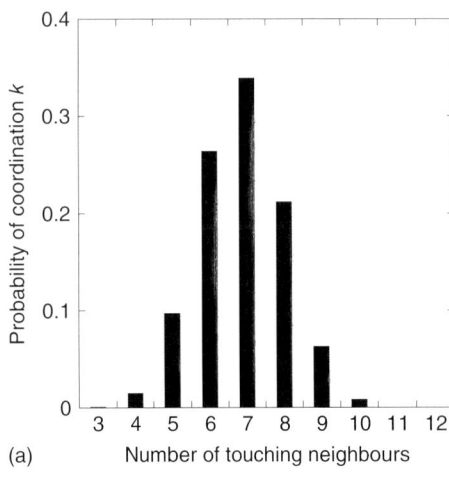
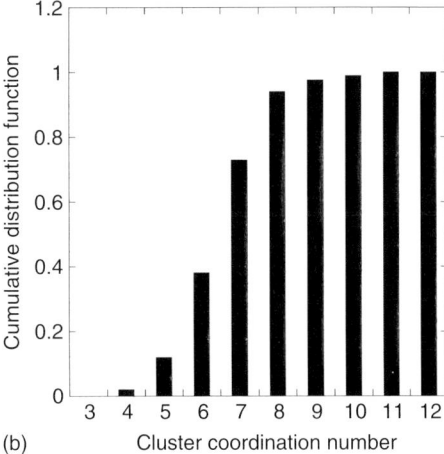

Figure 1.15 (a and b) Coordination distribution function, $\Psi_F(k)$, of random clusters predicted by Equation 1.25 and its cumulative distribution function (CDF).

Table 1.4 Computed values of the $\Psi_F(k)$ function for the selected values of the parameter s as shown.

k	s = 21	s = 20	s = 19	s = 18	s = 17
3	0.000748	0	0	0	0
4	0.015157	0.003930	0	0	0
5	0.097003	0.050298	0.015718	0	0
6	0.264060	0.205380	0.128360	0.049774	0
7	0.339510	0.352080	0.330080	0.255980	0.127990
8	0.212190	0.275070	0.343830	0.399970	0.399970
9	0.062872	0.097801	0.152810	0.237020	0.355530
10	0.008084	0.014670	0.027507	0.053329	0.106660
11	0.000367	0.000762	0.001667	0.003879	0.009696
12	0	0	0	0	0.000162
Sum =	1.000000	1.000000	1.000000	1.000000	1.000000

k values lower than $\bar{k}$, and vice versa, so that the full range of clusters is used, allowing for density fluctuations across the body of spheres.

In probability theory and statistics, the hypergeometric distribution (Equation 1.25) is a discrete probability distribution that describes the probability of successes in draws without replacement from a finite population of size containing a maximum of successes Figure 1.16.

The CDF of X will be discontinuous at the points k_i and constant in between:

$$\mathrm{CDF}(k) = \sum_{k_i \leq k} P(X = k_i) = \sum \Psi_F(k) \qquad (1.28)$$

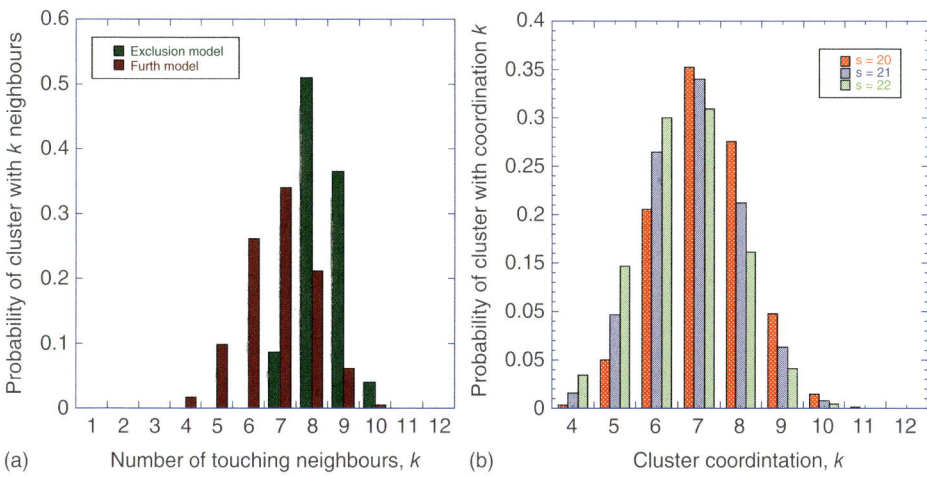

Figure 1.16 (a) Comparison of the distributions, Ψ_B and Ψ_F for randomly created clusters of equal-sized spheres. (b) Comparison of probabilities for clusters based on Equation 1.25 for s values as indicated.

For two-sized spheres, the result will depend on the relative size and whether smaller spheres touch a larger sphere or the other way round (Table 1.1).

1.3.4
Configuration of $(1 + k)$ Clusters

1.3.4.1 Regular Clusters
In the theory of crystallography, the symmetry elements of the unit cell describe the configuration of a given cluster. For example,

- hcp configuration with at least one axis of sixfold rotational symmetry (Figure 1.17a)
- fcc configuration with at least six axes of fourfold rotational symmetry and eight axes of threefold rotational symmetry (Figure 1.17b)
- bcc configuration with similar symmetry elements to that for fcc.

(a) (b)

Figure 1.17 (a and b) Atomic arrangements in clusters of crystalline form: hcp (3+6+3) layers and fcc (4+4+4) layers, showing the complete cluster and its components.

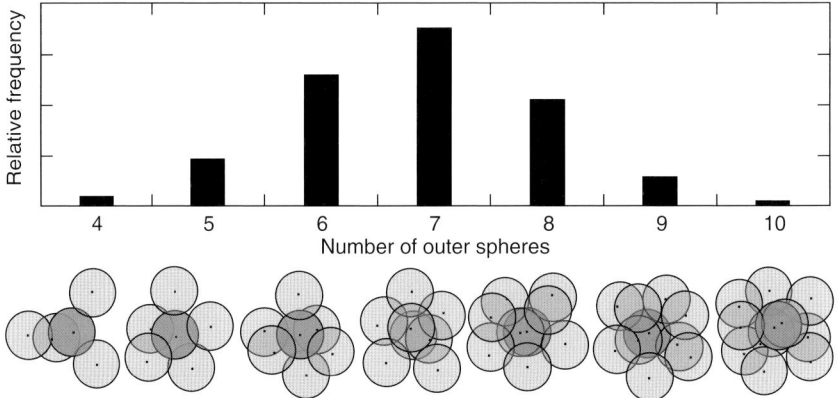

Figure 1.18 Examples of irregular clusters of equal-sized spheres with $k = 4$ to $k = 10$. Above, their relative frequency when randomly created by the Fürth method.

Therefore, the configuration of regular clusters is defined entirely by the symmetry elements of the corresponding crystal, although a primary cluster is a subset of the unit cell, and may have fewer symmetry elements than the unit cell itself.

Most importantly, the configurations of all crystalline clusters (unit cells) of a given form are exactly the same, although the size and chemistry may differ.

1.3.4.2 Irregular Clusters

In contrast to crystalline clusters, clusters created by a random process are different from each other even in the same (solid) substance. These variations are in terms of the number of touching neighbours, k, in the disposition of the outer spheres and their orientation in space. In theory, there will be an infinite variety of clusters and no two clusters alike. We refer to the positioning of the outer spheres as the *configuration* of the cluster. The orientation of the cluster in space is not important at this stage. Their appearance and relative number is shown in Figure 1.18.

For two clusters ($\ell = 1$ and $\ell = 2$) to be identical, we should necessarily have $k_1 = k_2, \zeta_1 = \zeta_2$ and $\Omega_1 = \Omega_2$ for the coordination number, the closing vector and spatial orientation, respectively.

Each cluster has nine possible values of k: 4, …,12. If the distribution of k is equiprobable, then the probability that two clusters have the same number of contact points (the same value of k) is $\sum_{k=4}^{k=12} (1/9)^2$.

For any given distribution $\Psi(k)$, we can write more generally that the probability for two clusters to have the same value of k is $\sum_{k=4}^{k=12} [\Psi_k]^2 = \frac{1}{9}c$, where $1 \leq c \leq 9$ is a constant. Next, we find that the probability of the two closing vectors to be the same as zero as ζ is a continuous variable. The same applies to Ω. Consequently, the probability of finding two identical clusters, if taken at random from a large (infinite) collection of random clusters, is zero.

Now, we turn our attention to the definition of the configuration of random clusters, which can be specified either individually for any given cluster or collectively for all clusters by appropriate statistics. The configuration of a cluster can be defined in two ways:

- by radial vector construction
- by spherical harmonics

1.3.4.3 Closing Vector Based on Radial Vector Polygon

This new method was conceived specifically to describe the configuration of random clusters and is capable of defining the configuration of an individual cluster, as well as describing the statistics of a large collection of random clusters (To *et al.*, 2006). In that sense, it provides a quantitative measure of a cluster's configuration in the absence of any regularity and symmetry elements. As an example, we take a cluster comprising nine outer spheres as shown in Figure 1.19a. For the given cluster, we identify the contact points between the inner and outer spheres by radial vectors, $\vec{R}_j, j = 1,...,k$, drawn from the centre of the inner sphere to the contact points on its surface as shown in Figure 1.19b. The contact points bisect the distances from the centre of the inner sphere to the centres of its touching neighbours.

Note that the positions of the contact points on the surface of the inner sphere are sufficient to define the configuration of the whole cluster uniquely and adequately. From now on, we focus on the properties of the inner sphere alone, bearing in mind that the sphere with its contact points represents the whole cluster.

For the purpose of analysis, we may arbitrarily number the contact points and align the contact point numbered 1 with the vertical axis of the external reference system, as shown in Figure 1.19b.

Then, the quantitative measure of the configuration of the cluster is defined in terms of a vector polygon with vertices R_j:

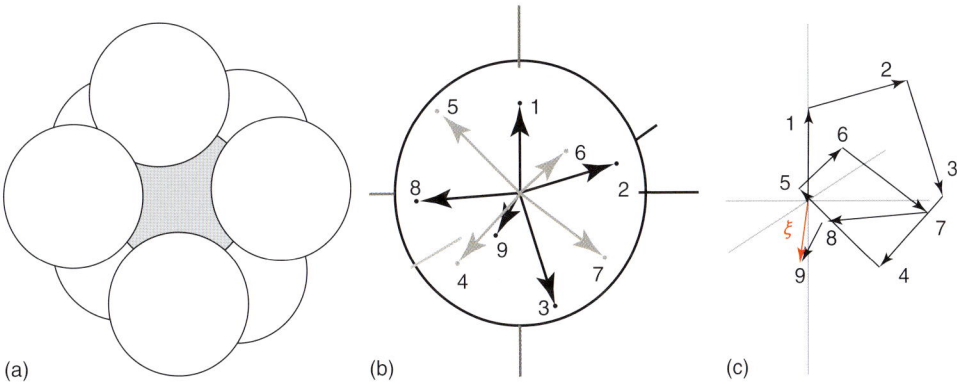

Figure 1.19 (a) An irregular (1 + 9) cluster. (b) The inner sphere of the irregular cluster, showing contact points and radial vectors. (c) A vector polygon constructed from the radial vectors.

$$\vec{\zeta} = \sum_{j=1}^{\kappa(x_\ell)} \vec{R}_j \tag{1.29}$$

where x_ℓ represents the ℓ'th sphere and κ is its corresponding number of its touching contacts. The sum represents a chain of the radial vectors and $\vec{\zeta}$ is the closing vector of the three-dimensional vector polygon.

The same vector construction can be applied to any and all clusters; therefore, the configurations of all clusters can be characterized by this method.

As the contact points are randomly positioned on the sphere, the vector polygon represents a *random walk* in three-dimensional space. A random walk is a sequence of random steps with independent and identically distributed increments. In mathematical analysis, there are two important questions relating to random walks:

- What is the end-to-end distance of a random walk consisting of n number of steps?
- What is the most probable end-to-end distance of such a random walk if the specific steps are not defined, but the number of the steps is known?

The answer to the first question is given by Equation 1.29. The answer to the second question is of statistical nature and is considered below.

Random walks

Random walk is an action involving consecutive jumps in space from point to point. The end of one step is the beginning of the next step, thus forming a continuous chain of steps or links. Each link can be represented as a vector of length, $\vec{a}$. The interest is to predict mathematically the so-called end-to-end distance after n steps.

Let the position of the walker be denoted by $\mathbf{R}_n$.

If the direction of each link is random and independent of the direction of the previous link, then the random walk is called *Gaussian random walk*.

The term random walk was first introduced into scientific literature by Karl Pearson in 1905 in relation to the spread of malaria by mosquitos. Random walks can explain the observed behaviour of processes in fields such as ecology, economics, psychology, computer science, physics, chemistry, and biology.

Consider a simple random walk of equal steps as shown in Figure 1.20. Let the length of the step be $a = |\vec{a}|$. After n number of steps, the end-to-end vector distance of the walk, denoted as $\vec{L}$, is given by

$$\vec{L} = \sum_{i=1}^{n} \vec{a}_i \tag{1.30}$$

The square of the magnitude of the end-to-end distance is equal to

$$|\vec{L}|^2 = \sum_{i=1}^{n} a_i \sum_{j=1}^{n} a_j = \sum_{i=1}^{n} a_{ii}^2 + 2 \sum_{j=1}^{n} a_i a_j \quad j \neq i \tag{1.31}$$

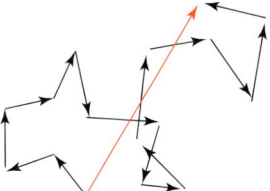

Figure 1.20 Random walk and its end-to-end distance.

This quantity is always positive regardless of direction. It may assume values in the range from 0 to $n\,a$. However, the statistical quantity, known as *the mean square end-to-end distance*, gives the most probable distance:

$$\langle L^2 \rangle^{1/2} = a\sqrt{n} \qquad (1.32)$$

where the <> brackets indicate statistical average.

The second term in Equation 1.31 is equal to $a^2 \times \cos(\angle_{ij})$, where $\angle_{ij}$ is the angle between two adjacent step vectors. If the orientation of the steps is completely random and uncorrelated (the so-called freely jointed chain), then the statistical average $\sum \cos(\angle_{ij}) = 0$, and therefore the second term is equal to zero, leaving only the first term.

Categories of random walks:

   Gaussian random walk (GRW) – carried out in unlimited space, with the direction of each step independent of previous steps. $\langle L^2 \rangle \sim n$.
   Self-avoiding random walk (SARW) – similar to GRW, but previously occupied points cannot be occupied again and each point has excluded space around it that cannot be occupied by other steps. Consequently, the mean square end-to-end distance becomes larger: $\langle L^2 \rangle \sim n^{6/5}$.
   Self-avoiding space limited random walk (SASLRW) – similar to SARW, but the space for executing random walk is limited (finite), and therefore, the walk must terminate when the available space is used up. No general analytical solution exists. A possible relationship is $\langle L^2 \rangle \sim n^{6/5} \exp\left(-\frac{\gamma}{1-n/n_{max}}\right)$, where γ and n_{max} are constants specific to the random walk.

We also define a function, ζ, equal to the magnitude of the vector, $\vec{\zeta}$:

$$\zeta(k) = \zeta = \|\vec{\zeta}\| \qquad (1.33)$$

Then, ζ is a summary statistic of any particular local cluster.

It should suffice to state that for any number and arrangement of contact points, that is for any number of random clusters, the ζ-function assumes values in a finite interval:

$$[0 \leq \zeta \leq \zeta_{max}(\kappa(x_\ell))] \qquad (1.34)$$

For each value of the number $k = 4, \ldots, 12$, there is a unique maximum value of the ζ-function $= \zeta_{max}(k)$. This value corresponds to an arrangement in which

Table 1.5 Calculated values of $\zeta_{max}(k)$, and $\zeta_{max}^{fixed}(k)$. For $k = 7$ & 11 by interpolation. Penultimate row for icosahedral. Last row for fcc and hcp.

k	$\zeta_{max}/(R)$	$\zeta_{max}^{fixed}/(R)$
1	1	Not applicable
2	$\sqrt{3}$	Not applicable
3	$\sqrt{2}\sqrt{3}$	Not applicable
4	$\frac{5}{3}\sqrt{3}$	$\sqrt{3}$
5	$\sqrt{3}\sqrt{3}$	$\sqrt{2}\sqrt{3}$
6	$\frac{4}{3}\sqrt{2}\sqrt{3}$	$\frac{5}{3}\sqrt{3}$
7	$\sqrt{3}\sqrt{3}$	$\sqrt{3}\sqrt{3}$
8	$2\sqrt{2}$	$2\sqrt{2}$
9	$\sqrt{3}\sqrt{2}$	$\sqrt{3}\sqrt{2}$
10	$\sqrt{3}$	$\sqrt{3}$
11	1	1
12$_{ico}$	≈ 0	≈ 0
12$_{fcc,hcp}$	0	0

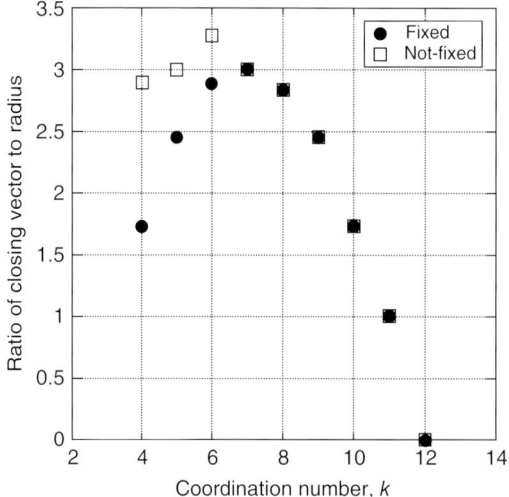

Figure 1.21 A plot of the closing vector values for fixed and non-fixed spheres from Table 1.5.

the outer spheres are rolled on the surface of the inner sphere towards the south pole into a close packed arrangement. The values of $\zeta_{max}(k)$ have been calculated according to this approach and are listed in Table 1.5. A plot of the closing vectors for the two conditions, not-fixed and fixed, is shown in Figure 1.21.

If the further condition is imposed requiring that the inner sphere be fixed by the outer spheres, then the values of $\zeta_{max}(k)$ change correspondingly to $\zeta_{fix}(k)$, as given in Table 1.5, third column. These values were calculated assuming that one of the outer spheres moves to a position diametrically opposite another touching sphere, so that for $k = 4$ we have only two radii (separated by an angle $\frac{1}{3}\pi$) contributing to the value of $\zeta_{\text{fix}}(4)$, for $k = 5$ only three radii, and so on. (The actual values must be arbitrarily smaller as the sphere has to be on the other hemisphere, however imperceptibly). Consequently, for fixed spheres, condition 1.34 is finally expressed as

$$[0 \leq \zeta \leq \zeta_{\text{fix}}(\kappa(x_\ell))]. \tag{1.35}$$

A vector polygon formed by a chain of radial vectors according to Equation 1.29 can be considered in terms of a random walk in space. We note that the steps of the walk are of the same size. If successive steps are represented by R_j, then the mean square displacement of a random walk is given by

$$\langle \zeta_k^2 \rangle = \sum_{j=1}^{k} \sum_{i=1}^{k} \langle R_j \cdot R_i \rangle = k\, A(j, i), \tag{1.36}$$

with a correlation function defined as $A(j, i) = \langle R_j \cdot R_i \rangle / |R|^2$. The probability for the end-to-end vector being of magnitude lying in the range $(\zeta, \zeta + d\zeta)$, equivalently, of lying in a shell of radius ζ and thickness $d\zeta$, is given by

$$P_k(\zeta)\, d\zeta = 4\pi \zeta^2 p_k(\zeta)\, d\zeta, \tag{1.37}$$

subject to the normalizing condition that

$$\int_0^{\zeta_{\text{fix}}(k)} P_k(\zeta)\, d\zeta = \int_0^{\zeta_{\text{fix}}(k)} 4\pi \zeta^2\, p_k(\zeta)\, d\zeta = 1, \tag{1.38}$$

where $P_k(\zeta)$ is a density function for the random variable, ζ and $p_k(\zeta)$ is a density function relating to spherical shells. In Equation 1.37, $P_k(\zeta)$ is related to the function $\Phi_k(v_1, \ldots, v_k)$ by

$$P_k(\zeta) = \int_{|v_1 + \ldots + v_k| = \zeta} \Phi(v_1, \cdots, v_k)\, dv_1 \ldots dv_k. \tag{1.39}$$

Well-known solutions to Equation 1.37 exist for random unrestricted walks. However, as the spheres are non-intersecting, the consecutive steps are subject to condition (3.1.2), leading to self-avoiding random walks (SARWs). Furthermore, the SARWs are limited by the finite space around the sphere's surface. We have shown earlier by geometrical calculations that the maximum end-to-end distance, that is the asymptotic limit for such SARWs decreases towards zero for $k \to 12$. The additional fixed sphere requirement limits the configurations of each cluster for $4 \leq k \leq 6$, such that $\zeta_{\max}$ cannot be reached and must be replaced by $\zeta_{\max}^{\text{fixed}} \leq \zeta_{\max}$. We note that $\zeta_{\max}^{\text{fixed}} = \zeta_{\max}$ for $k = 7$ to 12. There is approximate symmetry in the values of $\zeta_{\max}$ with respect to $k = 6$ and in the values of $\zeta_{\max}^{\text{fixed}}$ with respect to $k = 7$.

The function $P_k(\zeta)$ assumes a special and unique form for all clusters ($k = 4 \ldots 12$) for which $\zeta = 0$. For example, for equidistant angular configurations of points, $P_k(\zeta = 0) = 1$ by definition. Furthermore, clusters derived from Bravais lattices with $k = 6, 8$, and 12 (a subset of clusters with equidistant configurations) also have $\zeta = 0$, and $P_k(\zeta) = 1$.

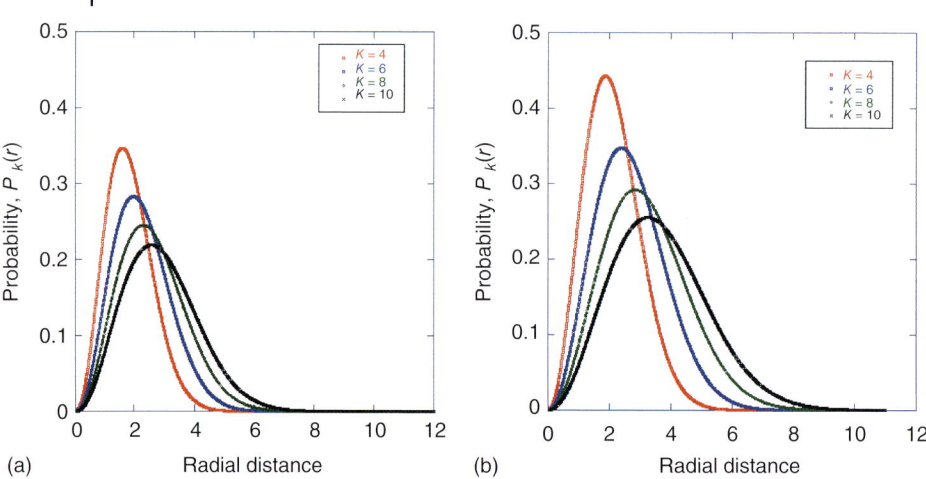

Figure 1.22 Probability functions for the random variable, ζ, for clusters with the values of k as shown. (a) Gaussian random walk and (b) self-avoiding random walk. In the latter case, peaks for SARW occur at larger values of k.

Table 1.6 Most probable random walk distances from Figure (1.22)

k	Gaussian	SARW	Ratio of SARW/Gauss
4	1.64	1.87	1.14
6	2.00	2.40	1.20
8	2.33	2.87	1.23
10	2.60	3.27	1.25

Therefore, for all regular clusters, $\zeta = 0$. For irregular clusters $\zeta \geq 0$, and the probability function, $0 \leq P_k(\zeta) < 1$. The variation of $P(\zeta)$ with k for irregular clusters is shown for GWR and SARW in Figure 1.22. The peak positions for the probabilities as a function of k have been determined from the graphs and are given in Table 1.6. The data shows how the end-to-end distance increases with k for the SARW relative to the Gaussian walk.

1.3.4.4 Physical Meaning of the Closing Vector, ζ

The geometry of an irregular cluster is adequately specified by the coordination number, k, and the magnitude of the closing vector, ζ. As stated before, the orientation of the cluster in space is not important, unless the cluster is placed in a physical force field, which may be magnetic, electric, elastic, or of any other nature. Then, the field will interact with the cluster through its geometrical and constitutional properties. The strength and direction of the interaction will be in proportion to the magnitude and direction of the closing vector, ζ.

Consider calculating a physical property of individual clusters. Let the vector polygon of Equation 1.29 be a piece-wise smooth curve C in $\mathcal{R}^3$. Let $\vec{p}_j$ be the polarity vector associated with each pair of spheres in the cluster identified with the radial vector, $\vec{R}_j$, that is the polarity of a pair comprising the inner and one outer sphere. Then, the polarity of the cluster is simply,

$$\vec{P}_{cluster} = \sum \vec{p}_j = c \times \sum_{j=1}^{\kappa(x_\ell)} \vec{R}_j = c \times \vec{\zeta}, \tag{1.40}$$

where c is a material characteristic constant. Let a charge of q Coulombs be associated with each pair of spheres. Then, the cluster placed in an electric field $\vec{E}$ experiences a force acting on it that can be calculated using Green's Theorem:

$$\vec{F} = \oint_C \vec{E}(x) \cdot d\vec{p}(x) \tag{1.41}$$

In materials science, the *coercivity* (also called the *coercive field* or *coercive force*) of a ferromagnetic material is the intensity of the applied magnetic field required to reduce the magnetization of that material to zero after the magnetization of the sample has been driven to saturation. Materials with high coercivity are called *hard ferromagnetic materials* and are used to make permanent magnets. Permanent magnets find application in electric motors, magnetic recording media (e.g. hard drives, floppy disks or magnetic tape) and magnetic separation. A material with a low coercivity is said to be soft and may be used in microwave devices, magnetic shielding, transformers or recording heads.

Other physical and mechanical properties can be related to the ζ function in a similar way.

1.3.4.5 Spherical Harmonics

Another way to describe the configuration of a cluster is by means of spherical harmonics, for which a general equation is as follows:

$$g_{\mu,\nu}(\phi,\theta) = \sum_{l=0}^{L} \sum_{m=-l}^{l} a_l^m(\mu\nu) \, Y_l^m(\phi,\theta) \tag{1.42}$$

where $a_\ell^m(i,j)$ are the spherical harmonics coefficients and $Y_\ell^m(\phi,\theta)$ are corresponding Bessel functions.

Spherical harmonics is an extension of Fourier series into three dimensions in spherical coordinates. The trigonometric functions in two dimensions represent the fundamental modes of a vibrating string, the spherical harmonics represent the fundamental modes of vibration of a sphere in much the same way. Fourier series decomposes periodic functions into the sum of simple oscillating functions, that is sines and cosines (or complex exponentials), and equivalent summation applies to spherical harmonics, as can be seen from Equation 1.42.

Wieder and Fuess (1997) proposed a modified Debye equation to consider the local structural order as follows:

$$I(\theta) = \sum_{i=1}^{N} \sum_{j=1}^{N} f_i f_j \left(\frac{n_{ij}}{d_{ij}} \right), \tag{1.43}$$

where:

$$n_{ij} = \int_{\phi=0}^{2\pi} \int_{\theta=0}^{\theta=\pi} \exp(i2\pi s(\alpha,\beta,\theta) r_{ij}(\phi,\theta)) \cdot h_{ij}(\phi,\theta) \sin\theta d\theta d\phi \quad (1.44)$$

$$d_{ij} = \int_{\phi=0}^{2\pi} \int_{\theta=0}^{\theta=\pi} h_{ij}(\phi,\theta) \sin\theta d\theta d\phi. \quad (1.45)$$

It can be noted that for $h_{ij}(\phi,\theta) = 1$, that is for spherical symmetry, the original Debye scattering Equation 1.42 is recovered. For the case, $h_{ij}(\phi,\theta) \neq 1$, the orientation function can be expressed in terms of spherical harmonics:

$$h_{ij}(\phi,\theta) = \sum_{\ell=0}^{L} \sum_{m=-\ell}^{+\ell} a_{\ell}^{m}(i,j) \, Y_{\ell}^{m}(\phi,\theta) \quad (1.46)$$

The probability function, $P(k)$ of Equation 1.37, has both a descriptive character and a predictive role. It describes the distribution of configurations for all clusters, as well as indicating the most probable configuration for each cluster. In that sense, it is also more appropriate for describing configurations of random clusters in ideal amorphous solids than spherical harmonics. The Q6 component of spherical harmonics has a range of different values for regular configurations, but it gives a non-discriminating value of zero for random packings. Only two-dimensional plots of spherical harmonics differentiate between various configurations. The function, $P(k)$, however, gives a more comprehensive statistical description of the range of random configurations around the most probable arrangement with a clear differentiation between clusters with different coordinations (k values) and independent of the choice of frame of reference.

1.4
Geometry of Sphere Packings

Packing of spheres is an assemblage (aggregate) of spheres of either finite or infinite extent. It is an object that can be studied *per se* in terms of its geometry and mathematical properties, and also it is a useful model of atomic arrangements in solids and liquids. The packing is a created object, that is either imagined or physically put together in some manner. This can be done in one two ways:

- by following a strict rule (or a set of self-consistent rules),
- in an undefined, or only partially defined way.

Packings in the second category are typically those based on so-called random pouring of spheres into containers and any variations thereof or on molecular dynamics computations. These methods simulate structures that resemble those of real solids, with particular packing arrangements and imperfections not present in perfect solids which are free of any imperfections by definition. Such models are ubiquitous and will be considered later.

Models in the first category are deterministic. If the geometrical rule of packing and the size of the spheres are known, then the packing can be created repeatedly and reproducibly following the rule, always with the same predictable result. This means that the created packing is repeatable and verifiable, which is a requirement for modelling of ideal structures, such as crystalline, quasi-crystalline and ideal amorphous atomic arrangements.

Packing of spheres involves a volume, V containing N number of hard spheres. For simplicity, assume all spheres to have the same diameter. The number density of such an assembly is,

$$\rho = N/V. \tag{1.47}$$

The total volume can be divided into

- volume occupied by all spheres, $V_o = N \cdot (4/3)\pi r^3 < V$
- unoccupied volume, $V_u = 1 - V_o$.

Naturally, for $N = 0$, $V_o = 0$, $V_u = V$ and $\rho = 0$. As N increases, the number density and the corresponding volume fraction, $0 \leq V_o/V < 1$, increases simultaneously. Under the condition of sphere impenetrability, the volume fraction is always less than one.

The maximum value of the volume fraction depends on both the arrangement of the spheres and the shape of the volume V. If the volume is concave at all points of its surface, and if the number of spheres that are packed into it is sufficiently large, the effect of its shape becomes negligible. Then, the geometrical arrangement of the spheres is the only deciding factor with regard to the maximum numerical density or volume fraction. The largest value of the volume fraction is found in sphere packings representing hexagonal close packed or fcc crystallographic arrangements. Any other arrangements result in lower maximum packing fractions, as described in the following.

We use a symbolic notation for a general packing of spheres. Let S_a represent a sphere of diameter, a, and let Ω be a set of points in Cartesian space representing the centres of N such spheres,

$$\Omega = \{x_1, x_2, \ldots, x_N\}. \tag{1.48}$$

Then,

$$S_a + \Omega \tag{1.49}$$

is a packing of spheres.

We first note some general properties:

1) Scale invariance:
   if $S_a + \Omega$ is a packing, then so is $S_{ca} + c\Omega$ for every positive c.
2) Homogeneity of point field:
   Let $N = [xn]$ be a point field. The point field is homogeneous if $N = [xn]$ and $N = [xn + x]$ have the same distribution for all $x \in En$. Then the density is independent of position and selection of the field. Therefore, packing fraction should be the same regardless of variation in atomic sizes in multi-atomic glassy alloys, that is random packing independent of the size of Voronoi cells.

1.4.1
Fixed and Loose Packings

Consider a cluster of spheres in packing comprising one inner sphere in the centre and k-number of outer spheres touching it. The number and arrangement of the outer spheres define a fundamental characteristic of the inner sphere in such a cluster; it can be in one of two distinctly different arrangements:

- Fixed spheres packing - A sphere is in a *fixed* position, when, for $4 \leq k \leq 9$, no more than $(k-1)$ of the outer spheres are located on one hemisphere; for $k > 9$, the sphere is always in a fixed immovable position. This condition implies that the touching spheres form a cage around the inner sphere. In $\mathbb{R}^3$, a minimum of four outer spheres is needed to fix the inner sphere. Otherwise, the inner sphere is *loose* and can move (Stachurski, 2003).
- Loose spheres packing - A sphere is always in a *loose* position, when, for $4 \leq k \leq 9$, all contacts are on one hemisphere only; the sphere may be confined but it is not fixed and can move to other loose positions in its vicinity. A packing containing loose spheres is called *loose packing*.

Any arrangement of outer spheres on a cluster with $k < 4$ cannot form a locking cage around the inner sphere, which gives reason to the lower limiting value, $k_{\min} = 4$, of touching neighbours in arrangements of spheres guaranteed to represent a perfect solid body. A perfect solid body is one in which all spheres are in fixed positions, disallowing any translational movement of any sphere. Frictionless rotations are allowed as they have no effect on solidity.

We shall now consider three types of packings, distinguished entirely by the geometrical arrangement of the spheres.

1.4.2
Ordered Packing

For ordered (crystalline) packing, the set containing the coordinates of the sphere centres possesses translational regularity in their values, and therefore, the position of every sphere in the ordered structure can be predicted using the formula:

$$\vec{r}_{\text{sphere}} = S(r_0) + u \cdot \vec{a} + v \cdot \vec{b} + w \cdot \vec{c} \tag{1.50}$$

If $\vec{a}, \vec{b}, \vec{c}$ are linearly independent vectors in Cartesian space, then the set,

$$\Lambda = \left\{ \sum_{i=1}^{3} z_i \vec{a}_i : z_i \in Z \right\} \tag{1.51}$$

is called a *lattice*, where Z is a set of all integers (Z is from German 'Zahlen' meaning numbers), and $(\vec{a}, \vec{b}, \vec{c})$ is the basis for Λ.

We call

$$S_a + \Lambda \tag{1.52}$$

an ordered (translative) packing of S_a on the lattice Λ, if

$$(\text{int}(S_a) + x_1) \cap (\text{int}(S_a) + \mathbf{x}_2) = 0, \qquad (1.53)$$

whenever $\mathbf{x}_1$ and $\mathbf{x}_2$ are distinct points in Λ. In the theory of crystallography, there are 14 distinct Λ-sets corresponding to 14 Bravais crystal lattices.

Ordered packing of spheres is evidenced by regularity of the arrangement. The most direct evidence is provided by testing for elements of symmetry displayed by the set. In particular, one can imagine a plane passing though such a packing on which many of the sphere's centres lie in some arranged ordered pattern. There is always an imaginary plane, passing through the set, which contains a large subset (more than three) of the points. This is the basis for crystallographic (translational) and for quasi-crystalline patterns (rotational regularity in the latter case).

We now require some special properties of the set Λ:

1) For any point, $x_\ell \in \Lambda$, its nearest neighbours are always equidistant from that point, that is $|x_{\ell j} - x_\ell| = a$ for $j = 1, \ldots, \kappa(x_\ell)$, where $\kappa(x_\ell)$ is the number of nearest neighbours. *This requirement implies that spheres are non-intersecting, that all spheres have the same diameter, and that nearest neighbours are touching*.
2) The number of nearest neighbours, $\kappa(x_\ell)$, is bounded: $[4 \leq \kappa(x_\ell) \leq 12]$. The lower limit of 4 is imposed by the condition that all spheres are fixed, and the upper limit of 12 is established as an absolute limit by the Kepler theorem.

1.4.3
Disordered Packing

Degree of disorder is defined by physicists and materials scientists as follows: Let p be the probability that a lattice position is occupied by a foreign replacement. The degree of disorder is expressed as

$$\begin{aligned} D &= \frac{\text{actual value of } p - \text{value of } p \text{ for complete disorder}}{\text{value of } p \text{ for complete disorder} - \text{value of } p \text{ for complete order}} \\ &= \frac{p-r}{1-r} \end{aligned} \qquad (1.54)$$

When order is complete, p is unity and D is therefore unity. When disorder is complete (whatever that means) and the arrangement of replacements quite random, only a fraction of r of the replacements will be in the positions of order. Therefore $p = r$, and $D = 0$.

This packing implies that it is disordered from the regular packing (by the introduction of imperfections usually referred to in crystallography as defects). Since in principle disorder can be reversed, disordered packing is deemed to be structurally derived from the ordered packing and therefore different from random packing.

1.4.4
Random Packing

We note that a random process is a repeating process whose outcomes follow no describable deterministic pattern but follow a probability distribution.

There exist three mechanisms responsible for random behaviour in systems:

1) Randomness coming from the initial conditions (chaos). Example: the motion of the end point of the so-called double pendulum is chaotic.
2) Randomness coming from the environment (Brownian motion). Example: (a) the motion of a gas molecule or (b) the tossing of a coin.
3) Randomness intrinsically generated by the system. Typically shows statistical randomness whilst being generated by an entirely deterministic causal process. This is also called *pseudo-randomness*. Example: (a) generation of random numbers by computer and (b) random packing of spheres.

The formation of real amorphous materials comes from randomness of density fluctuations in the liquid (Brownian motions). However, the construction of a random packing of spheres is generated by rules, and therefore, it is a pseudo-random process.

Random packing is fundamentally different from crystalline and quasi-crystalline packing and also different from disordered packing, and we represent it symbolically by the notation:

$$S_a + X \tag{1.55}$$

Random packing is characterized by irregularity of *all* point-to-point distances in the set X. An algorithmically random sequence (or random sequence) is an infinite sequence of binary digits that appears random to any algorithm. The definition applies equally well to sequences on any finite set of characters. The positions of the sphere centres cannot be described by an equation of the type (Equation 1.50). Instead, the distances between the points are described by an appropriate statistical density distribution function.

1.4.5
Random Sequential Addition of Hard Spheres

Consider a volume, V, into which hard spheres of the same diameter, d, are added sequentially, one at a time. The placing of each sphere inside the volume is at randomly chosen positions, with the constraint that spheres must not overlap (consequently, each addition must be preceded by a test of the available space). At any stage of this process, the packing is described by,

$$S_a + X_{RSA}. \tag{1.56}$$

If random sequential addition (RSA) of spheres always starts with V empty, and if it is repeated with ever-increasing V, and stopped when the same average density ρ is achieved, then in the limit $N, V \to \infty$ there is a unique limiting density, ρ^*, called the *saturation density*. With random sequential addition, it is obvious that a saturation density ρ^* is always less than the highest density of the ordered packing.

The process of random sequential packing can be described as follows. Let $v_{ex} = (4/3)\pi d^3$ be the volume from which one sphere excludes the center of another. Let $w_{ex} = w(\mathbf{r}_1, \mathbf{r}_2)$ be the combined volume from which the center of a third sphere is excluded by two spheres, one centred at $\mathbf{r}_1$ and the other at $\mathbf{r}_2$. We note that w_{ex} is not constant but varies from a minimum value of $(19/12)\pi d^3$ when the volumes, v_{ex}, overlap at $|\mathbf{r}_1 - \mathbf{r}_2| = d$ to a maximum value of $(8/3)\pi d^3$ when the centres of the two spheres are sufficiently far apart, $|\mathbf{r}_1 - \mathbf{r}_2| \geq 2d$.

Let $\mathbf{r}_1, \mathbf{r}_2, \mathbf{r}_3$ be an allowable configuration of the centres of the three spheres with the probability,

$$P(\mathbf{r}_1, \mathbf{r}_2, \mathbf{r}_3) dv_1\, dv_2\, dv_3 \tag{1.57}$$

where one of the spheres has its center in the infinitesimal volume element dv_1 at $\mathbf{r}_1$, one in dv_2 at $\mathbf{r}_2$ and one in dv_3 at $\mathbf{r}_3$. Then, from the definition of random sequential addition,

$$P(\mathbf{r}_1, \mathbf{r}_2, \mathbf{r}_3) = 2V^{-1}(V - v_{ex})^{-1}[(V - w_{12})^{-1} + (V - w_{23})^{-1} + (V - w_{31})^{-1}] \tag{1.58}$$

where $w_{ij} = w(\mathbf{r}_i, \mathbf{r}_j)$.

This probability density is not uniform in the space of allowable configurations but depends on the configuration through the w_{ij}'s. The greater the extent to which the exclusion volumes associated with the spheres i and j overlap, the smaller are w_{ij} and $(V - w_{ij})^{-1}$, and also the smaller is P. Thus, the probability, P, is biased in favour of those triplet configurations in which pairs of spheres are so distant that their associated exclusion spheres do not overlap.

By contrast, a uniform probability density in the space of allowable configurations, such as would characterize an equilibrium distribution, would for this case of three spheres be given by,

$$P = 6V^{-1}(V - v_{ex})^{-1}(V - \langle w_{ij} \rangle)^{-1}, \tag{1.59}$$

where

$$\langle w \rangle = (V - \omega)^{-1} \int w_{12} dv_2 \tag{1.60}$$

the integration being extended over all positions $\mathbf{r}_2$ in V of the center of a second sphere, which are allowed when a first sphere is fixed with its centre at $\mathbf{r}_1$. The constant $\langle w \rangle$ and the functions w_{ij} may be readily expressed in terms of the sphere diameter d, but this is not pursued here.

Sequential addition produces a configuration in which there remain no gaps as large as the sphere diameter, and there need not be any spheres in contact in this configuration. It belongs to the loose spheres packing category. Sequential random addition can be considered as an approximation to the equilibrium distribution in a model of hard sphere fluid.

1.4.6
Random Closed Packing of Spheres

Closed packing belongs to the fixed spheres packing category. The essential element of this random closed packing is that each new sphere that is added must rest on three, and only three spheres that have unequal spacings between them, $BC \neq CD \neq DB$, as shown in Figure 1.23.

In any group of four nearest-neighbour spheres selected from such a packing, the centres of three non-touching spheres must form an irregular triangle, as shown in Figure 1.23. The sphere, A, is touching the three spheres, B,C,D, to form touching contacts. Each triangle, ADC, ADB and ABC, is an isosceles triangle but with different angles: $\angle DAC \neq \angle CAB \neq \angle BAD$.

It is recalled from Section 1.3 that irregular clusters, constructed by the Fürth method or by the blocking method, consists of one inner sphere and k outer spheres randomly positioned. The outer spheres provide $k+2$ sites for the addition of spheres, which satisfy the condition for any combination of its three adjacent spheres that,

$$(\text{two sphere diameters}) > DC > DB > BC > (\text{one sphere diameter})$$

Now, consider an irregular cluster such as shown in Figure 1.24. Any three adjacent outer spheres provide a site for addition of another sphere. The added sphere, together with the underlying three spheres, forms a group of four spheres similar to that described earlier and depicted in Figure 1.23. Such addition of spheres onto available three-sphere sites will result in growth of the cluster. At the completion of the first growth stage, the original cluster will be increased in size by the addition of the second layer of spheres. All spheres in the second layer will have unequal distances between them and from the central inner sphere.

Next, the process of identifying and filling in of the possible sites can be repeated again, and again, according to the rules listed in the following. The above-mentioned process can be applied for as many times as required, until the

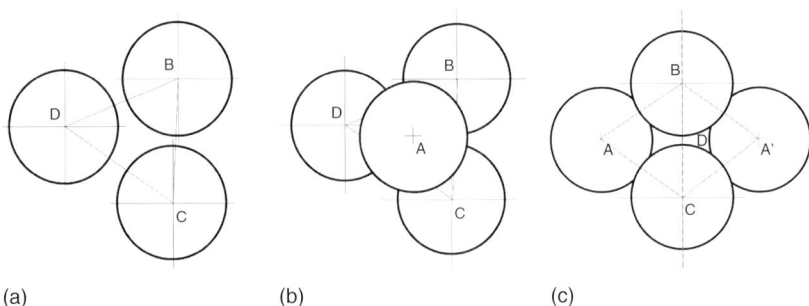

Figure 1.23 (a) Equal size spheres B, C and D, separated by unequal distances (DC > DB > BC), forming an irregular triangle. (b) Sphere A touches each of the underlying spheres B, C and D, and therefore, its centre is equidistant from each. (c) A five sphere subcluster; plane passing through BCD is a plane of mirror symmetry.

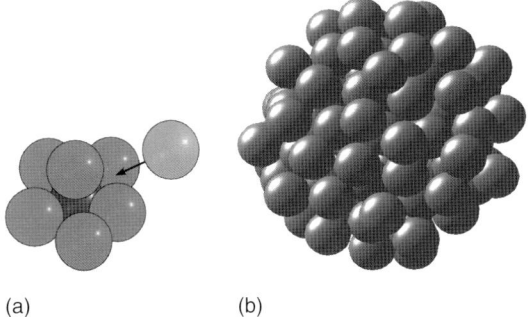

(a) (b)

Figure 1.24 (a) A primary cluster with a sphere added to a triangular site of an irregular shape formed by three adjacent outer spheres. (b) A cluster with several layers of spheres, added by the same method.

desired size of the cluster is reached. The geometrical construction follows a set of rules so that the outcomes are always reproducible and verifiable.

Rules for sphere addition to $S_a + X_{RCP}$ packing

1) For a given cluster of spheres, identify and index all possible triangular sites on its surface, based on three adjacent non-touching spheres.
2) Find (compute) the centres for spheres that can be potentially added to the sites
3) Sort all sites in order from the closest to the furthest in terms of the distance to the centre of the inner sphere;
4) Add a sphere to the site of the closest distance and then add a sphere to the next site in terms of the closest distance from the centre;
5) Ascertain that the second sphere does not overlap; reject the second sphere if there is an overlap; and
6) Continue the process (steps 4–5) until all allowable sites are filled in (with every completion of the loop another coordination shell is added).

This packing of spheres is represented by the notation:

$$S_a + X_{RCP} \tag{1.61}$$

This is a pseudo-random packing because it always results in the same packing, that is the same set X_{RCP} for the same initial configuration of the random cluster. However, the positions of the spheres are randomly distributed in space, without any short-, medium- or long-range order. Furthermore, the packing is different (different sphere positions) if the initial configuration of the random cluster is different. The random close packing possesses the following properties:

1) *The hard sphere condition* (Section 1.2); if x_i and x_j are the coordinates of two adjacent spheres, the $(x_i^2 - x_j^2)^{1/2} \geq D$ is true for all spheres without exception.
2) *The fixed sphere requirements* (Section 1.3); all spheres are in fixed positions.

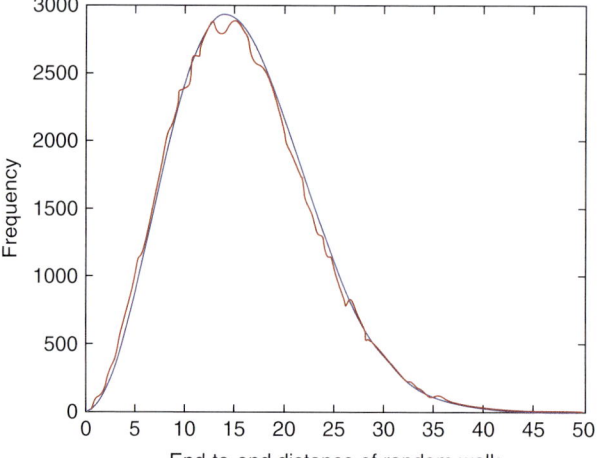

Figure 1.25 Probability distribution for random walks along contacting spheres in a random packing. The dithering curve is from experimental results, the smooth curve corresponds to theoretical Gaussian distribution. The agreement between the two curves is very close.

3) *Isotropy of a point field*; If $R(\theta)$ represents rotational transformation, and if $N = xn$ and $R(\theta)N = R(\theta)xn$ have the same distribution for any angle θ, then the field is said to be isotropic.
4) *Motion invariance*; If a field is isotropic and homogeneous, then it is said to be motion invariant.

In Section 1.3, radial vectors emanating from the centres of all spheres to the contact points on their surfaces were defined. The construction, when carried out for all spheres in a packing, results in a three-dimensional network spanning the whole body. The sum of all vectors in all spheres must tend to zero as the size of the body increases to infinity:

$$\sum_{i=1}^{N} \vec{R} \to 0, \quad \text{as} \quad N \to \infty, \quad \text{summation over all vectors} \quad (1.62)$$

proof: $\sum \vec{R} = \sum \langle R\cos\Theta \rangle = R \sum \langle \cos\Theta \rangle = 0$. For a finite system (Equation 1.62) becomes $\sum \vec{R} \to 0$ as $N \to \infty$.

A large simulation cell with more than 2×10^6 spheres, comprising approximately 60 coordination shells, was used for random walk experiments. Random walks of 300 steps were carried out from one touching sphere to another touching sphere. No limitations imposed by the cell boundary were encountered. All random walks were started at random from the innermost 10^4 spheres. To obtain reasonable statistical distribution, 10^5 such random walks were executed, with the result shown in Figure 1.25. The wavering curve represents the experimentally determined distribution; the smooth curve is drawn in accordance with Gaussian statistics of random walk (Equation 1.39). The agreement is remarkably good, confirming that the size of the cell is sufficient to carry out undisturbed random walks along the spheres. However, the random walk element, although necessary to identify this type of amorphous solid, is not a measure sensitive enough to detect the presence of packing imperfections, and therefore not sensitive enough to distinguish ideal amorphous solids from non-ideal amorphous solids.

Random walks and Markov chains

A Markov chain is a stochastic process with the Markov property on a finite or countable state space. The term *Markov chain* refers to the sequence (or chain) of states such a process moves through. Usually, a Markov chain is defined for a discrete set of times (i.e. a discrete-time Markov chain).

Often, random walks are assumed to be Markov chains. A discrete-time random process involves a system which is in a certain state at each step, with the state changing randomly between steps. The steps are often thought of as moments in time, but they can refer to physical distance or any other discrete measurement; formally, the steps are the integers or natural numbers, and the random process is a mapping of these two states. The Markov property states that the conditional probability distribution for the system at the next step (and in fact at all future steps) depends only on the current state of the system and not additionally on the state of the system at previous steps.

1.4.7
Neighbours by Voronoi Tessellation

Now, we can increase the size of the primitive cluster to include additional layers of spheres around it, as shown schematically by the circles for two dimensions in Figure 1.26, and consider neighbours of the A sphere.

We assign spheres into discrete neighbouring positions by the Voronoi tessellation method when they do not touch the inner central sphere. This method is used to allocate spheres into the so-called *nearest neighbours* (touching and not touching the inner sphere) and to separate them from further positioned,

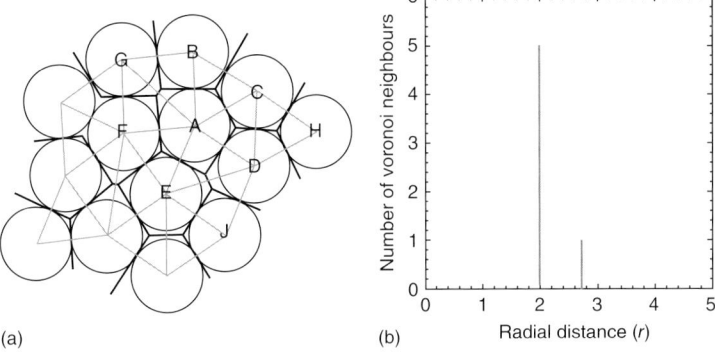

(a)   (b)

Figure 1.26 (a) A cluster of circles and construction of Voronoi polygons around circles in two dimensions. (b) Histogram of Voronoi nearest neighbours around circle A.

more distant neighbours. In this method, a Voronoi polyhedron is constructed for each sphere according to a unique mathematical procedure (Voronoi, 1908; Brostow et al., 1978; Tsumuraya et al., 1993). Then, by definition, two spheres are Voronoi neighbours if they share a common face of their respective polyhedra.

Formally, the Dirichlet–Voronoi polyhedron is constructed around each sphere by space partition, as defined in the following equation:

$$V_V(x_{ij}) = \{x : \|x, x_i\| \leq \|x, x_j\| \text{ for all } x_j \in X_{\text{IAS}}\} \tag{1.63}$$

For the present purposes, we give an operational definition, and for perspicuity, we limit it to two dimensions. Consider a set of points, each representing centres of circles (Figure 1.26). First, we draw links between neighbouring points. Then we take each link and produce a line perpendicular to it and passing through the point equidistant from terminal points. Such bisectors produce polygons around the points. For each point, the smallest polygon so constructed is the Voronoi polygon. It delineates the space that belongs to that particle. Extension to three dimensions is straightforward, in which polygons become polyhedra. The number of faces of a Voronoi polyhedron is equal to the number of all h-neighbours.

One can always, in an unambiguous way, divide up a structure, made of points, into (usually irregular) tetrahedra. This is done using first the Voronoi (or Dirichlet) decomposition of space into individual cells which contain the regions of space, closer to a given point than to any other one. In generic cases, the Voronoi cells have three faces sharing a vertex of the cell. Then, connecting the original points of the set whenever their associated Voronoi cells share a face, defines a unique decomposition of the space into tetrahedra. This simplicial decomposition is equivalent, in three dimensions, to a point set triangulation in two dimensions. This procedure also provides the best way to define the coordination number in dense structure: it is equal to the number of faces of the Voronoi cell. In a topological sense, the Voronoi cell and the coordination polyhedra are dual. In a tetrahedral division of space, the set of vertices closest to a given site forms its first coordination shell, which is a triangulated polyhedron (a deltahedron).

According to Figure 1.26a, circles B to F are the Voronoi coordination shell neighbours to sphere A (by touching) and circle G is also a neighbour, although not touching. Spheres H and J are not Voronoi neighbours; they do not share a common face with A. The criterion for the first Voronoi coordination neighbour is that the distance between centres is within the range: $D \leq h \leq D\sqrt{2}$. In three dimensions, the criterion is $D \leq h \leq D\sqrt{8/3} = 1.633\,D$ (Lee et al., 2010). Faces tangential to the sphere/circle separate D-neighbours and contain respective contact points. Therefore, the Voronoi tessellation method allows a greater number of spheres to be nearest neighbours than just the touching neighbours – significantly so for three-dimensional clusters. This is clearly evident in Figure 1.27b which shows the distributions for both touching and Voronoi nearest neighbours obtained from a random packing of spheres of the same size.

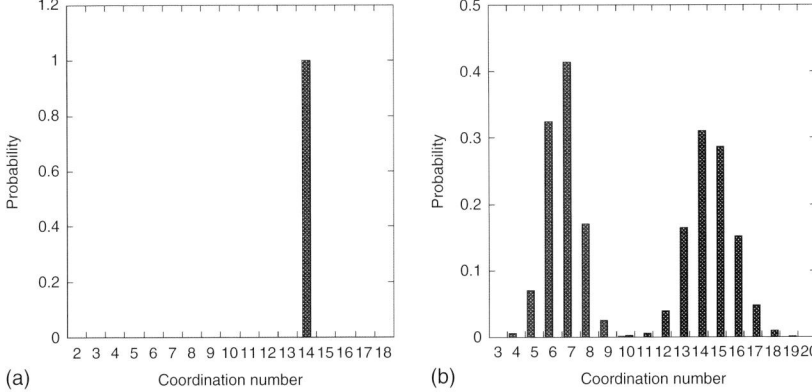

Figure 1.27 Histograms of neighbours according to Voronoi method: (a) ordered (bcc) cluster and (b) statistics for random clusters with touching neighbours (on left) and Voronoi neighbours (on right).

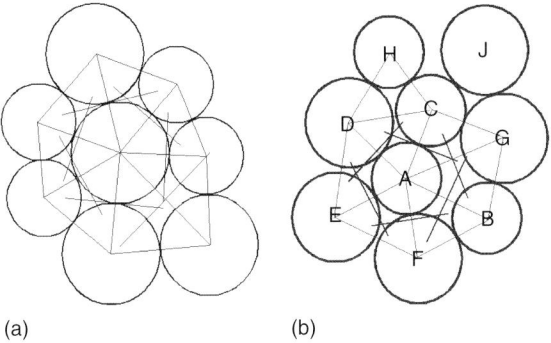

Figure 1.28 Voronoi tessellation applied to a cluster with two sizes of spheres. Notice that the bisectors between large and small spheres fall within the larger sphere: (a) large inner spheres and (b) small inner sphere.

For disordered, irregular and random packing, the Voronoi polyhedra are irregular and vary in size and shape. For regular clusters and ordered packings, particularly on Bravais lattices, each Voronoi polyhedron is the same size and shape for each and every sphere (of the same size). Thus, the coordination number distributions, as seen in Figure 1.27b for the random packing, degenerate into a single valued function for ordered clusters. For example, for a bcc structure every primary cluster has 8 touching neighbours plus 6 Voronoi neighbours, giving all together 14 Voronoi nearest neighbours, as seen in Figure 1.27a. Crystallographers use the term *Wigner-Seitz cells* for the Voronoi construction (Wigner and Seitz, 1933).

Voronoi tessellation does not work straightforwardly for spheres of different sizes. This is clear from the diagrams shown in Figure 1.28, where it can be observed that the bisector lines do not go through the contact points, but cut into

Figure 1.29 Hryhoryi (Georgyi) Voronoi, Ukrainian mathematician (1868–1908) (from Wikipedia).

the larger sphere. The problem does not present itself in cases where packing simulation is carried out by molecular dynamics. In such cases, atoms are force centres and not spheres with clearly defined radii; and therefore, unwittingly, the problem is neglected. Consequently, what are nearest neighbours by this method for multi-sized spheres has to be redefined and weighted Voronoi tessellation applied.

Hryhoryi Voronoi, Ukrainian mathematician whose portrait is shown in Figure 1.29 (1868–1908). A conference dedicated to Voronoi diagrams is held every two years and published in the International Journal of Voronoi Diagram Research Center, Hanyang University, Seoul, Korea.

The mathematics of the polyhedra in question begins with the work of Johann Peter Dirichlet (Dirichlet, 1850), followed by the work of Gregoryi (pronounced Hrehoryi) Voronoi, an Ukrainian mathematician, who was a student of A. Markov (of Markov chains) in Kiev, later working in Warsaw in the beginning of the twentieth century (Voronoi, 1908, 1909). This is why the polyhedra are most often named after him.

The application of Voronoi tessellation to describe the amorphous structures can be traced to the early work by Bernal and his coworkers on the structure of liquids. One of the earliest publications of this type appeared in 1970 by Finney (1970a), in which early computations of Voronoi polyhedra were carried out on random close packed structures, using main-frame computers. The relationship between the distribution of Voronoi volumes and the radial distribution function was established for these simple monoatomic structures in a subsequent publication (Finney, 1970b).

Brostow *et al.* developed the first exact computational method of Voronoi tessellation (Brostow *et al.*, 1978). Definitions of indirect, degenerate and quasi-direct neighbours of a given particle (point) have been introduced at the same time. The method relies on constructing first the so-called direct polyhedra, relatively simple, with a considerable saving of time. Shortly after, Tanemura *et al.* (1983) published an efficient algorithm for computing the Voronoi tessellation, and these two papers must be considered to be the seminal works on this subject, including the definition of amorphous structure in solid materials and in liquids (Medvedev *et al.*, 1990).

The earliest Voronoi diagrams for polymers have been published by Theodorou and Suter (1985). The tessellation was computed for static, equilibrated structures, with the results describing mainly the Fürth distribution. Pfister and Stachurski

(2002) used the concept of Voronoi polyhedron around a polymer chain to postulate a deformation mechanism in amorphous polymers. For inorganic glasses, Voronoi tessellation approach was used by Tsumuraya and coworkers (1993). In addition, Watanabe and Tsumuraya (1987) studied liquid sodium by molecular dynamics and applied the Voronoi approach to distinguish glass formation from crystallization and in the distinction of amorphous solids and liquids.

The Voronoi diagram has also been used to define the fundamental difference between an amorphous solid and a liquid in terms of percolations of two kinds of structures (Medvedev et al., 1990). Atomic level strain was connected to the Voronoi diagram (Mott et al., 1992) and plasticity (Stachurski and Brostow, 2001).

It is important to stress that Voronoi polygons are an important concept in other branches of science, in particular that of surface visualization, as evidenced by two examples of publications by De Floriani et al. (1985), and Hoff et al., (1999). The selected publications quoted here contain numerous references to other publications related to this topic.

Many computational packages have Voronoi routine built-in. For example, Materials Studio©, MatLab©, Mathematica©, Maple© and others.

1.4.8
Neighbours by Coordination Shell

A more general method for assigning spheres into positions relative to a central sphere is based on the radial distance from the centre. Given a central sphere (or a point), the number of spheres enclosed within each concentric shell, $r_{shell} \leq r \leq r_{shell} + dr$, represents the neighbours in that shell, as illustrated in Figure 1.30. In other words, the number of sphere centres in a shell of thickness, dr, at a distance, r_{shell}, from the origin is given by:

$$n(r) = \frac{N}{V} \int_{r_{shell}}^{r_{shell}+dr} g(r) 4\pi r^2 dr \tag{1.64}$$

where N represents the total number of spheres in the volume V extended over the whole body, and therefore, N/V is the numerical density of the body. The factor $4\pi r^2$ represents the surface area of the shell, and $g(r)$ is the *radial particle density distribution function*.

The radial density distribution function, frequently called the pair distribution function (PDF), describes the distribution of distances between pairs of particles contained within a given volume. Mathematically, if a and b are two particles in a fluid, the PDF of b with respect to a, denoted by $g_{ab}(\vec{r})$, is the probability of finding the particle b at the distance $\vec{r}$ from a, with a taken as the origin of coordinates.

In statistical mechanics, the radial distribution function (or pair correlation function), $g(r)$ in a system of particles (atoms, molecules, colloids, etc.), describes how density varies as a function of distance from a reference particle. In simplest terms, it is a measure of the probability of finding a particle at a distance of r away from a given reference particle, relative to that for an ideal gas.

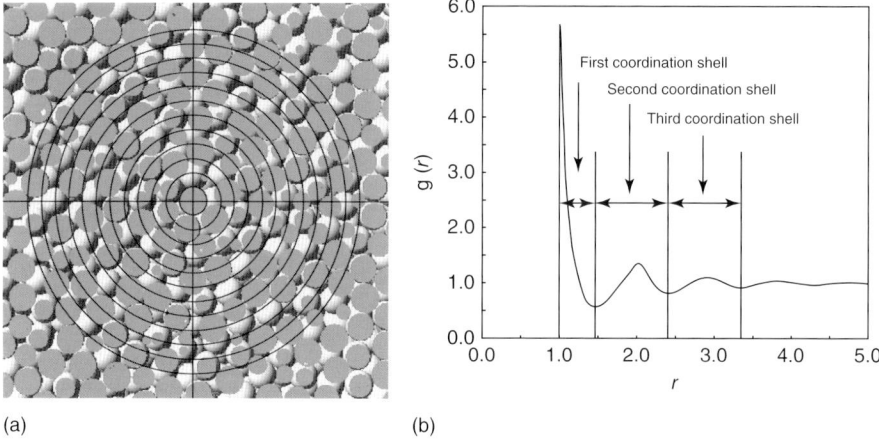

Figure 1.30 (a) Cross section through randomly packed spheres of the same size with concentric shells drawn in. (b) Radial distribution function for the packing on the left plotted as a function of inter-sphere separation. The first three coordination shells are delineated by the vertical blue lines.

In computer simulations, the counting of individual spheres in a packing of spheres is made by means of the following formula, which is valid for any value of r:

$$g(r) = \frac{1}{\rho} \frac{n(\Delta r)}{4\pi r^2 \Delta r} \tag{1.65}$$

So derived $g(r)$ is a stepwise discrete function, although it is usually drawn as a smooth curve. It has a value of zero up to a distance $r_{min} = 2R$ at which it rises sharply to a maximum value. For hard sphere packings, the initial slope at r_{min} is infinity.

From its maximum value, the function decreases and oscillates, showing a number of peaks that diminish in height with increasing r. In the limit of $r \to \infty$, $g(r) \to 1$ because $n(r)/(4\pi r^2 \Delta r) \to \rho$ (Figure 1.30).

By this method, there will be first shell neighbours at distances corresponding to the first peak, second shell neighbours corresponding to the second peak, third shell neighbours and so on, as shown in Figure 1.30. Accordingly, the first coordination number is given by

$$n_1 = 4\pi\rho \int_{r0}^{r1} r^2 g(r)\, dr \tag{1.66}$$

The second coordination number is defined similarly:

$$n_2 = 4\pi\rho \int_{r1}^{r2} r^2 g(r)\, dr \tag{1.67}$$

We note that the smaller Δr in Equation 1.65 the finer the resolution of the PDF. This is clearly evident when comparing the PDFs in Figure 1.31a and b.

The initial slope of PDF ought to be vertical as in Figure 1.30. In Figure 1.31a, the slope is less than $+\infty$ because of the existence of the smaller spheres (atoms)

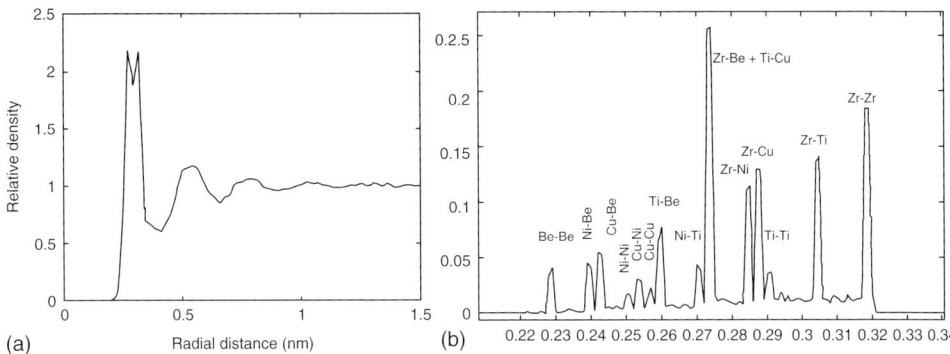

Figure 1.31 PDF for Zr-Ti-Cu-Ni-Be metallic glass computed using Equation 1.65. (a) With $\Delta r = 0.23$ nm and (b) with $\Delta r = 0.001$ nm. Note the change of horizontal scale.

at separation distances between 0.23 and 0.27 nm as seen in Figure 1.31b, whereas the double peak corresponds to the presence of Zr–Be and Ti–Cu pairs that happen to have the same separation. Indeed, Figure 1.31b shows the component peaks of the double peak in Figure 1.31a. All pairs under the first peak belong to the first coordination shell in accordance with the definition in Figure 1.30. All pairs under the second peak (between 0.33 and 0.70 nm) belong to the second coordination shell and so no.

1.4.8.1 Pair Distribution Function

The PDF describes the distribution of distances between pairs of particles. In relation to Figure 1.32, $g_{AB}(\vec{r})$ gives the probability of finding the particle B at a distance $\vec{r}$ from particle A which is chosen to be at the origin. If the system of particles is homogeneous and isotropic, then there is an equal probability of finding particle B at any position $\vec{r}$, and the PDF becomes centre-symmetric, so that $p(\vec{r}) = 1/V$.

However, the probability of finding *pairs of particles* at given positions is not uniform. Then, the PDF is obtained by the two-body probability density function:

$$g(\vec{r},\vec{r}) = p(\vec{r},\vec{r}) \, V^2 \, \frac{N-1}{N} \approx p(\vec{r},\vec{r}) \, V^2 \qquad (1.68)$$

For any particulate system, with N large enough to allow for a meaningful statistical analysis, $N - 1 \sim N$.

The simplest possible PDF for non-dense randomly packed spheres is

$$g(r) = \begin{cases} 0, & \text{if } r < d, \\ 1, & \text{if } r \geq d. \end{cases} \qquad (1.69)$$

For dense packing, pairs of spheres are separated by $r = ah$, where a is any real number starting from 1. Then the PDF can be represented by a set of delta functions:

$$g(r) = \sum_i \delta\,(r - ia) \qquad (1.70)$$

(a)   (b)

Figure 1.32 Probability of contacts on mixing spheres of different types. (a) Spheres separated into same type regions. (b) Spheres mixed at random.

The radial PDF is defined to be independent of orientation. In statistical mechanics, it is given by the expression:

$$g_{AB}(r) = \frac{1}{N_A N_B} \sum_{i=1}^{N_A} \sum_{i=1}^{N_B} \langle \delta(|\mathbf{r}_{ij}| - r) \rangle \tag{1.71}$$

1.4.8.2 The Probability of Contacts

Probability of contacts between atoms is frequently evaluated in condensed matter physics. For instance, to a first approximation the internal energy, U_{int}, of a thermodynamic system comprising two types of atoms is predicted to be:

$$U_{int} = N/2 \left[(c_A u_{AA} + c_B u_{BB}) - c_A c_B (u_{AA} + u_{BB} - 2u_{AB}) \right] \tag{1.72}$$

where $c_A = N_A/N$ is the concentration of A-type atoms with specific interaction energy of u_{AA}, and c_B is the concentration of B-type atoms with specific interaction energy u_{BB}, respectively. The last term in Equation 1.72 expresses the enthalpy of mixing, which turns out to be a quadratic function of the probability of contacts between atoms A and B; this probability is expressed by the product of concentrations, $c_i c_j = c_A c_B$, and it is given by the statistical formula:

$$P_{ij} = \delta \cdot c_i \cdot c_j \tag{1.73}$$

where i and j stand for A and B, respectively, and δ has a value of 2 when $i = j$, and a value of 1 when $i \neq j$. The expression is verified for variety of sphere sizes, except for situations when one sphere is so small as to fit into an interstitial position between the large spheres.

1.4.8.3 Contact Configuration Function

Edwards and coworkers (2002) postulate a statistical mechanics approach to jammed (Torquato et al., 2000) configurations. In particular, they develop the concept of entropy of micro-canonical ensemble of jammed configurations expressed as follows:

$$S = \log \int \delta(V - W(\varsigma)) \Theta(\varsigma) \, d\varsigma \tag{1.74}$$

where δ is the Heaviside function, V the volume of the body of N spheres (analogous to the internal energy of a system), ς the collective coordinates of contact point positions in the static packing and $W(\varsigma)$ a volume function expressed in terms of contact variables (analogous to the Hamiltonian of a system in statistical mechanics). According to Edwards, under the condition, $V = W$, the system is said to be 'jammed', and the function $\Theta(\varsigma)$ ensures that all spheres are in fixed (locked) positions.

It is conjectured that it may not be possible to obtain an IAS (ideal amorphous solid) by any experimental method, just as it is impossible to grow an ideal, completely defect-free crystal. Certainly, all experimental attempts at random packings of spheres suffer at least three limitations that prevent even a close approximation to IAS. First, the force of gravity: its effect during the dynamic stage of pouring causes some clusters of spheres to arrange into locally close packed configurations (fcc and hcp) or to rearrange into these at a later stage by tapping or vibration. Second, friction: it limits the freedom to form random configurations and it is likely to be responsible for loose spheres in the structure. Third, inhomogeneity: this results from the directionality of pouring of the spheres into a container and also from edge effects because of the presence of the container walls. All three limitations can be avoided by computer simulation of round cells, as described herein.

1.5 Ideal Amorphous Solid (IAS)

We call a body consisting of randomly packed spheres *an* ideal amorphous solid (IAS) if it is constructed by following precisely the packing rules of Section 1.4 and therefore contains absolutely no imperfections of packing of any kind. The IAS is defined with the same purpose as one defines an *ideal crystal* in the theory of crystallography. The term *ideal* denotes perfection of structure, theoretical and hypothetical. It is meant to provide a universal model for representing and understanding the structure of amorphous solids, in a way analogous to crystallography, with the following advantages:

1) It uniquely defines the structure of the amorphous glassy solid by specifying the atomic arrangement, atomic radii, concentrations and chemical elements. In particular:
   a. all spheres are in fixed positions (i.e. perfect solidity), and the positions of the spheres are specified by the expression:

   $$S_{a,b,\ldots m} + X_{IAS} \tag{1.75}$$

   b. coordination distribution function, $\Psi(k)$
   c. configuration distribution function, $\Phi_k(\zeta)$
   d. free volume $= 0$
2) The IAS represents the base-line model for random atomic arrangement in amorphous solids and can be used to predict:
   a. the *ideal* radial distribution function
   b. the *ideal* coordination distribution

c. the *ideal* configuration distribution
d. the *ideal* structure factor
e. the *ideal* X-ray scattering patterns
f. the *ideal* physical properties
g. and so on.

The centres of any three adjacent but non-touching spheres form triangles of unequal sides, due to (i) random positions and/or (ii) different sphere radii. There is not a single incidence of four adjacent spheres that are co-planar, in direct contrast to any of the crystallographic Bravais lattices where the multiple centres of spheres are co-planar. Consequently, there is no translational symmetry in this structure.

The condition of solidity and rigidity is implicit in the concept of an ideal amorphous solid, and the same is true for an ideal crystalline solid. The rigidity of every sphere is achieved with the conditions stipulated in Section 1.3. The geometrical construction of IAS lends itself readily to description and analysis by Voronoi tessellation and associated Delaunay simplexes.

The coordination numbers, based on touching neighbours, will be range bound to $k_{min} \leq k \leq k_{max}$, where $k_{min} = 4$, and the value of k_{max} will depend on the composition of the IAS and the specific cluster selected. In arrangements composed of different sized spheres, one can use the values for k_{max} listed as a function of the ratio r_0/r_i, relying on calculations from Clare and Kepert (1986).

The statistical functions, $P_k(\zeta)$ and $\Psi(k)$ defined in Section 1.3, characterize the randomness of the packing arrangement. These functions provide a novel way to describe the structure of randomly packed spheres. Additional structural measures are described in Chapter 2, as well as by:

- X-ray scattering (described in the next chapter)
- Pair distribution function (described earlier in Section (2.1))

X-ray scattering is a direct corroboration for the conjecture that atomic arrangement in amorphous solids is random. The conjecture is based on the fact that the scattering property is a Huygens-style sum of amplitudes from all points of scattering and Fourier transforms. For instance, the local density at point, **r**, is related to the average density multiplied by the radial distribution function, $\bar{\rho} \times g(r)$; the radial distribution function can be derived from the X-ray scattering measurements, and the scattered intensity is computed from the Debye equation. All of these relations are well established (Torquato, 2002).

That no short-range order (SRO) exists in the IAS random packing of spheres can be reasoned as follows. As all spheres are identical, variation in packing arrangement is the only source of differentiation. Then each cluster can be described by (at least) three statistics: (1) the number of contact points, k (2) the configuration of the nearest neighbours of the inner sphere, described by ζ, (3) the orientation of the cluster in space, described by some measure, Ω.

For two clusters (say, 1 and 2) to be identical, it is necessary that $k_1 = k_2$, $\zeta_1 = \zeta_2$, and $\Omega_1 = \Omega_2$.

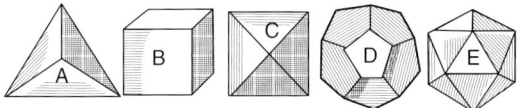

Figure 1.33 The five ideal Greek solids.

Each cluster has nine possible values for $k = 4,...,12$. If the distribution for k is equiprobable, then the probability that two clusters have the same number of contact points (the same value of k) is simply 1/9. For any given distribution $\Psi(k)$, as shown in Figure 1.16, we can write more generally that the probability for two clusters to have the same value of k is given by $\sum_{k=4}^{12} [\Psi_k]^2 = C/9$, where $1 < C < 9$ is some known constant if Ψ_k is known.

For the other two variables, we note that two independent copies of random variables with continuous distributions agree with probability 0. Therefore, the probability that two or more neighbouring clusters have identical coordination and configuration and are oriented in precisely the same way, is zero. Repeatability of an ordered or regular pattern in adjacent clusters extending along any line is zero. Consequently, it can be stated with generality that the IAS body has no short-, medium- or long-range translational and/or orientational order.

Such an ideal amorphous solid has no free volume, $v_f = 0$. From this condition, it can be deduced that such an ideal amorphous solid would possess Frenkel and Orowan theoretical strength and that its coefficient of diffusivity would be equal to zero, $D_{IAS} = 0$. These properties are also true of a perfect single crystal as defined by crystallography.

The IAS is an isotropic body when considered at a sufficiently large size. For real amorphous materials, this size is of the order of 10–100 nm (see the definition of representative volume element in Chapter 2).

The ancient Greek philosophers also identified and documented the so-called ideal solids, known as *Platonic solids*, shown in Figure 1.33. They are *regular* polyhedra, also called *ideal* because each, without exception, is made up of identical faces; the tetrahedron has four identical equal-sided triangles, the cube has six identical squares faces, and so on. In other words, a regular polyhedron has all faces as regular polygons, identical in both shape and size. These solids have resulted from a meticulous study of the properties of geometrical shapes. However, it may be imagined that these ideal shapes have also given ideas as to the appearance of mineral single crystals. In addition, they were used for the size of heavenly spheres of the universe (see books on Copernicus or Kepler).

1.6
Construction of an Ideal Amorphous Solid Class I

The construction of the IAS model is based on the hard sphere packing principles described in this chapter. It can be carried out using the IAS software, which is available as freeware and can be downloaded from the web site: http://users.cecs.anu.edu.au/~u9300839/index.php

The IAS software is written in MatLab, and it remains the copyright of the authors as indicated in the User Instructions included with a complete download. The download software includes all the packages necessary to run the software. The first part of the program creates a virtual round cell of randomly packed spheres according to the IAS rules described in this book. The second part analyses the packing and provides results for

- packing fraction
- Voronoi packing fraction
- pair contact probability
- radial distribution (PDF) by centre statistics
- radial distribution (PDF) by centre statistics (fine graded)
- coordination distribution statistics by contact and by Voronoi method
- random walk vector magnitude
- Voronoi volume statistics
- equally divided points on central sphere
- IAS visualization
- X-ray scattering

The starting point of construction of an IAS model is the creation of an irregular primary cluster, as shown in Figure 1.35, by any of the two methods described in this chapter. This function is in-built into the IAS software. Entering a call for the IAS-GUI opens up the following dialog as shown in Figure 1.34.

To enter data in IAS graphical user interface (GUI) click in the field [At. symbol], then a new window with a table of chemical elements will open. In the table of

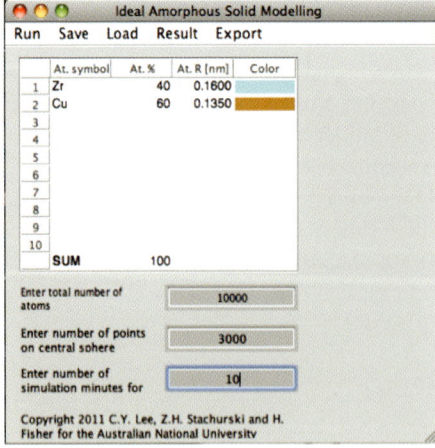

Figure 1.34 (a) The IAS-GUI dialog at the start of the program.

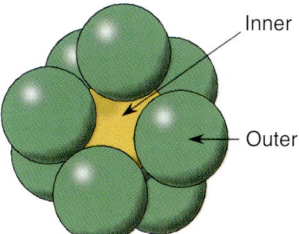

Figure 1.35 Distinction between inner and outer spheres in a primary cluster.

elements, click on the desired element. The element symbol, its atomic radius, and its colour will appear automatically in the IAS-GUI.

Repeat the above-mentioned actions for additional elements, as required.

Finally, assign atomic percentage(s), in the field(s) – [At. %], for each element such that the SUM of [At. %] adds up to 100.

Note: The radius of each element can be changed by clicking on the corresponding field and entering the data manually. The colour of each element can be changed by clicking on the corresponding field and selecting it from the colour chart.

In computational terms, division of the sphere's surface into equally spaced points results in a list, which may be denoted as $\mathbf{L}[\mathbf{x}_n, p]$, where p are the point sequential numbers and $\mathbf{x}_n$ the points' corresponding coordinates, x_p, y_p, z_p. For all $1 \leq p \leq n, x_p^2 + y_p^2 + z_p^2 = R^2$, where R is the radius of the inner sphere. The number of chosen points depends on the required outcome. The larger the number, the greater the division and the finer the random positioning of the outer spheres. The smaller the number, the lower the memory requirements and faster execution of the program. It has been found by experience that $n = 3000$ points is a minimum workable number for this purpose.

Once the list has been compiled, the program choses at random a point from the list. A random number, $\mathcal{R}$, is generated in the range $0 \leq \mathcal{R} \leq 1$, which is multiplied by n, so that a point, $p = \text{Integer}[n \cdot \mathcal{R}]$, on the list is chosen. As k points will be finally selected, the first can have id number, $k = 1$, assigned to it.

For each subsequent point ($k > 1$), a check must be made to ensure no overlap of spheres, see Figure 1.36. The geometric condition is for an angular separation to avoid overlap onto each other's excluded space. This condition is correct for the case of monoatomic size, that is $|\mathbf{R}_\ell| = |\mathbf{R}_k|$ for all $1 \leq \ell \leq k$. If the condition is satisfied, then the $k = \ell$ point is selected and its coordinates stored in memory. This loop process is continued until no further points can be added to the inner sphere. The end result is a list of k points with their respective coordinates, all satisfying the non-overlap condition.

For the case of spheres of different radii, the lower limit, $\pi/3$, will be smaller or larger, depending on the actual sphere sizes involved, which is resolved geometrically.

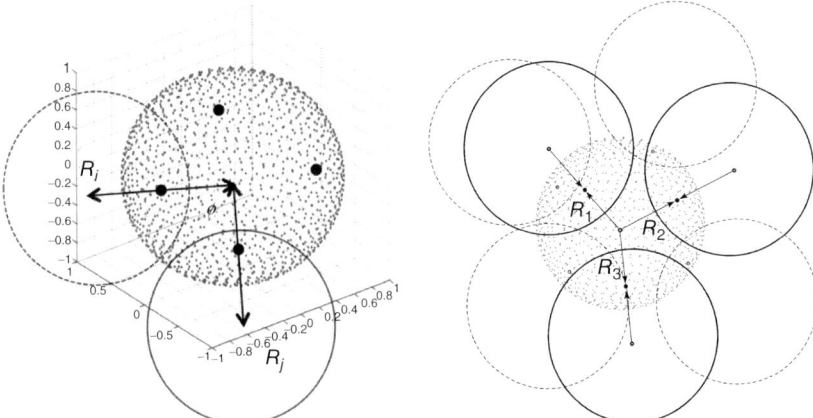

Figure 1.36 Selected points are tested for overlap.

The end of this process is reached when the blocking function, $\hat{b} = 1$. This means that none of the remaining free space around the inner sphere is of sufficiently large extent to allow the addition of another sphere onto it. In principle, this is a straightforward and easily understood condition. In practice, it is not so simple because there is no analytical function describing the blocking function, so it cannot be coded analytically. Instead, an algorithm is made to delete p-points from the original **L**-list until none are left; then the program is halted. The deletion of the p-points occurs for every successful and every unsuccessful point selection. In the former case, all points lying within the spherical cup around the selected point and extending over the conical angle, $\phi = 2\pi/3$ are deleted. In the latter case, only the selected point is deleted. As n is finite, the execution of the program is finite and relatively quick.

To complete the construction of the primary cluster, a sphere is placed in contact with the inner sphere at the chosen k points. Consequently, the distance between centres of the inner and outer spheres is $2R$ for a monoatomic model or $R_m + R_{m+1}$ for a model involving m atomic sizes.

With the primary cluster completed, the following important observations can be made.

- The centres of any two outer spheres of the same size make an isosceles triangle with the centre of the inner sphere.
- For a multi-size case, the maximum number of spheres that can be made to touch the inner sphere depends on the specific case. However, it can be stated categorically that $\hat{k}$ for that model is always less that k_{max}.
- For any three adjacent outer spheres in a given shell, the distances between their centres are unequal. This means that the centres of any three adjacent spheres form *irregular triangles*. Consequently, the seed cluster is an irregular cluster (Figure 1.37). This is a fundamental requirement for the construction of an ideal amorphous solid model.

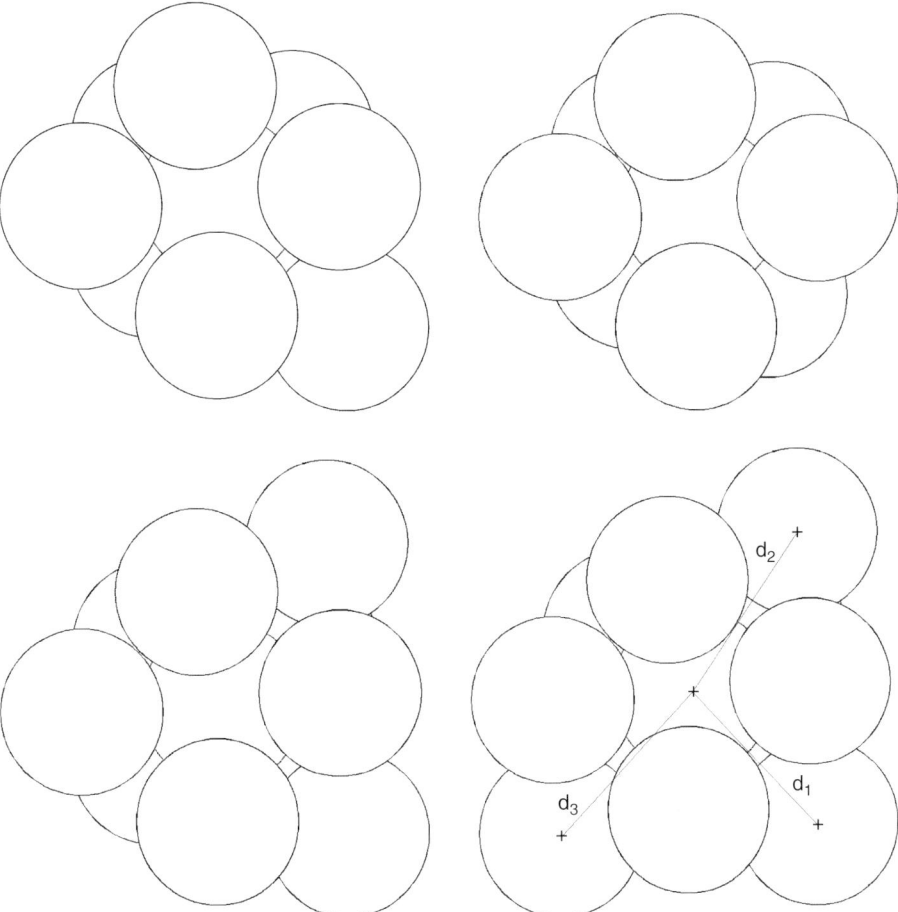

Figure 1.37 Second layer growth of an irregular primary cluster with initial $k = 8$. Three additions are shown with $d_1 \neq d_2 \neq d_3$.

Growth of the primary cluster is achieved by addition of spheres to three-sphere sites. As shown in Figure 1.37, an irregular primary cluster with $k = 8$ has the first sphere in second shell added onto three-sphere site with the closest distance to the inner sphere; then a second sphere is added onto a three-adjacent-spheres site; then a third sphere is added ($d_1 < d_2 < d_3$), and so on, until the shell is filled in.

In the appropriate field of the GUI, enter the total number of atoms (to be simulated): any number between 1 and 10^6 is acceptable. Figure 1.38 (below) shows a previously simulated round cell comprising 10^6 atoms in a metallic glass composed of 65% Mg, 25% Cu and 10% Gd atoms.

Next, enter the number of points on the central sphere: typically 3000, as shown in Figure 1.39. This is the number of computed equidistant points on the surface of the first inner sphere. The outer spheres will be added by the program onto

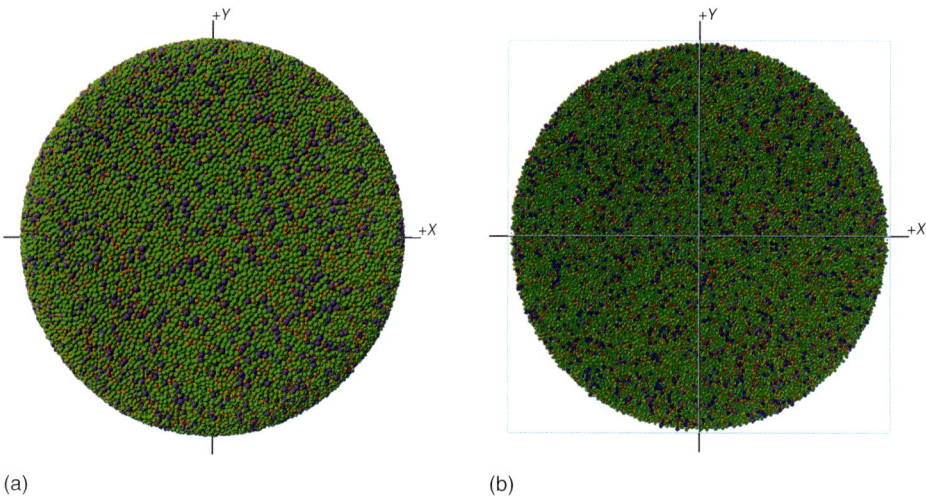

Figure 1.38 (a and b) A view of a round cell and its cross section, comprising 10^6 spheres of three different sizes.

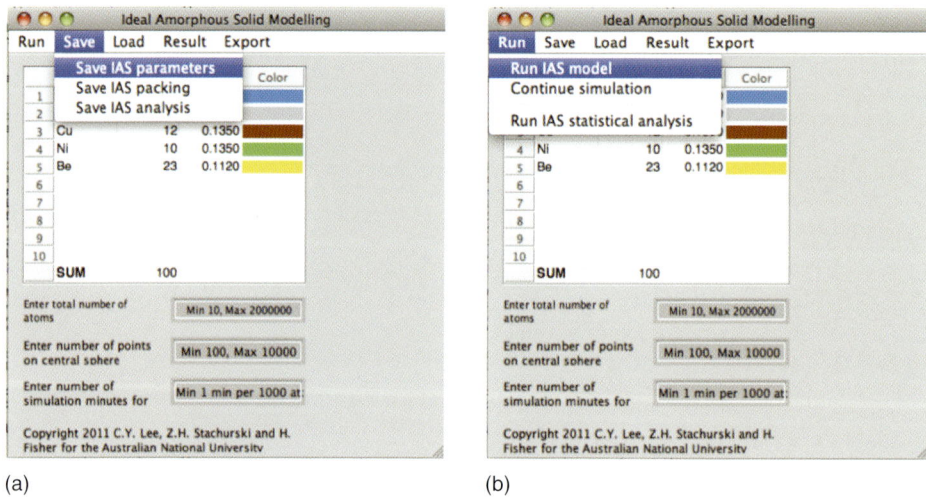

Figure 1.39 (a and b) Two views of the IAS main dialog.

points randomly chosen on the surface of the central sphere as described earlier. Consequently, the outer spheres will form an irregular cluster. The program will accept any number between 1 and 10^5.

Enter the number of simulation minutes. Suggested numbers are 2 for 10^4 spheres, 20 for 10^5 spheres and 200 for 10^6 spheres.

Finally, save the IAS parameters (Figure 1.39). The software directs the file to the 'IAS parameters directory', to be found in the IAS software folder. The outcome of this process is a three-dimensional geometrical pattern of randomly packed

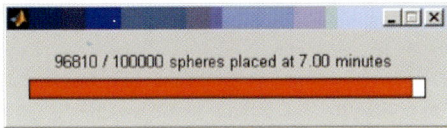

Figure 1.40 A wait bar will open during simulation.

spheres, called a *round cell* if it is of finite dimensions or an IAS if it is of infinite extent. The essential information about the round cell is stored in a matrix, $M[x_N, m]$, where N is the total number of atoms/spheres in the model, the vectors x_N define the positions of the sphere's centres with respect to the origin and m identifies the corresponding type of the sphere, usually by its radius. Example of a round cell and its cross section is shown in Figure 1.38.

Go to the RUN menu and select 'Run IAS model' as shown in Figure 1.39. A wait bar will open and indicate progress of the simulation (Figure 1.40).

1.7 Elementary Theory of Amorphousness

1.7.1 Background

A theory is a collection of logically interrelated ideas on and descriptions of a particular topic. The ideas may include axiomatic statements and subsequent deductions that form a self-consistent framework of knowledge on a given subject. Therefore, a theory involves logical analysis and reasoning from assumed premises; it puts forward a hypothesis to be maintained or disproved. Furthermore, a theory is used to explain data and generate hypotheses that can be tested by research.

A personal knowledge is an *understanding* of the studied topic. It involves *experience* (skills and familiarity) and requires *holding in memory*:

1) an amount of information (i.e. a number of facts)
2) a concept of relationship(s) between the elements (parts) of the information
3) an ability to manipulate the concepts and facts to derive and present reasonable and logical arguments

The level (depth) of understanding, and therefore the level of knowledge, depends on the amounts of (1) and (2) and the ability in (3).

An overview of the field of solid state, and of its corresponding theories, can be summarized as shown in the table below. There are three main divisions of the solid state: crystalline, quasi-crystalline and amorphous. Each area has a corresponding geometrical theory that defines the ideal solids in terms of perfect packing of spheres or (for the case of quasi-crystalline state) other geometrical objects.

	Solid-state fields	
Crystalline	**Quasi-crystalline**	**Amorphous**
Theory of crystallography Ideal crystalline solids	Theory of quasi-crystallography Penrose tiles	Theory of amorphousness Ideal amorphous solids

This representation can be followed by associated theories of solids containing imperfections.

	Solid state fields	
Crystalline	**Quasi-crystalline**	**Amorphous**
Theory of crystal defects	Theory of imperfections	Theory of imperfections

Certainly, theory of crystal defects and theory of crystallography are well developed. Presently, this cannot be said of the other two fields.

There are several methods to test a theory. However, the foremost kind of testing is to verify the predictions of the new theory. If it is verified, then the theory has, for the time being, passed the test; there are no reasons to discard it. The amount of empirical information conveyed by a theory, or its empirical content, increases with its degree of falsifiability. A theory is falsifiable if there exists at least one class of non-empty basic statements which are forbidden by it. According to Karl Popper, the larger that content of the class the more falsifiable is the theory. It also means that the theory says more about the world of experience than a theory with fewer falsifiable statements.

A special random packing that corresponds to the packing in ideal amorphous solid is characterized by properties of and limitations imposed on the set X_{IAS}, summarized from Section 1.4, and shown below:

1.7.2
The Axioms

A1. Choose any two spheres in the IAS packing and draw a straight line passing through the centres of the two spheres.
Then, the probability of any other sphere in the packing having its centre lying on that line is zero.

A2. A straight line of arbitrary direction passing through an IAS will be divided by the atoms/spheres that it cuts through (or is tangent to) into irregular

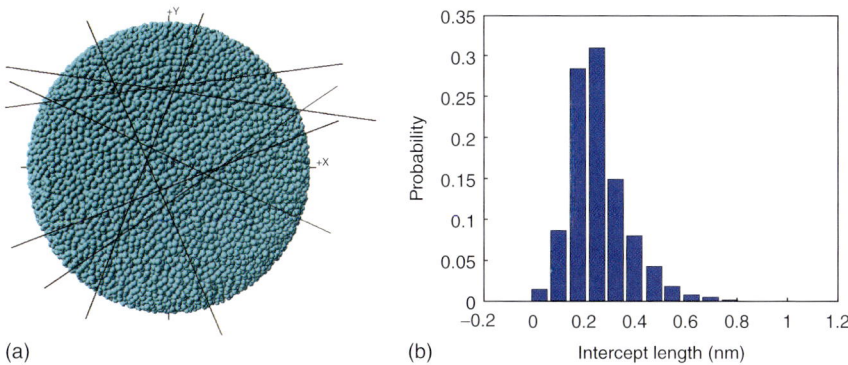

Figure 1.41 Illustration of the axioms A1 and A2: (a) straight lines cutting through the IAS model at random orientations and (b) distribution of irregular line segments in a monoatomic IAS. The graph is a summary statistics of segments from 10^3 lines.

length segments, in contrast to crystalline solids where interatomic distances are always regular (providing the line has a rational slope).

A3. For any IAS, the statistical distribution of lengths of the irregular line segments will be the same regardless of the direction of the line as a consequence of the solid being homogeneous and isotropic.

Figure 1.41 is an illustration of the concept expressed by the last two axioms earlier. The round cell, shown on the left, contains 10^6 sphere of the same size, and the graph on the right contains the corresponding statistical data.

1.7.3
Conjecture

It is conjectured that there exists an ideal random packing of spheres, constructed in accordance with the above-mentioned rules and possessing the properties and complying with the limitations given in Tables 1.7 and 1.8. Then,

$$X_{IAS} = X_{RCP} \tag{1.76}$$

Table 1.7 Global properties of the set X.

	Property	Object	Requirement
P1	scale invariance	if $S_a + X$ is a packing	$S_{ca} + cX$ is also a packing
P2	homogeneity	let $N = [xn]$ be a point field and $N = [xn + x]$ is a point field	both fields must have the same distribution for all $x \in En$
P3	isotropy	if $N = [xn]$, and $R(\theta)N = R(\theta)[xn]$	have the same distributions then the field is isotropic

Table 1.8 Limitations imposed on the set X.

	Limitation	Object	Requirement
L1	solidity	all spheres in *fixed* positions	touching hard spheres $\|x_{\ell j} - x_\ell\| = a$
L2	coordination, k	for any point $x_\ell \in X$	$4 \leq k \leq 12$ with probability $\Psi(k)$
L3	configuration, ζ	for any cluster centred on $x_\ell \in X$	$0 \leq \zeta \leq \zeta_{max}$ $\Phi_k(\zeta)$

Such an ideal random packing of spheres is a model of atomic arrangements in an ideal amorphous solid, as described in this book.

1.7.4
The Rules

Structures of any kind are purposefully created by following carefully designed rules of construction. The rules for creating a model of an ideal (theoretical) amorphous solid should be considered as analogous to the rules for creating a perfect (ideal) crystalline arrangement. However, the resultant amorphous structure must contain none of the elements of symmetry present in crystalline arrangements, such as mirror, rotation or glide. Therefore, to create an amorphous packing of spheres, the rules must not enforce any ordering beyond the nearest neighbour and therefore can apply only to adjacent spheres. Such basic rules for amorphous packing of spheres can be expressed as follows:

R1. Every sphere must be in *fixed* position so that the resultant structure is solid. A sphere is in fixed position when for $4 \leq k \leq 9$ touching contacts with neighbouring spheres, no more than $(k-1)$ are located on one hemisphere; for $k > 9$, a sphere is always in a fixed position.

R2. Three adjacent non-touching spheres must form an *irregular* triangle, so that no two triangles have the same (identical) shape.

The first rule requires that all spheres in the amorphous packing must have touching contacts and that the number of contacts for any individual sphere in a monoatomic structure is limited to the range of $4 \leq k \leq 12$. This packing of spheres models a perfectly rigid solid as no sphere in the body can be moved in any direction.

The second rule implies that in each cluster there are adjacent but non-touching spheres, with distributed distances between them limited to the range: $2r < D_{ij} << 2\sqrt{3}\, r$. A consequence of this packing arrangement is that no four adjacent spheres (touching or not) lie in a plane. It ensures that none of the symmetry elements, characteristic of crystalline arrangements, are present in the amorphous body.

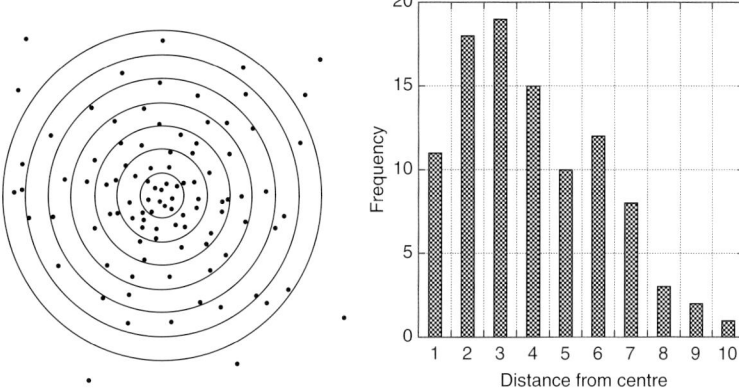

Figure 1.42 Target with randomly distributed shots and a corresponding Gaussian plot.

It transpires that, under the above rules, there are no solutions for random packing of spheres in one or two dimensions. In three dimensions, the IAS proposed by To *et al.* (2006) is a solution for a perfect random packing of spheres obeying the rules and axioms listed earlier.

A final point to be made here is that an ideal amorphous structure is not the same and not equal to a thermodynamic ideal structure. The former is purely a geometrical construct, the latter is particular state of physical matter.

1.7.5
Statistical Correspondence

A geometrical theory of amorphousness can be postulated on the basis of statistical correspondence, two related axioms and two rules, as described in the following. The condition of correspondence in statistical geometry limits the reconstruction of events from their descriptive distributions. For example, given a distribution of shots fired at a target, one can confirm by analysis that it is Gaussian (Figure 1.42), but from the given distribution, it is not possible to reconstruct the precise positions of the shots. A schematic representation of this relationship is shown in the diagram below:

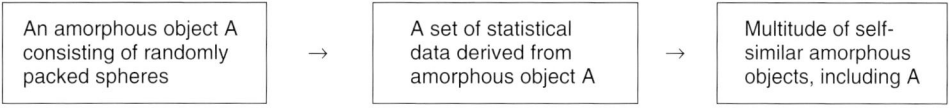

This is in contradistinction to an ordered structure, which can always be reconstructed identically from its unit cell and symmetry elements.

Analysis of the packing of spheres, $S + X$, gives information that is uniquely derived from the packing. However, the reverse relationship is not unique; from the given information for a random packing, one can recreate not one but a multitude of packings (multitude of sets X) of the same characteristics and properties, including the one identical to the original.

1.8
Classes of Ideal Amorphous Solids

At this stage of theoretical development, three separate classes of IAS can be distinguished:

- IAS Class I – represented by random packing of spheres as described in the previous section.
- IAS Class II – represented by random packing of spheres with local SRO.
- IAS Class III – represented by random three-dimensional networks without any local order.

1.8.1
IAS Class I: Random Close Packing of Individual Atoms

The random packing of hard spheres in fixed configurations, as described in the previous sections, represents packing of individual atoms with no preference for, and no bonding to, nearest neighbours. It is proposed that IAS Class I is a good model suited particularly to metallic glasses. It is the most general, least restricted IAS model from which other classes can be derived. Its only correlations are limited to the touching neighbours. Otherwise, this class has no short-range, no medium-range and long-range correlations present.

1.8.2
IAS Class II: Random Close Packing of Linear Model Chains

The IAS Class II has similar characteristics of the IAS Class I, with the addition of SRO between specific nearest neighbours that are bonded. This means that the bonded neighbours do not separate when flowing or in the liquid state and therefore represent linear, chain-like random atomic arrangements found in organic and inorganic glasses.

Consider a simulation cell containing N randomly close packed equal-sized touching spheres, that is IAS Class I (where N is a large number). Convert it to Class II by making a chain of touching spheres by the following process: Chose one sphere as the starting point and link it with one of its touching neighbours. Next, link the second sphere at random with one of its touching neighbour, except the one already joined to. Continue in this manner the three-dimensional SARW, linking spheres until a chain of $(n - 1)$ links is made.

Next, chose another unattached starting sphere, and following the above-mentioned process construct another chain of n spheres. Repeat the process until (N/n) chains are formed. As $N >> n$, there are many ways to form the

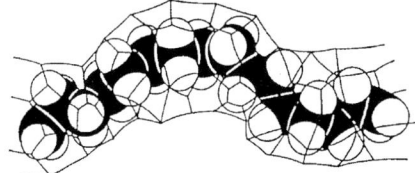

Figure 1.43 Representation of a segment of polyethylene macromolecular chain with C and H atoms to scale and a schematic drawing of Voronoi tessellations.

chains, and therefore, there exists a finite probability that it is possible to form (N/n) chains without violating the SARW or creating cross-links, and without any 'free' sphere(s) left over. This cell of linear model chains (macromolecules) constitutes an ideal polymeric amorphous solid of Class II. Such a simplification of the molecular chains has been used in the science of polymers which can be found in relevant textbooks. This approach has been used for the results shown in Figure 1.25. If the monomers in the chains are represented by spheres, then IAS Class II is indistinguishable from IAS Class I.

However, more accurate representation of macromolecular chains requires modelling of atoms to scale, for example as shown in Figure 1.43 for polyethylene. The H atoms are drawn in towards the C-atom in each monomer due to strong covalent bonding, and therefore representing CH_2 group as one sphere is possible, but usually quite inaccurate.

The degrees of freedom of a chain in IAS-II are defined by free rotation around each link and are constrained by (i) the fixed distance between linked spheres and (ii) the angle between neighbouring links being limited to $60° \leq \Theta \leq 180°$. Therefore, rotations and translations are coupled. In a chain segment in which the first and the last spheres are in fixed positions, we find that

1) in a segment containing three or four spheres, all are in fixed positions imposed by their bonds, regardless of other contact points,
2) in segments longer than four spheres, the middle spheres acquire loose position properties, unless fixed by contacts with other surrounding spheres,
3) for chains with fixed valency angle (typically $\Theta = 109°$), sufficient kinematic freedom appears in segments of seven or more links.

The kinematics of chains is a subject of its own in the field of mechanics. The analysis of linkages requires co-ordinate transformations around a closed chain which must satisfy the following relationship for each and every configuration:

$$[T_{n,1}] \ldots [T_{34}][T_{23}][T_{12}] = [1] \qquad (1.77)$$

where $[1]$ is identity matrix and $[T_{i,j+1}]$ transforms coordinates of a point P in i system to its coordinates in $i+1$ system. Suffice it to say that analysis of a model chain based on polyethylene with seven links can be shown to satisfy the condition for general deformation, providing the end of the chain P is confined to move in limited space.

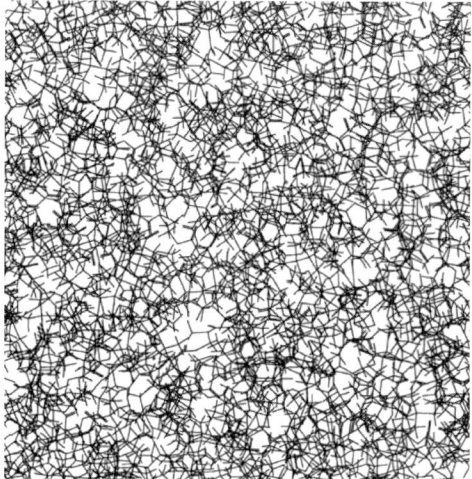

Figure 1.44 Randomness of bonds in a computer-simulated cell of cross-linked polymer.

1.8.3
IAS Class III: Random Close Packing of Three-Dimensionally cross-Linked Chains

Presently, there are no models of ideal packing of chains of this kind based on packing of spheres. The structure of cross-linked polymers is described by other methods; one way of representation of the randomness of the structure is by displaying the skeleton of bonds only, omitting the atoms, as shown in Figure 1.44.

1.9
Imperfections in IAS

Perfection of structure is defined firstly by the definition of imperfections that may occur, and secondly by the strict requirement of the absence of those imperfections, namely no blemishes, flaws or defects of any kind, for which we shall use the collective term *flaws* (in contrast to 'defects' used in crystalline solids). For example, a gas in which the particle velocities conform precisely to the Maxwell distribution is referred to as an *ideal gas*; any deviation from this distribution is an imperfection and its behaviour will deviate from that of the ideal gas law. In the same sense, in an ideal amorphous solid, the arrangement of spheres conform precisely to distributions consistent with the functions $\Psi(\kappa(x_\ell))$ and $\Phi_k(\zeta)$; any deviation is then an 'imperfection'. Two general categories of flaws in the random amorphous structure can be defined: (i) geometrical and (ii) statistical, as described in the following.

The classical science of crystallography describes ideal crystalline solids by emphasizing their perfection, which is defined (by common usage) as the physical state of total absence of imperfections. Then, defects in crystalline solids are defined with regard to the perfection of the ideal crystalline structures. Defects introduce disorder in ideal crystalline solids and alter their behaviour. Many properties of real crystalline solids (both natural and synthetic materials) are governed directly by the presence of defects in their structure. For example,

- Diffusion is governed by the density of vacancies or interstitials (point defects) (Flynn, 1972)
- Screw dislocations in zeolites can confine diffusion to one direction only (line defects) (Walker et al., 2004).
- Plastic deformation is understood mainly in terms of dislocation glide (line defects) (Cottrell, 1965) or grain boundary sliding (surface defects) 1975.
- Semiconductors are created by doping crystals (introducing defects) to modify their electronic structure (point defects) (Turek, 1997).

The significance of these relationships is reinforced when realizing that no diffusion is possible without defects that crystals reach their theoretical strength if no dislocations are present and that semiconductors become insulators if doping is removed.

1.9.1
Geometrical (local) Flaws

We distinguish three kinds of geometrical flaws that can occur in the IAS-I structure:

1) spheres of different sizes (as impurities)
2) loose spheres
3) vacancies/holes
4) compositional inhomogeneities.

In an ideal amorphous solid composed of spheres of the same size, any sphere of a different size must be considered as a geometrical (and structural) flaw, as would be the case in ideal crystalline solids (substitutional or interstitial defect). On a local level, such a different sphere may have its contact number outside the allowable range, depending on the size difference. On a larger scale, the effect of its presence would be evident (if in sufficient concentration) in the radial distribution function or equivalently in the structure factor. Loose spheres possess limited freedom of movement within the cage created by its neighbours. A quantitative measure of the movement of any sphere (labelled, say i) is given by the value of the displacement (Equation 2.43), where u_i, v_i and w_i are components of the allowable movement (in a Cartesian frame of reference) relative to its neighbours. By definition, for a fixed sphere, $J_\ell = |\mathbf{J}_\ell| \equiv 0$; for a loose sphere, $0 < J_\ell < a$ because here we do not allow a vacancy. The definition of IAS requires that $\mathbf{J}_\ell \equiv 0$ for $i = 1, \ldots, N$, that is all spheres are in fixed positions.

A *vacancy* is defined as an empty hole (unoccupied space) with its minimum dimension equal to or greater than the diameter of a sphere ($l_{min} \geq a$). It is considered as a serious flaw having a significant effect on the overall density of IAS. The condition that all spheres are fixed is sufficient to disallow the occurrence of vacancies in the IAS-I. An exception is the special case of clusters of 13 touching spheres with either fcc ($k = 4 + 4 + 4$) or hcp ($k = 3 + 6 + 3$) packing arrangement. In these cases, removal of the central sphere would result in a vacancy with the surrounding spheres remaining in fixed (jammed) positions (Torquato et al., 2000). The occurrence of such clusters is considered to be a statistical flaw in the IAS structure, as described later.

The existence of vacancies in crystalline solids is well proved; the existence of vacancies in real amorphous solids is a matter for debate. It is conjectured that in amorphous solids vacancies occur in the form of free volume distributed throughout the volume rather than concentrated at some specific points as vacancies. This proposition has support from the arguments presented in Section (1.4.5). In the IAS model for packing of spheres, individual clusters vary in their coordination number and packing fraction (or density) even though the free volume = 0. In real solids, interatomic forces are active and such density variations result in stored elastic strain energy, which will provide the thermodynamic driving force for (i) a certain degree of relaxation of density variations and (ii) free volume to become > 0. The supposition is that free volume is trapped whilst cooling from the liquid phase. It is well known that densification of glasses can be achieved by annealing close to the glass transition.

1.9.2
Statistical (global) Flaws

It is possible to specify statistical flaws under two categories:

1) flaws associated with the $\Psi(k)$ function; and
2) flaws associated with the $P_k(\zeta)$ function.

With regard to the first category, a sphere with a contact number $k < 4$ constitutes a flaw in the ideal amorphous structure. It is both a statistical flaw with respect to the $\Psi(\kappa(x_\ell))$ function in that it violates the allowable range of values in the variable and also a geometrical flaw in that it violates the requirement that all spheres be fixed. At the other end of the spectrum, we note that in Euclidean space no sphere can have $k > 12$. Such flaws cannot occur in any packing of spheres of identical diameter (physical impossibility). An important, and more likely, source of statistical flaws, is the departure from the *ideal* distribution of contacts predicted by the $\Psi(\kappa(x_\ell))$ function. In a special packing, when the distribution becomes single valued (e.g. $k = k_{bcc} = 8$ for all atoms, and the half peak width, $\omega_\Psi = 0$), the body will acquire regular structure, which may be perfectly ordered, or disordered, but it is no longer amorphous.

With regard to the second category, any arrangement of contact points on a sphere which corresponds to any of the five regular arrangements (Greek solids) is classified as a flaw because it represents an ordered (crystalline) structure and a departure from the prescribed distribution of contact points. More generally, any arrangement of contact points on a sphere which corresponds to equidistant angular configurations of points or any of the Bravais lattices is classified as a flaw. Specifically, clusters of 13 spheres with hcp or fcc arrangement are considered as serious flaws in IAS-I, in that these increase drastically the average density of the body. It should be noted that an icosahedral arrangement of 13 spheres ($k = 1 + 5 + 5 + 1$) is not a flaw because in this case the touching neighbours can be moved around the inner sphere (whilst remaining in contact with it). Hence, $0 \leq \zeta \leq d$ (where d is a constant), and the width of the distribution, $w_p > 0$. Furthermore, removal of the inner sphere from an icosahedral cluster of 13 touching spheres (symmetrical or not) will make the outer spheres loose. Therefore, it is neither a special case nor a statistical flaw.

1.9.3
The Effect of Flaws on the Density of IAS

The transition from liquid- to solid-like behaviour of randomly packed spheres is considered to occur at a packing fraction close to 0.49 (Reiss and Hammerich, 1986; Kopsias, 1998). This represents the lower limit for "solid" random packing of spheres. At the other end of the spectrum, the hcp or fcc arrangement of spheres represents the upper bound for sphere packing fraction at 0.74. Therefore, the atomic packing fraction p_{IAS} for IAS must lie within the range $0.49 < p_{IAS} < 0.74$. The packing density of the IAS model is in the range 0.63 ± 0.05. There is no precise and unique value for the packing fraction as it depends on the composition, range of atomic radii, the coordination and configuration of the starting cluster and possibly other factors.

In a given experimental random packing of spheres (non-IAS), there must be flaws of the type described earlier. This view is supported by the results for simulated cells in which an increasing number of vacancies and loose spheres was shown to lower overall density. On the other hand, higher density random packing models, approaching a packing fraction of 0.64, show distinct splitting of the second peak in the radial distribution function, a clear indication of the presence of fcc or bcc clusters or their fragments or alternatively a spiral connection of tetrahedra or local icosahedral packing.

Flaws vitiate the structure of the ideal amorphous solid, causing its density to vary from the hypothetical ideal value. Loose spheres and vacancies have the effect of lowering the density, whereas an undue presence of fcc and hcp clusters tends to increase density, both locally and globally. Data from two separate studies in which the properties of simulated cells comprising large numbers of randomly packed spheres were described are sketched in Figure 1.45. First, round cells comprising randomly packed spheres around a central nucleus were created by computer simulation. An algorithm was devised to add spheres to an existing surface, starting at the nucleus. Round cells of different densities were obtained in one of two ways: (i) specific parameters in the algorithm were adjusted to reduce optimum packing

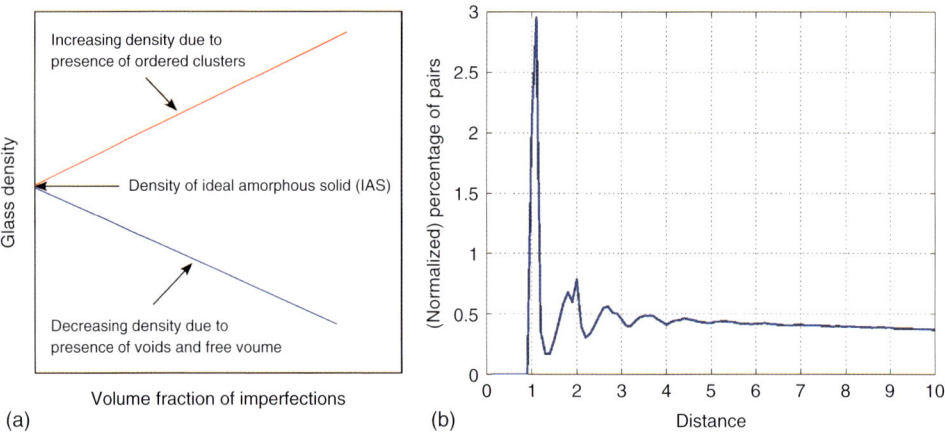

Figure 1.45 (a) The effect of packing imperfections on the density of IAS. (b) PDF showing split second peak due to packing imperfections (order).

and (ii) spheres were removed at random from a given cell of maximum density. As expected, global packing density decreases as the fraction of loose spheres increases. A line fitted to the data points, starting at 0.49 and 80% loose sphere content (solid–liquid transition point), and extrapolated to zero content of loose spheres, points to a value close to 0.63 ± 0.05 for the density of a packing containing none of these flaws. It is proposed that the round cell with the maximum packing density is a close approximation of IAS-I and that the most likely packing density of IAS-I is close to that value, measurably lower than the so-called typical value of 0.64.

In another publication, the so-called cylindrical cells were constructed by pouring a large number of hard spheres into a container whilst holding a stick in the middle of it; the stick was then removed, leaving a packing density of approximately 0.58. Higher densities, up to approximately 0.64, were achieved by gently tapping the sample. The packings were analysed by X-ray tomography and computational rendering of the spheres, so that geometrical properties of the packings could be measured. In particular, the results showed significant fractions of fcc and hcp clusters occurring in the samples. The data in Table 2 from Aste et al., (2004) are also plotted in Figure 1.45, illustrating the relationship between the packing density and the combined fractions of fcc and hcp clusters. In this case, the global packing density of samples increases with the number of high density clusters which are considered as both geometrical and statistical flaws in the IAS-I. This is a remarkable observation in that we are accustomed to thinking of defects as always decreasing the density of crystalline materials, whereas a flaw in an ideal amorphous solid evidently can have either positive or negative effect on density, depending on the nature of the flaw. It is supposed that the two red square points at 5% flaw content, but well below 0.6 packing fraction, must correspond to cylindrical cells that retained the large number of loose spheres and vacancies expected from the method of construction, with a small content of fcc and hcp clusters formed by gentle tapping. It is assumed that in these cells the presence of loose spheres and vacancies is the dominant flaw type causing overall reduction in packing density.

Now, consider the effect of short- and long-range 'order' as factors in density variations. By definition, SRO in solids refers to ordered packing of atoms in small volumes, ranging over one or two atomic distances. Ordered packing in a region (say, convex) means that one or more symmetry elements should be found in the arrangement of atoms confined to within that region. For any random cluster to exhibit SRO in itself, it would have to possess a minimum of one statistical geometry element. Examples of SRO in real materials include

- Guinier–Preston (GP) zones in Al–Cu alloys
- cationic short-range order in crystalline ionic conductor
- i-phase nano-crystals in Zr-based bulk metallic glass (BMG).

The GP zones form in the shape of a disk with translational symmetry of $AlCu_2$ atomic groups along its basal plane and rotational symmetry perpendicular to its base. The second example refers to supercell structure within a Hollandite crystal consisting of TiO_6 and MgO_6 octahedra within a tetragonal unit cell. The third example represents near spherical particles with icosahedral configuration (quasi-crystalline) appearing in the five element metallic alloy with random amorphous atomic packing. Is short-range 'order' possible in the random packing of spheres, and specifically, in an ideal amorphous solid Class I. We can disqualify compositional variation as all spheres are identical, leaving variation in packing arrangement as the only source of differentiation. Each cluster can be described by (at least) three statistics:

- the number of contact points k;
- the configuration of the nearest neighbours of the inner sphere, described by ζ; and
- the orientation of the cluster in space, described by the measure Ω.

We propose that if a monoatomic amorphous solid is an IAS then its structure must conform without exception to two statistical properties, subject to invariance under rotation:

- $\Phi_k(v_1, \ldots, v_k), k = 4, \ldots, 12$, is a family of joint probability density functions, on k unit vectors $v_1, \ldots, v_k$ in $\mathbb{R}^3$, of the configurations of clusters of spheres.
- $\Psi(\kappa(x_\ell)), k = 4, \ldots, 12$, a probability distribution for the number of nearest neighbours (coordination number).
- In geometrical terms, amorphous solids are fundamentally different from crystalline solids in that they cannot be constructed by the crystallographic method of translation of the basis along the lattice points. They do not possess any regular lattice.
- A random structure is not merely a disordered structure. A disordered crystal structure can be gradually restored to its fully ordered state by removal of its defects. In principle, an amorphous structure with flaws can be rearranged to its ideal (perfectly random) state.

- The IAS-I is an irregular assemblage of spheres containing no imperfections. By definition, it shows the absence of preferred local configurations and local flaws. It is of infinite extent (no edge effects) and shows no orientation. It is homogeneously random (it is ergodic with respect to detail of local patterns).
- Flaws in amorphous solids can increase or decrease average body density, depending on the type of flaw.
- The mean coordination number of spheres is rational in an ideal crystalline structure and hypothesized to be irrational in an ideal amorphous solid.

1.9.4
Short and Medium Range Order

We have concluded in Section 1.5 that the IAS packing has no short-range, medium-range, nor long range order. However, all real amorphous materials have flawed IAS packing, and encompass deviations from ideal amorphous packing that give rise to local short and medium range arrangements of atoms. Short range order, limited to individual primary clusters, is likely to form during the cooling process if the interatomic interactions overcome the thermal fluctuations entropic effect. Medium range order may form during annealing by diffusion and atomic rearrangement, usually around the clusters with short range order. Density variations within the glass structure expedite this process. The latter process is similar to recrystallisation observed in heavily strained polycrystalline metals.

Short and medium range order means that the local arrangement of atoms in their immediate neighbourhood acquires either some geometrical symmetry elements or recurring chemical stoichiometry, or both. For as long as these remain below detectable levels, they will be referred to as flaws in the amorphous packing. However, above some specific critical volume fraction the presence of these arrangements will begin to affect the properties of the amorphous material, and in that case will be described as nano-crystals or nano-particles.

References

Aste, T., Saadatfar, M., Sakellariou, A. and Senden, T.J. (2004) Investigating the geometrical structure of disordered sphere packings. *Physica A*, **339**, 16–23.

Barlow, W. (1883) Probable nature of the internal symmetry of crystals. *Nature (London)*, **29** (738), 186–188.

Bernal, J.D. (1959) A geometrical approach to the structure of liquids. *Nature*, **183** (4655), 141–147, doi: 10.1038/183141a0.

Brostow, W., Dussault, J.-P. and Fox, B.L. (1978) Construction of voronoi polyhedra. *Journal of Computational Physics*, **29/1**, 81–92.

Clare, B.W. and Kepert, D.L. (1986) The closest packing of equal circles on a sphere. *Proceedings of the Royal Society of London Series A: Mathematical and Physical Sciences*, **405** (1829), 329–344.

Conway, J.N. and Sloane, N.J.A. (1998) *Sphere Packings, Lattices and Groups*, 3rd edn, Springer-Verlag, New York.

Cottrell, A. (1965) *Dislocations and Plastic Flow in Crystals*, Clarendon Press, Oxford, New York.

De-Floriani, I., Falcidieno, B. and Pienovi, C. (1985) Delaunay-based representation of surfaces defined over arbitrary shaped domains. *Graphics and Image Processes*, **32**, 127–140.

Dirichlet, G.L. (1850) Über die Reduktion der Positiven Quadratischen Formen mit. Drei Unbestimmten Ganzen Zahlen. *Journal für die Reine und Angewandte Mathematik*, **40**, 216.

Edwards, S.F. and Grinev, D.V. (2002) Granular materials: towards the statistical mechanics of jammed configurations. *Advances in Physics*, **51**, 1669-1684.

Finney, J.L. (1970a) Random packings and the structure of simple liquids. i. the geometry of random close packing. *Proceedings of the Royal Society of London Series A: Mathematical and Physical Sciences*, **319** (1539), 479–493.

Finney, J.L. (1970b) Random packings and the structure of simple liquids. ii. the molecular geometry of simple liquids. *Proceedings of the Royal Society of London Series A: Mathematical and Physical Sciences*, **319** (1539), 495–507.

Flynn, C.P. (1972) *Point Defects and Diffusion*, Clarendon Press, Oxford.

Furth, R. (1964) A new approach to the statistical thermodynamics of liquids. *Proc. R. Soc. Edinburgh*, **66**, 232.

Hales, T.C. (1998) The Kepler conjecture. *arXiv preprint math/9811078, http://arxiv.org/abs/math/9811078* (accessed 25 June 2014).

Haüy, R.J. (1821) *Traité Élémentaire De Physique*, Bachelier, Paris.

Hoff, K.E., Culver, T., Keyser, J., Lin, M. and Manocha, D. (1999) Fast computation of generalised voronoi diagrams using graphics hardware. *SIGGRAPH99 Conference Proceedings*, ACM, p. 277.

Kepler, J. (1611) *De Nive Sexangula*, Elsevier.

Kopsias, N.P. and Theodorou, D.N. (1998) Elementary structural transitions in the amorphous lennard-jones solid using multidimensional transition-state theory. *Journal of Chemical Physics*, **109** (19), 8573–8582.

Lee, Ch.-Y., Stachurski, Z.H. and Welberry, T.R. (2010) The geometry, topology and structure of amorphous solids. *Acta Materialia*, **58** (2), 615–625.

Leopardi, P. (2006) Equally spaced points on a sphere. *Electronic Transactions on Numerical Analysis*, **25**, 309–327.

Medvedev, N.N., Geiger, A. and Brostow, W. (1990) Distinguishing liquids from amorphous solids: percolation analysis on the voronoi network. *J. Chem. Phys.*, **93** (11), 8337–8342.

Mott, P.H., Argon, A.S. and Suter, U.W. (1992) The atomic strain tensor. *Journal of Computational Physics*, **101**, 140.

Pfister, L.A. and Stachurski, Z.H. (2002) Micromechanics of stress relaxation in amorphous glassy pmma part ii: application of the rt model. *Polymer*, **43**, 7419–7427.

Reiss, H. and Hammerich, A.D. (1986) Hard spheres: scaled particle theory and exact relations on the existence and structure of the fluid/solid phase transition. *Journal of Physical Chemistry*, **90** (23):6252–6260, doi: 10.1021/j100281a037.

Stachurski, Z.H. (2003) Definition and properties of ideal amorphous solids. *Physical Review Letters*, **90** (15), 5502–5506.

Stachurski, Z.H. and Brostow, W. (2001) Methamorphosis of voronoi polyhedra as a definite measure of the elastico-dissipative atomic displacement transition. *Polimery*, **46** (5), 302–306.

Tanemura, K. (1983) A new algorithm for three-dimensional voronoi a new algorithm for three-dimensional voronoi tessellation. *Journal of Computational Physics*, **51**, 191–207.

Taylor, C.C.W. (1999) *The Atomists: Leucippus and Democritus*, University of Toronto Press.

Theodorou, D.N. and Suter, U.W. (1985) Atomistic modelling of mechanical properties of polymeric glasses. *Macromolecules*, **19** (1) 379–387.

To, L.Th., Daley, D.J. and Stachurski, Z.H. (2006) On the definition of an ideal amorphous solid. *Solid State Sciences*, **8**, 868–879.

Torquato, S. (2002) *Random Heterogenous Materials*, Springer-Verlag, New York.

Torquato, S., Truskett, T.M. and Debenedetti, P.G. (2000) Is random close packing of spheres well defined? *Physical Review Letters*, **84** (10), 2064–2067.

Tsumuraya, K., Ishibashi, K. and Kusunoki, K. (1993) Statistics of voronoi polyhedra in a model silicon glass. *Physical Review B*, **47** (14), 8552–8557.

Turek, I. (1997) *Electronic Structure of Disordered Alloys, Surfaces and Interfaces*, Kluwer Academic Publishers, Boston, MA.

Voronoi, G. (1908) Recherches sur les paralleloedresprimitifs. *Journal für die reine und angewandte Mathematik*, **134**, 198–287.

Voronoi, G. (1909) Nouvelles applications des paramètres continus à la théorie de formes quadratiques. *Journal für die reine und angewandte Mathematik*, **136**, 67.

Walker, A.M., Slater, B., Gale, J.D. and Wright, K. (2004) Predicting the structure of screw dislocations in nanoporous materials. *Nature Materials*, **3** (10), 715-720.

Watanabe, M.S. and Tsumaraya, K. (1987) Crystallisation and glass formation in liquid sodium: a molecular dynamics study. *Journal of Chemical Physics*, **87** (8), 4891.

Wieder, T. and Fuess, H. (1997) A generalized debye scattering equation. *Zeitschrift für Naturforschung*, **52a**, 386–392.

Wigner, E.P. and Seitz, F. (1933) On the constitution of metallic sodium. *Physical Review*, **43**, 804–810.

Gittus, J. (1975) Creep, Viscoelasticity and Creep Fracture of Solids. *Appl. Sci. Publ.*, London.

Zarzycki, J. (1991) *Glasses and the Vitreous State*, Cambridge University Press, Cambridge.

Books on Random Walks

Barber, M.N. and Ninham, B.W. (1970) *Random and Restricted Walks Theory and Application*, Gordon and Breach, New York.

Hughes, B.D. (1995) *Random Walks and Random Environments*, Clarendon Press, Oxford.

Rudnick, J. and Gaspari, G. (2004) *Elements of the Random Walk*, Cambridge University Press, Cambridge.

Books on Sphere Packings

Conway, J.N. and Sloane, N.J.A. (1998) *Sphere Packings, Lattices and Groups*, 3rd edn, Springer-Verlag, New York.

Daley, D.J. (2000) Packings and Approximate Packings of Spheres. Technical Report, National Institute for Statistical Sciences, grant No DMS- 9313013.

Sadoc, J.-F. and Mosseri, R. (1999) *Geometrical Frustration*, Cambridge University Press, Cambridge.

Zong, C.M. (1999) *Sphere Packings*, Springer-Verlag, New York, ISBN: 978-0-387-22780-1.

Books on Crystal Imperfections

Flynn, C.P. (1972) *Point Defects and Diffusion*, Clarendon Press, Oxford.

Kelly, A. and Groves, G.W. (1973) *Crystallography and Crystal Defects*, Longman, London.

Nelson, D.R. (2002) *Defects and Geometry in Condensed Matter Physics*, Cambridge University Press.

A selected list of books on topics covered in this chapter.

Books on Crystallography

Buerger, M.J. (1956) *Elementary Crystallography*, John Wiley & Sons, Inc., New York.

Friedel, G. (1964) *Leçons de Crystallographie*, Blanchard, Paris.

Books on Glasses

Doremus, R.H. (1973) *Glass Science*, John Wiley & Sons, Inc., New York.

2
Characteristics of Sphere Packings

2.1
Geometrical Properties

2.1.1
The Coordination Distribution Function, $\Psi(k)$

Consider a large finite spherical region, say $\mathcal{S}$, in a random packing of spheres of infinite extent. Let N denote the number of spheres with centres in $\mathcal{S}$, and $\mathbf{C}_k$ the number of such spheres with given coordination number k, where $k = 4, \ldots, 12$. Then

$$\sum_{k=4}^{12} \mathbf{C}_k = N \tag{2.1}$$

We could as easily regard $\mathcal{S}$ as being one of the sequence of such regions, indexed by N say, and similarly $\mathbf{C}_k(N)$. Then the fraction of spheres with coordination number k we would regard as satisfying

$$\Psi(k) = \lim_{N \to \infty} \frac{\mathbf{C}_k}{N} = \Psi(\kappa_\ell) \tag{2.2}$$

that is, the coordination number distribution $\{\Psi(k)\}$ coincides with the nearest-neighbour distribution and can be regarded as an asymptotic property of the solid packing.

$$\Psi(k) = \frac{\mathbf{C}_k}{N} \tag{2.3}$$

Suppose we regard clusters of touching spheres as being the building blocks of amorphous solids, analogous to a unit cell of a crystal. A *local* cluster of $(1 + k)$ impenetrable spheres, all of the same diameter d, comprises an inner sphere, with centre x_ℓ say, which is touched by exactly k outer spheres with centres at $x_{\ell j}$, $j = 1, \ldots, k$. Subject to this contact requirement, the local cluster can have any configuration, regular or otherwise. Where needed we denote the number of contact spheres in the local cluster with centre x_ℓ by the function $\kappa(x_\ell)$. Any sphere can belong to several clusters. We note that $\Psi(k)$ of Equation 2.4 is directly related to the probability distribution for the nearest neighbours, as defined earlier:

$$\Psi(k) = \Psi(\kappa_\ell) \qquad (2.4)$$

Examples of the functions for ordered and random packings are shown in Figure 1.27.

2.1.2
Tetrahedricity

When the network of radial vectors is combined with the Voronoi polyhedra, each Voronoi polytope becomes internally divided into tetrahedra. In other words, each polytope is decomposed into a simplicial complex, or a union of simplices, satisfying certain properties. A polytope is a geometric object with flat sides. A polygon is a polytope in two dimensions, a polyhedron is a polytope in three dimensions. In geometry, generalization of the notion of a triangle is a *simplex*, and in three dimensions, it is a tetrahedron.

Some simplicial atomic configurations have specific physical significance. Thus, simplices with large circumradii correspond to low-density configurations (see Figure 2.1).

Within the circumsphere of any Delaunay simplex, there are no other atoms (centres). Hence, if a circumradius exceeds an average value, the simplicial configuration circumscribes a larger volume than the average for a given system. On the contrary, simplicial atomic configurations with the minimum circumradius correspond to the most dense arrangements. For example, there are a large number of close to (but not quite) equilateral tetrahedral configurations in models of amorphous phases of spherical atoms. Such configurations play an important role in the amorphous structure.

To measure the irregularity of tetrahedra, Hiwatari et al. (1984), proposed a formula of the tetrahedral configurations, based on the Minkowski measure (see the following text):

$$\delta = \sum_{i=1}^{6} \frac{|\ell_i - \overline{\ell}|}{6\overline{\ell}} \qquad (2.5)$$

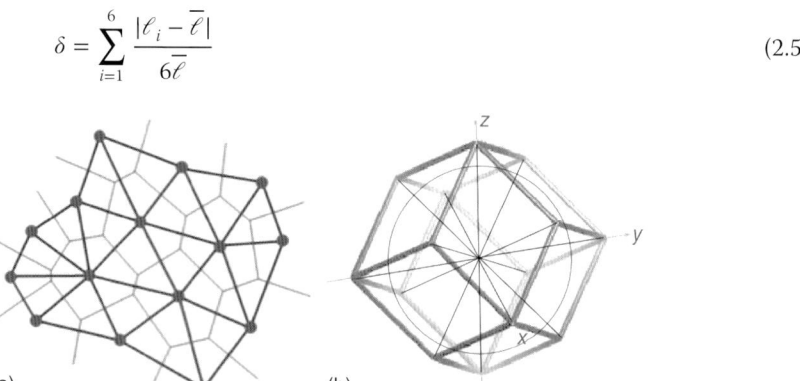

Figure 2.1 Delaunay tessellation. On the basis of (a) the central points in two dimensions and (b) the centre of inscribed sphere inside Voronoi polyhedron.

where ℓ_i is the length of the ith edge and $\overline{\ell}$ is the average length of a given polygon.

Another formula, based on Euclidean 2-norm measure, was proposed by Medvedev et al. (1990). The Γ parameter measures the irregularity of the tetrahedron formed by four adjacent spheres, in which l_i and l_j are the lengths of the edges and $\langle l \rangle$ is the mean edge length.

$$\Gamma = \sum_{i>j} \frac{(l_i - l_j)^2}{15\langle l \rangle^2} \tag{2.6}$$

$\Gamma = 0$ corresponds to a regular tetrahedron.

$\Gamma > 0$ corresponds to distortions from the regular shape.

It is found that the maximally jammed random packings of spheres do contain a fraction of tetrahedra with $\Gamma = 0$, which manifests itself in the so-called split second peak in the radial distribution function (Torquato, 2002). However, as the fraction of these imperfections in random packing decreases, the evidence of their presence becomes statistically insignificant and possibly overlooked. Nevertheless, they are *not perfect* random packings.

In the quenched liquid phase, there is a cluster with a low local density, spanning the whole of the specimen (in this respect, it can be considered macroscopic). The atomic arrangement is such that in each simplicial configuration belonging to that cluster, atomic mobility is possible. It seems only natural that any specimen containing such a cluster is a fluid. The cluster has the capability of macroscopic structural reorganizations such as shear flow inherent to fluid systems.

In the quenched phase, the liquid-like cluster is absent. Instead, there is a different percolative cluster consisting of atomic configurations close in form to the regular tetrahedron. This second type of a cluster may play the role of a backbone, imparting mechanical rigidity to the specimen. Indeed, the regular atomic tetrahedral configuration in the system is more stable than any other (distorted) configurations. The cluster involving several contiguous tetrahedral configurations (decahedrons or icosahedrons) should be more stable than a single tetrahedron: the atoms involved are less labile because of the geometrical rigidity of the cluster and have lower local energy than other atomic configurations of this system.

Percolative clusters appear to exist in all glass phase models of spherical atoms and hard spheres. They always involve a high fraction of almost regular tetrahedral configurations and a high fraction of pentagonal faces on the Voronoi polyhedra and the splitting of the second peak of the radial distribution function $g(r)$. They occur mainly due to the appearance of a large fraction of clusters of regular tetrahedral configurations with contiguous faces.

In analytic geometry, the distance between two points of the xy-plane can be found using the distance formula. In the Euclidean space, the distance between two points is usually given by the Euclidean distance (2-norm distance). The 2-norm distance is the Euclidean distance (a generalization of the Pythagorean theorem):

$$d = \sqrt{(x_1 - x_2)^2 + (y_1 - y_2)^2 + (z_1 - z_2)^2} \tag{2.7}$$

Other distances, based on other norms, are also used instead. For example, the Minkowski distance of order p is defined as

$$\text{Distance} = \left(\sum_{i=1}^{n} |x_i - y_i|^p \right)^{1/p} \tag{2.8}$$

for a point $(x_1, x_2, \ldots, x_n)$, and a point $(y_1, y_2, \ldots, y_n)$,

2.1.3
Voronoi Polyhedra Notation

A *polyhedron* is a solid object whose surface is made up of a number of flat faces, which themselves are bordered by straight lines.

The Euler polyhedra theorem states that

$$V - E + F = 2 \tag{2.9}$$

where V is the number of vertices, E is the number of edges and F is the number of faces in any convex polyhedron. The formula is true for all convex polyhedra, including the result from Voronoi tessellation. A convex polyhedron is one in which a line joining any two points in the solid is completely contained within the solid or any plane section.

Voronoi tessellation of a packing results in a Voronoi diagram, which is a three-dimensional network of convex touching polyhedra, completely filling space. The polyhedra share common faces, vertices and edges. With some thought, it can be deduced that each edge is shared by three faces; hence, a face, F_i of i edges, contributes $\frac{1}{3}i$ edges to the network, thus

$$E = \frac{1}{3} \sum_i i F_i, \quad \text{for} \quad 3 \leq i \leq n \tag{2.10}$$

Also, because four edges meet at one vertex and each edge is associated with two vertices, we can write a relationship between the number of edges and vertices:

$$V = \frac{1}{2} E \tag{2.11}$$

Substitution of the above relations into the Euler formula (2.9), and carrying out the summation over i, results in a generalized Euler equation for such a network:

$$3F_3 + 2F_4 - F_5 - F_7 - 2F_8 - \ldots = 12 \tag{2.12}$$

where F_3 is the number of faces with three edges, F_4 the number of faces with four edges, F_5 the number of faces with five edges and so on. This relation is obeyed by every network and indicates the exact topology of polyhedra which can occur. Note that F_6 is absent from the equation (the F_6 terms in Equation 2.12 cancel out). Six-sided faces do occur, but the number of faces with six sides can be changed without changing the number of any other n-edged faces. This means that six-sided faces have no effect on the topology.

To a large extent, the topology of most commonly occurring polyhedra is effectively determined by the numbers of F_3, F_4 and F_5, because the number of seven-edged faces and higher are small compared to the numbers of the first three. From this assumption, one can derive the criterion

$$3F_3 + 2F_4 + F_5 \geq 12 \tag{2.13}$$

For more complex polyhedra, where the values of F_7, F_8, F_9 and so on may be significant, criterion (2.13) is extended to

$$3F_3 + 2F_4 + F_5 \geq 14 \tag{2.14}$$

On the basis of the above analysis, topology of polyhedral types indicates that

- The most frequently occurring types are 13-, 14- and 15-edged polyhedra.
- Polyhedra can be described by a set of digits, corresponding to the number of F_3, F_4, F_5, F_6, F_7 an so on.

Molecular dynamic computer studies of model glasses show that the polyhedra with 14-edged faces are most abundant and 5-edged faces are most common, as shown in Table 2.1.

2.1.4
Topology of Clusters

We take clusters with a limit on k, such that, $4 \leq k \leq 12$, and require that the inner sphere is immobilized, that is, fixed by the outer spheres. The k outer spheres can be placed on the inner sphere in two ways: (i) in a regular arrangement or (ii) in an irregular arrangement.

The topology of clusters can be considered in terms of the triangles made by joining the centres of neighbouring spheres. The number and arrangement of the outer spheres defines the topology of the cluster. In that sense, topology is equivalent to coordination plus configuration.

The geometry and topology of clusters of atoms have been considered by Frank and Kasper (1959) in order to describe the complex atomic arrangements in

Table 2.1 Most frequent irregular polyhedra satisfying criterion (2.13) as described by the topological notation, derived from (Finney, 1970).

F_3	F_4	F_5	F_7	F_8	$\sum F_i$	$3F_3 + 2F_4 + F_5$
0	1	10	2	0	13	12
0	3	6	4	0	13	12
0	5	3	6	1	15	13
1	3	4	5	1	14	13
0	2	8	4	0	14	12
0	4	4	6	0	14	12
1	1	8	3	1	14	13
2	2	4	4	2	14	14

crystal structures of transition metal alloys. They chose clusters with coordination shells that contain triangular faces only. In classifying the topology of clusters, they distinguished three cases, depending on the shape of the triangles formed by three adjacent atoms:

> Case 1 Any three adjacent atoms make equilateral triangles with the centre (inner) atom (hexagonal close packed (hcp) and face-centred cubic (fcc) in Figure 1.17a and b).
> Case 2 Coordination shell atoms make equilateral triangles; the shell atoms make isosceles triangles with the centre (icosahedral in Figure 2.2a).
> Case 3 Coordination shell atoms make irregular triangles only; the shell atoms make isosceles triangles with the centre atom (in Figure 2.2b).

2.1.4.1 Ordered Clusters

In the first category, any four adjacent spheres are always touching as they are derived from the ordered, close-packed crystalline structures (fcc and hcp). Therefore, all angles between any three spheres are equal to $\pi/3$, and the triangle is isosceles. In the second case, the triangle formed by any two outer spheres with the inner sphere is equilateral.

The atomic packing in crystals is regular, which implies exactly the same distances between atoms when viewed along straight line passing through the crystal, especially along any crystallographic direction. This implies that volume, packing fraction and density of each unit cell (or Wigner–Seitz cell) are precisely the same. For a perfect single crystal, this situation can be represented by the graph shown in Figure 2.3a, drawn for an fcc crystal structure. The graph, represented by a single

(a)   (b)

Figure 2.2 Clusters of same-size spheres. (a) Icosahedral cluster (1+5+5+1) (outer spheres symmetrically arranged but not touching) and (b) an irregular cluster without any symmetry.

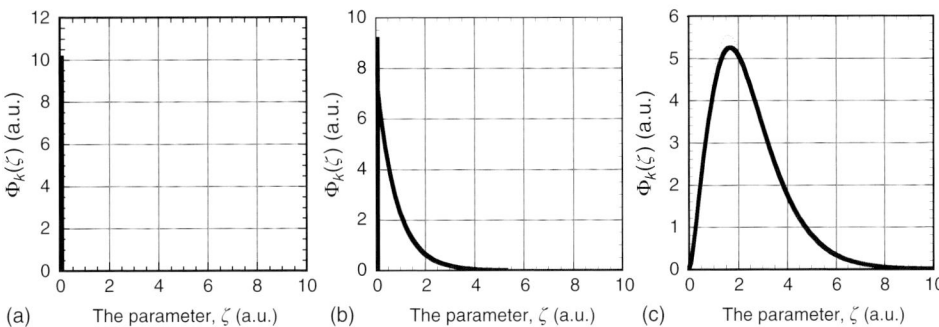

Figure 2.3 Distribution functions that clearly differentiate between perfect crystal, disordered crystal and an amorphous random structure.

line peak at the packing fraction of $p_f = 0.74$, signifies that the probability of this arrangement is precisely one for each and all Wigner–Seitz cells in that crystal.

Crystalline clusters with a maximum number of touching neighbours can occur in three different ways:

- hcp with outer spheres arranged in three layers of 3-6-3 pattern, as shown in Figure 1.17,
- fcc with outer spheres arranged in three in layers of 4-4-4 pattern, also shown in Figure 1.17,
- icosahedron with outer spheres arranged in four layers of 1-5-5-1 pattern has h-neighbours with h slightly larger than D amongst the outer spheres, shown in Figure 2.2.

The fcc and hcp arrangements are unique in that each sphere is in a fixed position relative to its neighbours so that every fcc or hcp cluster is identical in the number of outer spheres (coordination number) but differ in the configuration (arrangement) of the cluster. In the icosahedral arrangement, there is a certain degree of freedom to move the outer spheres in that the outer spheres do not touch each other (but all touch the inner sphere). The interval between the centres of the outer spheres has been calculated as 1.051 of the sphere diameter (Mackay, 1962). Within the limited degree of freedom, the icosahedral clusters can have infinite number of configurations (arrangements), of which 12 will be regular with 5-fold symmetry. The reader is referred to texts in crystallography for further information about ordered clusters.

2.1.4.2 Irregular Clusters

The third Frank and Kasper case has irregular arrangement of the coordination atoms, and therefore, any three adjacent outer spheres form irregular triangles, as shown in Figure 2.2. However, any two outer spheres form isosceles triangle with the inner sphere.

In amorphous solids, the distances between nearest-neighbour atoms cannot be the same and must vary within a certain limited range. From this we infer that the Voronoi volumes around each atom must vary; therefore, packing fractions must vary too. This is shown schematically in Figure 2.3c for an irregular packing with an average packing fraction, $p_f \approx 0.625$. By the nature of this arrangement, the structure has a distribution of the volumes of Voronoi cells, and each cell has a unique non-repeatable configuration.

2.1.5
The Configuration Distribution Function, $\Phi_k(\zeta)$

The configuration of a packing of spheres is specified by the configuration distribution function. This function contains the statistics of all clusters in the form of the probability functions, $P_k(\zeta)$ for all $1 \leq k \leq 12$. Examples of these functions for global random walk (GRW) and self-avoiding random walk (SARW) are shown in Figures 1.22a and b.

Mathematicians are interested to find the asymptotic behaviour for correlated walks (where each step depends on previous steps) of the mean-square displacement which is not linear in the number of steps (k). The asymptotic behaviour for random walks on the surface of a sphere is in the self avoiding space limited random walk (SASLRW) category (see Section 1.3.4.3), as is evident from Figure 2.4. As $k \to 12$, $\zeta \to 0$ and $P_k(\zeta) \to 1$, with the width of the distribution reducing to zero.

The scalar quantity, ζ, defined by Equation 1.33, is chosen to be the random variable describing the configuration of clusters in the random packing. It is equal to the magnitude of the end-to-end vector of the SARW formed by the radial vectors, R_j for $j = 1$ to k_ℓ, for each ℓ-sphere, where ℓ is an index for spheres in the packing.

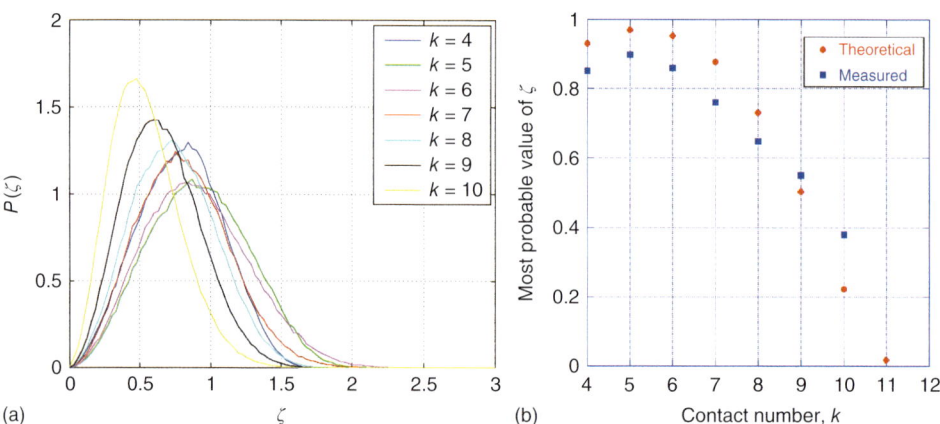

Figure 2.4 (a) Probability functions for the random variable, ζ, for clusters with the values of k as shown, derived from random packing of equal-sized spheres. (Reprinted with permission from To *et al.* (2006), Copyright (2006) by American Physical Society.) (b) The most probable value of $\zeta = \hat{\zeta}$ as a function of coordination number, k, predicted from Equation 2.15 and measured from simulations.

For a Gaussian random walk, the most probable value of $\zeta = \hat{\zeta}$ is proportional to $k^{1/2}$. However, $\hat{\zeta}$ varies with k in a way which is not in accordance with Gaussian random walk. This is because (i) the consecutive steps on the radial vectors, R_j for $j = 1$ to k_ℓ, are subject to hard sphere non-overlap condition leading to SARWs and (ii) the SARWs are limited in extent by the finite space around the inner spheres surface. It is conjectured that the self-avoiding limited random walk is described by the following equation:

$$\hat{\zeta}(k) = [(2R/3) k^{3/5}] \left[\exp\left(-\frac{\gamma}{1 - k/k_{\max}} \right) \right] \quad (2.15)$$

Analytical solutions for the probability generating function (1.37) for self-avoiding walks of this type are not available at present (Madras and Slade, 1993). Equation 2.15 is conjectural and must not be taken as validated. The problem of SARW on the surface of a sphere is of wider interest, for instance, in biology in modelling the transport of proteins on the surface of a spherical cell, in robotics in predicting the limiting envelope of movement of multi-arm robot, in mobile ad-hoc networks and so on.

Instead, computer simulations of random clusters were carried for each of $k = 4$ to 10, calculating the ζ value for each cluster and collecting the statistics. Clearly, for $k = 3$ (not simulated), all configurations would be in the loose category, whereas for $k \geq 8$, the probability of generating a loose configuration becomes insignificant. The accumulated statistical data for the ζ value for each cluster, excluding loose configurations, have been plotted in Figure 2.4. It can be seen that the effect of self-avoidance in the limited space on the surface of the sphere forces the most probable displacement of the walk to decrease with increasing k for $k > 7$, unlike that for the GRW (as shown in Figure 1.22a) or SARW (shown in Figure 1.22b). Indeed, for a close-packed regular configuration of the outer spheres (fcc or hcp), the random walk must return to the starting point ($\zeta = 0$). This is a distinguishing feature between a regular and an irregular packing of spheres. By definition, for all ideal crystalline structures, $P_k(\zeta) = 1$ for $\zeta = 0$ and $P_k(\zeta) = 0$ for $\zeta \neq 0$.

The shifting position of the peaks in Figure 2.4, at first moving away from $\zeta = 0$ (for $k = 4$ to 6) and then towards $\zeta = 0$ (for $k > 7$), is an effect expected from the geometrical limitations of the SASLRWs. A comparison of $\hat{\zeta}$ predicted by Equation 2.15 with the values measured from computer simulations is shown in Figure 1.22b. The trends are similar.

2.1.6
The Volume Fraction

Consider a sphere of maximum radius that can be fitted inside a polyhedron. Then the volume fraction of the polyhedron occupied by the sphere is given by

$$v_f(x_{ij}, X) = \frac{v(S_3)}{V_V(x_{ij})} \quad (2.16)$$

where $v(S_3)$ is the volume of the inscribed sphere and $V_v(x_{ij})$ is the volume of the corresponding Voronoi polyhedron. In geometry, an inscribed sphere or in-sphere of a convex polyhedron is the largest sphere that is contained wholly within the polyhedron and, therefore, tangent to the polyhedron's faces (Figure 2.5).

2.1.6.1 Regular Polyhedra
Regular polyhedra are special in that spheres inscribed into them will touch all of their faces. As the number of faces increases, so the volume fraction increases as shown in Table 2.2.

2.1.6.2 Irregular Polyhedra
As a rule, irregular polyhedra have only some faces tangent to the largest sphere that can be fitted in. In a random packing of spheres, the sizes of the polyhedra vary from sphere to sphere, and it is generally agreed that the distribution of their sizes is finite as shown in Figure 2.6.

The minimum Voronoi volume is for 12 contact points, defined by fcc and hcp. The maximum Voronoi volume, associated with four contact points, cannot be defined precisely, but it is estimated to be larger than the tetrahedral volume. This

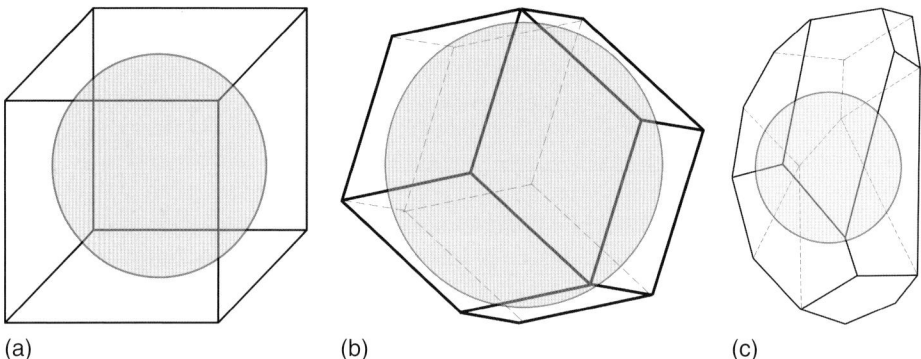

(a)   (b)   (c)

Figure 2.5 Examples of polyhedra with inscribed spheres. The spheres touch all faces in the regular polyhedra but not in the irregular case.

Table 2.2 Volume fractions for selected regular polyhedra.

Polyhedron of side, s	Number of faces	Volume	In-sphere radius	Volume fraction v_f
Tetrahedral	4	$\sqrt{2}\,s^3/12$	$s/2\sqrt{6}$	0.302
Simple cubic	6	s^3	$s/2$	0.524
bcc	8	$\sqrt{2}\,s^3/3$	$s/\sqrt{6}$	0.680
fcc	12	$s^3/4$	$s/20$	0.740
hcp	12	$s^3/12$	$s/12$	0.740

bcc, body-centered cubic.

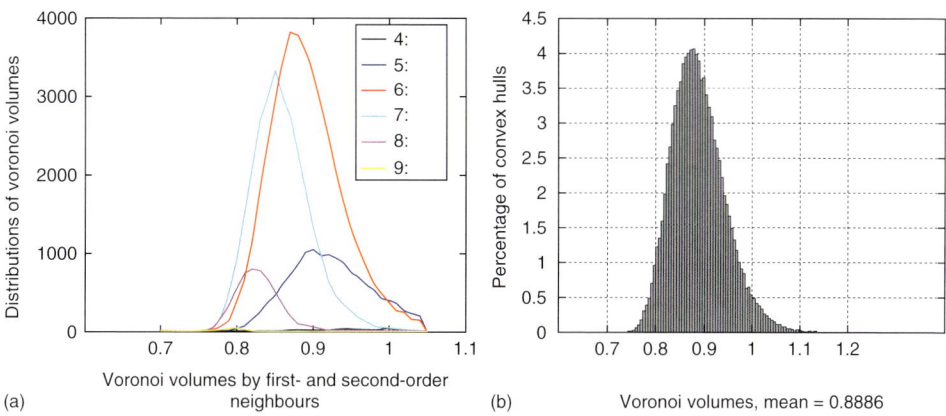

Figure 2.6 (a) The Voronoi data can be segregated into bins with specific coordination number. (b) Total Voronoi volume distribution for a random packing of spheres.

Table 2.3 Average volume fractions in random packing of spheres.

k	4	5	6	7	8	9	10	11	12
v_f	0.582	0.588	0.596	0.616	0.631	0.654	0.705	0.726	0.739

allows estimations of the bounding volume fractions for contact numbers $k = 4$ to 8, 10 and 12 to be carried out. Volume fractions for $k = 9$ and 11 can be estimated by interpolation. The average volume fraction for each bin will vary from a maximum of 0.739 ± 0.001 to a minimum of approximately 0.58 ± 0.02, as shown in Table 2.3.

The distribution of Voronoi volumes for irregular packing can be described to a good approximation by the following function:

$$f(V_V) = x^\alpha \exp\left[-\left(\frac{x}{\lambda}\right)^2\right], \quad x = (V_V - V_0) \tag{2.17}$$

where V_V is the Voronoi volume for an atom (or monomer), V_0 is the corresponding minimum volume extrapolated to 0 K, λ describes the width of the peak and is related to the average volume, $\overline{V}_V$. The exponent, α, and the constant, λ, are chosen to fit to a given distribution.

2.1.7
The Packing Fraction

Consider random packing of spheres inside a simulation cell, which will be referred to as the *Round Cell* (because of its spherical shape). The volume of a

Round Cell is equal to

$$V_{RC} = (4\pi/3)R_{RC}^3, \qquad (2.18)$$

whereas the volume occupied by N spheres is

$$V_{spheres} = N(4\pi/3)R^3, \qquad (2.19)$$

where R_{RC} is the radius of the Round Cell and R is the radius of a sphere.

The ratio of the volume occupied by the spheres to the volume of the Round Cell is called the *packing fraction*:

$$p_f = \frac{V_{spheres}}{V_{RC}}. \qquad (2.20)$$

The packing fraction is a specific case of the general definition of volume fraction (2.16)

2.1.7.1 The Average Packing Fraction for the Round Cell

The average packing fraction for a random packing of spheres can be predicted by Equation 2.21,

$$v_f = \sum_{k=4}^{12} \Psi(k) \times v_f(k) \qquad (2.21)$$

where $\Psi(k)$ is the probability of contact number (coordination = k) and v_f is the corresponding average packing fraction for that coordination.

The values for the average volume fraction v_f are given in Table 2.3, and the probability of coordination $\Psi(k)$ is obtained from Figure 1.12 for clusters of the blocking method, and from Figure 1.15 for clusters by the free volume method, and also shown in Table 2.4. Assume that the whole random packing involves distribution of clusters based on the blocking method as described in Section 1.3. Then the calculation using Equation 2.21 gives

$$p_f(\text{block}) = 0.641 \qquad (2.22)$$

for the average packing fraction. On the other hand, if the distribution of clusters is based on the free volume method, then the predicted average packing fraction is

$$p_f(\text{Furth}) = 0.612. \qquad (2.23)$$

Of course, for a given packing of spheres, the average packing fraction can be calculated directly using Equation 2.20. Chapter 3 contains descriptions of a number

Table 2.4 Probabilities for the blocking and Furth clusters.

k	4	5	6	7	8	9	10	11	12	Σ
Ψ_B	0	0	0.009	0.08	0.50	0.37	0.038	0.003	0	1.00
Ψ_F	0.015	0.097	0.265	0.339	0.213	0.063	0.008	0	0	1.00

of computer simulated random packing of spheres, both mono- and multi-sized, with average packing fraction falling usually in the range 0.63 ± 0.005.

2.1.7.2 The Local Packing Fraction
The local packing fraction of $S_3 + X$ around $S_3 + x_i$, is given by

$$\bar{p}_f = \frac{\sum \mu_{ij} p_f(x_{ij}, X)}{\sum \mu_{ij}}, \qquad (2.24)$$

where

$$\mu_{ij} = \frac{v(D(x_{ij}))}{m_i} \quad \text{for } m = \text{card}\{X_i\}. \qquad (2.25)$$

2.1.7.3 The Limits of Packing Fraction
There is an important distinction between the two types of packings, as shown in Figure 2.7. In regular ordered packing of spheres, the local packing fraction around each sphere is unique, the same for all spheres and equal to the average packing fraction for the whole packing, as shown in Figure 2.7a. By contrast, in a system comprising of randomly packed spheres, the local density will vary from sphere to sphere; there will be a distribution of packing densities as shown in Figure 2.7b.

In each case, the packing fraction is bound by limits

$$p_f(x_i, X)_{\min} \le p_f(x_i, X) \le p_f(x_i, X)_{\max} \qquad (2.26)$$

The theoretical and experimental upper bound for both ordered and random close-packed spheres of equal size is $p_f(x_i, X)_{\max} = \pi/\sqrt{18} = 0.7405$, so that,

$$[p_f(x_i, X)_{\max}]_{\text{crystalline}} = [p_f(x_i, X)_{\max}]_{\text{random}} \qquad (2.27)$$

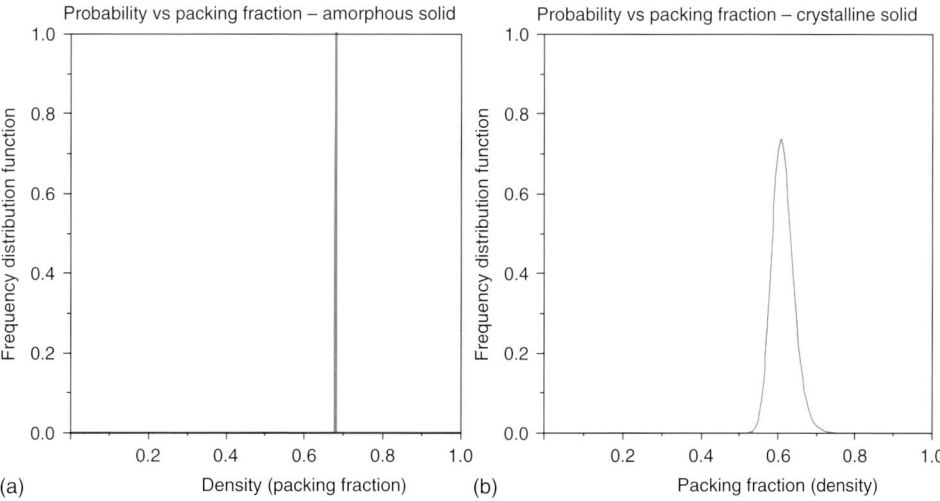

Figure 2.7 Distributions of local density (packing fraction) in crystalline and amorphous sphere packings.

However, this is not the case for the minimum value. The limit for the minimum local packing fraction for crystalline packings is less than that for random local packing (see Figure 2.8). For the crystalline case, the value of $[p_f(x_i, X)_{min}]$ is 0.302 (see Table 2.2). For the random packing case, the value of $[p_f(x_i, X)_{min}]_{random}$ is approximately 0.58 (see Figure 2.8), derived from measurement of the Voronoi volumes of individual clusters in closed packing of spheres. So that

$$[p_f(x_i, X)_{min}]_{crystalline} < [p_f(x_i, X)_{min}]_{random} \qquad (2.28)$$

For finite models of sphere packings, the boundary effect must taken into account when calculating the packing fraction (Figure 2.9). Spheres whose centres lie within the radius, $r \leq (R_{RC} - 2R)$, are wholly inside the simulation volume, whereas on the boundary of the Round Cell, some spheres will be partially in and partially out of the Round Cell (see Figure 2.10). This applies to spheres whose centres are within the shell of radius, $(R_{RC} - R) < r < (R_{RC} + R)$. The partial volumes of these spheres must be included so that the calculated packing fraction is accurate. A method for calculating the partial volumes is given in the following.

A method for calculating partial sphere volumes

A sphere whose centre is lying outside the boundary of the Round Cell but is intersected by it as shown in Figure 2.10. The partial volume enclosed between the two spheres must be included in the packing fraction.

It can be considered as the sum of the two spherical segments, one belonging to the Round Cell and the other belonging to the sphere.

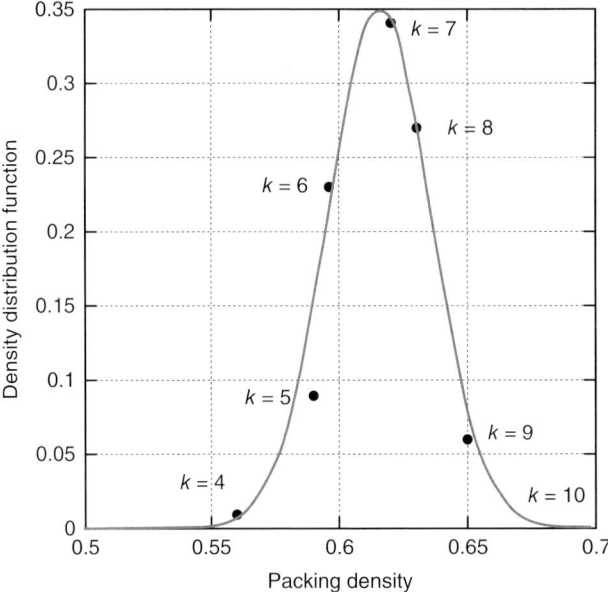

Figure 2.8 Volume fractions of regular and irregular polyhedra inferred from packing density and distribution of cluster's density.

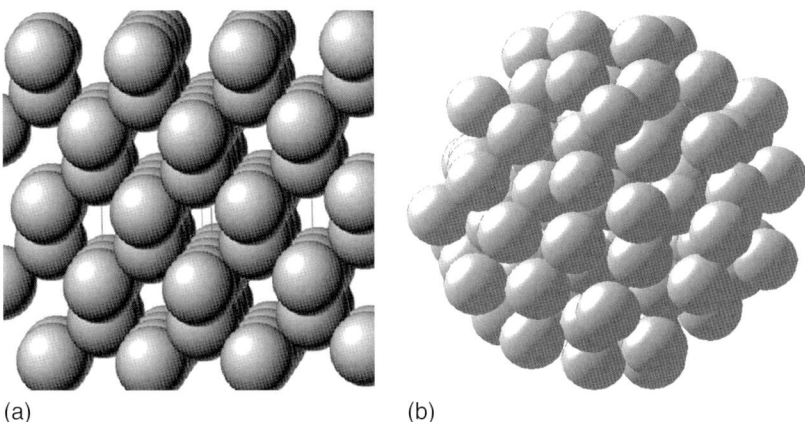

Figure 2.9 Example of low atomic packing fraction in Si crystal compared to higher packing fraction in random packing.

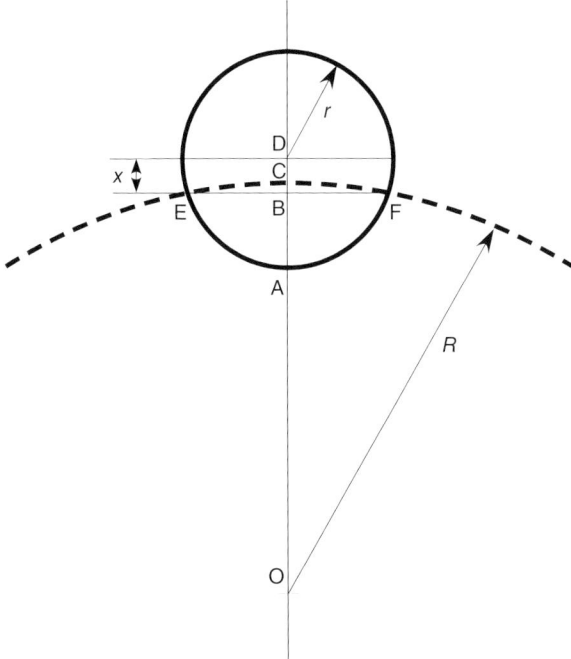

Figure 2.10 Geometrical relations between the radius of the Round Cell ($R_{RC} = R$) and the radius of the sphere ($R = r$) for the calculation of packing fraction in sphere packings.

For the purpose of this derivation, and in relation to the drawing in Figure 2.10, we use the following notation:

r is the radius of a sphere from the packing (small sphere),
R is the radius of the Round Cell (large sphere),
$h = AB$ is the height of the spherical cap belonging to the small sphere,
$H = BC$ is the height of the spherical cap belonging to the Round Cell.

Then the volume of the spherical segment AEBFA is given by

$$V_r = \frac{\pi h^2(3r - h)}{3}, \tag{2.29}$$

and if $h = 0$, then $V_r = 0$ as expected. The volume of the spherical segment BECFB is given by

$$V_R = \frac{\pi H^2(3R - H)}{3} \tag{2.30}$$

The following relationship exists for the two spherical segments:

$$r^2 = (r - h)^2 + a^2 \tag{2.31}$$

$$R^2 = (R - H)^2 + a^2 \tag{2.32}$$

From Equations 2.29, 2.30 and 2.32, we derive the following formulae for the partial volumes of the spherical segments:

$$V_r = \frac{\pi}{3}\left[(r - x) - \frac{r^2 - x^2}{2y}\right]^2 \left[2r + x + \frac{r^2 - x^2}{2y}\right] \tag{2.33}$$

$$V_R = \frac{\pi}{3}\left[\frac{r^2 - x^2}{2y}\right]^2 \left[3R - \frac{r^2 - x^2}{2y}\right] \tag{2.34}$$

where

$x = OD - R$
$y = R - x$

Note that for $x = r$, $V_r = 0$ and $V_R = 0$, as required.
Also, for $x = -r$, $V_r = 4\pi r^3/3$ and $V_R = 0$, as required.

Therefore, for each boundary sphere, the partial volume must be included in the calculation of the packing fraction is equal to the sum:

$$V_{total} = V_r + V_R \tag{2.35}$$

Application of the coordination function to granular materials The variation and distribution of porosity within a random assemblage of packed spheres (spherical particles) is of interest in defining local transport coefficients for flow of liquids and gases. In a macroscopic packing, porosity is delineated by the unoccupied spaces between the spheres. A random packing of spheres is characterized by bulk mean porosity or packing fraction as defined later. As the packing involves stochastic arrangement of a variety of clusters of different coordinations, the local variations in porosity occur over distances of few sphere diameters. Such local effects are important where high rates of heat and flow of gas or liquid are involved, as in heat exchanges or in liquid permeation chromatography (GPC or HPLC).

The local mean flow coefficient through a cluster will be determined by the mean velocity with respect to the completely free sphere and, therefore, dependent on the local coordination. If the bulk mean coefficient can be defined as $k_b = (1/N) \sum k_l$, then the local mean coefficient is given by

$$k_l = \frac{\int_0^{A_s} k_p dA_s}{\int_0^{A_s} dA_s} \tag{2.36}$$

where A_s is the total surface area of one particular sphere and k_p is the value of the coefficient at a point on the sphere.

2.1.8
Representative Volume Element

Consider a body created by random dense packing of N hard spheres of identical diameter, D, where N is sufficiently large for a meaningful statistical analysis. The body has a numerical density, $\rho_b = N/V$, which is conveniently expressed in terms of a packing fraction, p_f, as defined above. Let the body be divided into smaller parts. For subsequent divisions, the density of each part will remain the same to within an accepted tolerance, t.

$$\rho_b - t \leq \rho_{part} \leq \rho_b + t \tag{2.37}$$

Eventually, there will be a size of division below which the density will vary from one part to another by an amount greater than the accepted tolerance.

The *representative volume element* (RVE) of the randomly packed body is the smallest size division for which the density is still the same and equal to ρ_b to within the accepted tolerance (this translational invariance in the mean is a requirement imposed on all macroscopically large disordered systems). The size of RVE is defined by the mathematical probability statement:

$$P(|\bar{\mu} - \bar{\rho}_b| \leq t) \geq \alpha \tag{2.38}$$

where $\bar{\mu}$ is the mean density of RVE, $\bar{\rho}_b$ is the mean density of the body and the tolerance, t, and the confidence limit, α, are chosen in accordance with specific requirements.

In the limit, the largest variations in local densities, ρ_i, will be found when volumes around each sphere are expressed in terms of the corresponding Voronoi polyhedra. From the implicit assumptions, one can make the following observations:

1) Density fluctuations around individual clusters are in both positive and negative sense from the average density, ρ_b.
2) $\rho_{max} = \pi/\sqrt{18} \sim 0.74$ occurs in regular close-packed structures and provides the absolute limiting value on the right-hand side of the distribution.

2 Characteristics of Sphere Packings

3) For a symmetric distribution, the following is true $|\rho_{min} - \bar{\rho}| = |\rho_{max} - \bar{\rho}|$. Assuming an asymptotic average value, $\bar{\rho} = 0.63 \pm 0.05$, gives a value for the minimum density, $\rho_{min} \sim 0.54$.
4) For a non-symmetric distribution the minimum value of density (for any sphere inside the body) is less than that for the symmetric case, but greater than zero: $0 \leq \rho_{min} \leq 0.54$.

2.1.9 Density of Single Phase

Archimedes density is defined as the ratio of bulk mass over the total volume,

$$\rho = \frac{m}{V} \left(\frac{kg}{m^3}\right) \tag{2.39}$$

It should be noted that when referring to the density of a solid material, mass m refers to the mass of the solid and V to the corresponding occupied volume. Whilst, when referring to a particulate material, mass m is the mass of the material as if it were smoothly distributed throughout its volume V, disregarding the porosity (packing factor).

2.1.9.1 Density of Crystalline Solid

Density of a pure solid chemical element is always listed in relevant textbooks (and Wikipedia on the Internet) as the density of the equilibrium crystalline state of that element, usually at room temperature unless specified otherwise. This information must be borne in mind when considering the density, for example, of copper which has an fcc crystal structure, as shown in Tables 2.5 and 2.6.

The packing fraction for an fcc closed-packed structure of hard spheres is 0.740. The slight discrepancy between the two methods of calculated packing fractions (last two lines in Table 2.6) is due to truncation and rounding off.

2.1.9.2 Density of Amorphous Solid

The density of copper in the amorphous solid state at room temperature can be predicted by the ratio of packing fractions, that is, $(0.63/0.74) \times 8.96 = 7.63$ g cm^{-3}. This method is based on the following assumptions:

Table 2.5 Bulk (continuum) properties of copper (Cu).

Density of copper	8.96 g cm^{-3} = 8.96 × 10^3 kg m^{-3}
Relative atomic mass	63.6
Atomic number	29
Molar volume	63.6/8.96 = 7.09 cm^3
Avogadro number	6.02 × 10^{23}
Volume per atom	7.09 × 10^{-21}/6.02 × 10^{23} = 0.01178 nm^3 atom^{-1}

Table 2.6 Crystallographic properties of copper (Cu).

Unit cell size @ 295 K	0.361 nm = 361 pm
Unit cell volume	$(0.361 \text{ nm})^3 = 0.047 \text{ nm}^3$
Atomic radius	$\frac{1}{4} \times 0.361 \times \sqrt{2} = 0.128 \text{ nm}$
Atomic volume	$(4\pi/3)(0.128)^3 = 0.00878 \text{ nm}^3$
Atomic packing factor	$0.00878/0.01178 = 0.74$
Atomic packing factor	$4 \times 0.00878/0.047 = 0.74$

Table 2.7 Predicted amorphous density of selected elements.

Element	Crystalline packing fraction	Amorphous packing fraction	Crystalline density (g cm^{-3})	Amorphous density (g cm^{-3})
Cu	0.74	0.63	8.96	7.63
Fe	0.68	0.63	7.87	7.29
Si	0.34	0.63	2.33	4.32
C-diam	0.34	0.63	3.52	6.52

- The atoms of copper are represented by hard spheres of appropriate radius ($r_{Cu} = 0.128$ nm in Table 2.6).
- The radius of the copper atom remains the same when in the amorphous state.
- The random packing follows the rules described in Section 1.4.

The same method can be applied to other elements, in some cases with unexpected results, as shown in Table 2.7.

Calculated by this method, the densities of Si and diamond are greater in the amorphous state than in the crystalline state. It must be emphasized that the predictions are hypothetical, based on the assumptions listed above, unlikely to be correct because the electronic structure of these elements would prevent random packing of atoms as if they were free of directional bonding.

2.1.10
Density of a Composite

Consider a solid body comprising two phases, as shown in Figure 2.11. Let the continuous phase be called phase 1, and the particulate, phase 2.

The body has mass, m (kg) and occupying volume, V (m^3).

From the law of conservation of mass, we write

$$m_{total} = m_1 + m_2 \tag{2.40}$$

where m_1 and m_2 are the masses of phase 1 and 2, respectively.

Now, substitute $\rho \cdot V$ for m from the definition of density: $\rho = m/V$, and obtain

$$\rho_{total} V_{total} = \rho_1 V_1 + \rho_2 V_2 \tag{2.41}$$

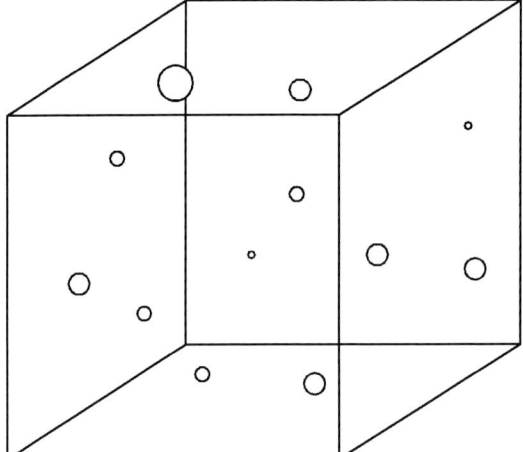

Figure 2.11 A body composed of two phases.

Division of both sides by V_{total} gives

$$\rho_{\text{total}} = v_1\rho_1 + v_2\rho_2 \quad \text{with} \quad v_1 + v_2 = 1 \tag{2.42}$$

where v_1 and v_2 are the volume fractions of phase 1 and 2, respectively. Equations 2.40 and 2.42 are frequently applied to a system such as shown in Figure 2.11. It can be applied to a system containing more than two phases.

Whilst Equation 2.40 is always correct, Equation 2.42 is correct only under certain conditions, which are

- the atomic volumes are additive
- the atomic volumes remain the same in different structural arrangements.

The mass and volume ought to be measured with as large a piece of single crystal of copper as can be found, of high purity, and suitably annealed to eliminate crystal defects, such as vacancies and grain boundaries. Then the density, as the ratio of mass to volume, would correspond to a nearly perfect (ideal) crystalline state of the substance.

2.1.11
Solidity of Packing

For perfect rigidity of a packing of hard spheres, the centre of a sphere, x_ℓ, must always lie inside a polyhedron defined by the nearest neighbours, $x_{\ell j}$, as its vertices. *This condition requires that with $\kappa(x_\ell) = k$: for $4 \leq k \leq 9$ no more than $(k-1)$ contacts must be located on one hemisphere in order to maintain the sphere in fixed position, for $k \geq 10$, the sphere is always in a fixed position.* This leads to *perfect solidity*, defined by the requirement that

$$\mathbf{J}_\ell = \left(\sqrt{\mathbf{u}^2 + \mathbf{v}^2 + \mathbf{w}^2}\right)_\ell \equiv 0, \quad \text{for all} \quad \ell = 1, \ldots, N, \tag{2.43}$$

where $\mathbf{J}_\ell$ represents the possible relative displacement of an ℓth sphere within the cage created by its neighbours and $\mathbf{u}, \mathbf{v}, \mathbf{w}$ are possible displacements if the sphere is loose, measured in the x, y, z directions, respectively.

Proof: As all spheres are fixed, therefore, $u_\ell^2 = 0$ and $v_\ell^2 = 0$ and $w_\ell^2 = 0$ for all ℓs, and therefore, $\mathbf{J} \equiv 0$ (rigid body).

For a packing of spheres to be representative of a solid, it must have the property of *solidity*. Solidity requires dense packing and imposes spatial constraints on the spheres. Rotations of spheres are admitted as they are inconsequential, but translations must be disallowed.

When all spheres in a packing are in fixed positions, we call it *closed* packing or rigid packing. However, when a packing contains spheres that are loose, then it is not rigid and will possess the flow property. An example of a packing containing loose spheres is in Figure 2.12, showing a distribution of possible displacements for many of its loose spheres.

This is not the same as close packing in crystallography which refers to hcp or fcc packing of spheres of the highest packing density. Incidentally, both hcp and fcc are also closed packings in the sense defined above.

We note that all clusters derived from crystallographic packings represent closed packings, providing they do not contain defects of any kind. All atoms occupy unique lattice sites or related sites governed by minimum in the interaction potential.

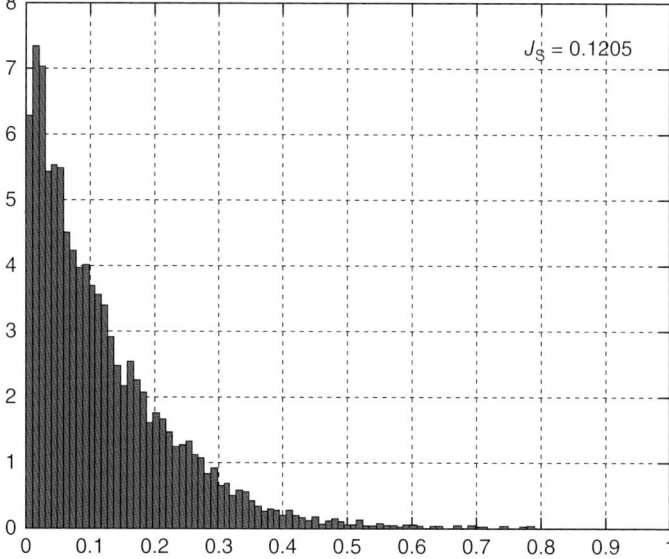

Figure 2.12 Distribution of $J(\ell)$s for a random packing with loose spheres. Vertical axis represents percent of loose spheres with given J_ℓ value. On the horizontal axis, the value of 1 corresponds to one sphere diameter.

It is necessary to point out that a body of densely packed spheres can remain solid and stable even if a number of spheres inside it are allowed to be loose (sometimes called *rattlers*). Such packings of spheres have been described by Hopkins et al. (2010).

2.2 X-ray Scattering

2.2.1 Introduction

At the beginning of the twentieth century the pioneering work of von Laue on X-ray diffraction of single crystals led rapidly to the determination of precise locations of atoms in crystalline minerals. Studies of space group symmetry provided theoretical basis for the definition and description of crystal structures, corroborated by the diffraction methods. The discovery of Bragg's law by W.L. Bragg and W.H. Bragg immediately allowed extension of the diffraction theory to polycrystalline materials. The pair distribution function (PDF) method, developed in the 1990s, resulted in precise determination of complex crystalline structure from the Fourier transformation of the measured powder X-ray diffraction data (Egami and Billinge, 2003) and analysis of diffuse scattering lead to understanding of defect structures (Welberry, 2004 Neder and Proffen, 2008).

For amorphous materials, the geometrical methods are less well developed. The positions of atoms are usually not known a priori and the vision of atomic arrangements is difficult to perceive clearly. However, as in crystalline solids, X-ray scattering pattern is a direct indication of atomic positions in the material.

2.2.2 Geometry of Diffraction and Scattering

X-ray measurements are usually carried out on an X-ray diffractometer, an instrument designed specifically for crystalline samples. In this instrument, the sample is placed in the X-ray beam and rotated continuously around an axis perpendicular to the beam, whilst the detector, receiving the diffracted beam, is rotated simultaneously at twice the rate, as shown in Figure 2.13a. The angle of the crystalline planes in the sample make with the incoming beam is referred to, by convention, as the *Bragg angle*, θ (from Bragg's law: $\lambda = 2\, d \sin \theta$). The diffracted intensity is measured by the detector and plotted as a function of the diffractometer angle, 2θ.

When measurements are carried out on amorphous samples, usually the same diffractometer instrument is used. As a result, the scattered intensity is also plotted against the diffractometer angle, 2θ, even though the angle does not have the diffraction meaning as in the case of the crystalline samples. In fact, the orientation of the amorphous sample with regard to the beam is immaterial as it

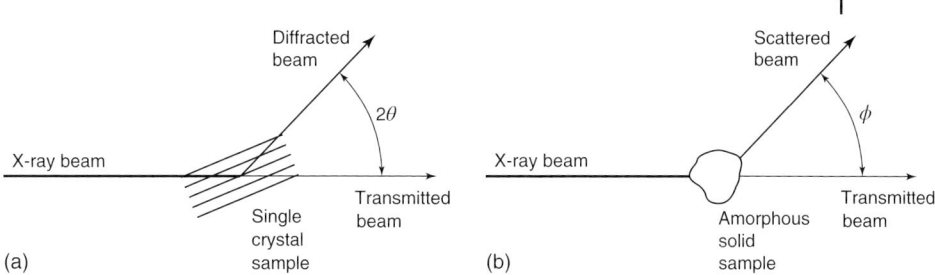

Figure 2.13 Definitions of the Bragg diffraction angle, 2θ, and the scattering angle, ϕ.

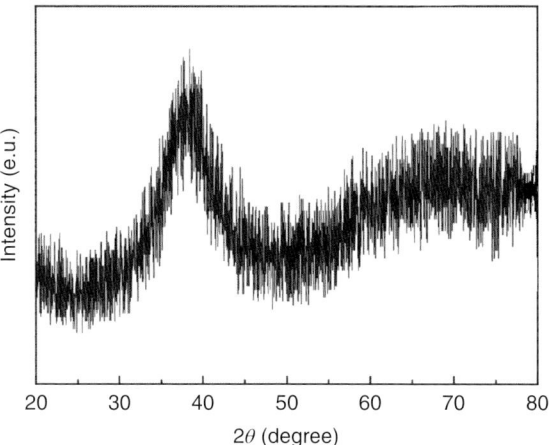

Figure 2.14 An example of a diffractometer recording.

does not have any crystalline planes. To distinguish between the two cases, the scattering angle for the amorphous sample should be denoted by ϕ, noting that instrumentally, $\phi = 2\theta$ (see Figure 2.13b). However, in practice and by tradition, the 2θ angle is used as shown in Figure 2.14.

An alternative to diffractometer is the flat plate camera. Its geometry is shown in Figure 2.15, and a recording obtained in such a measurement is also shown. The diffractometer recording is a radial profile of the intensity from flat camera recording.

2.2.3
Intensity of a Scattered Wave

Consider an X-ray beam directed onto a solid. The amplitude and phase of the scattered radiation in the whole of the surrounding space is represented by a wave

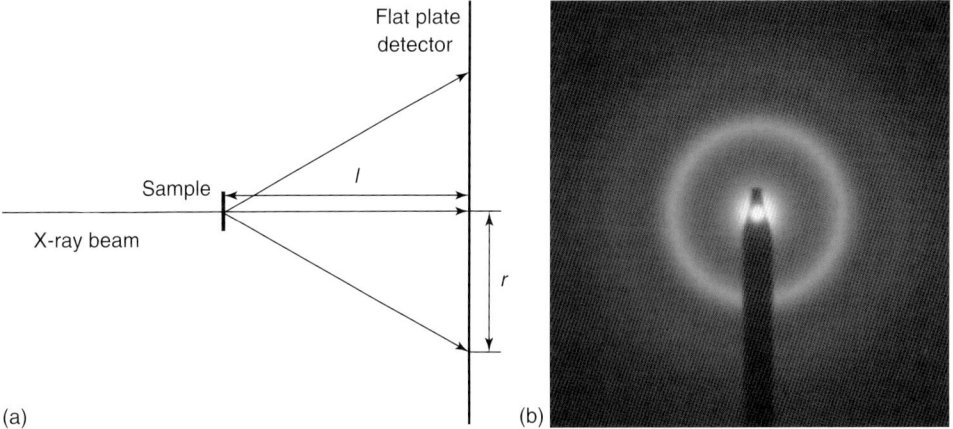

Figure 2.15 The side-on geometry of the flat plate camera and an example of a corresponding flat camera recording.

equation given in integral form by:

$$F(\vec{Q}) = \int_s \rho(\vec{r}) \exp(i\vec{r}\vec{Q}) \, dr, \quad \exp ix = \cos x + i \sin x \quad (2.44)$$

The term $\rho(\vec{r})$ represents the electron density as a function of the position $\vec{r}$ and $\exp(i\vec{r}\vec{Q})$ represents the phase of the scattered wave. We want to know the intensity of scattered radiation at some point P, positioned at a long distance, R relative to the size of the sample, at an angle ϕ with the x-axis (Figure 2.16). In general terms, the intensity of the beam is given by

$$I(Q) = F(\vec{Q})F^*(\vec{Q}) \quad (2.45)$$

where F^* is the complex conjugate of F.

If the positions of atoms in the solid are known to be at $\mathbf{r}_0, \mathbf{r}_1, \mathbf{r}_2, \ldots, \mathbf{r}_N$ (for example, from the set X in Equation 1.47), and if their corresponding atomic scattering factors, $f_1, f_2, \ldots, f_N$, are also known, then the total scattered intensity radiating from such a solid is given by the Debye equation:

$$I(s) = \sum_i^N \sum_j^N f_i(s)f_j(s) \frac{\sin(2\pi s r_{ij})}{2\pi s r_{ij}}, \quad \text{for } i \text{ and } j \text{ from 0 to } N \quad (2.46)$$

where $r_{ij} = |\mathbf{r}_j - \mathbf{r}_i|$ is the distance between scattering centres and $s = |(S - S_0)|/\lambda$ is the magnitude of the scattering vector. The phase difference between any two scattered rays is, $\delta = 2\pi s r_{ij}$, determined only by r_{ij} for a fixed angle of scattering, with an equal probability of phase difference having any value between 0 and 2π.

Spherical symmetry of the electron density can be considered as a time average over the course of an X-ray scattering experiment. Its Fourier transform will also have spherical symmetry, and consequently, it will be a function of the modulus of Q but not its direction. This Fourier transform is usually called the *atomic scattering factor*, given by

$$f(Q) = \int_0^\infty 4\pi r^2 \, \rho(r) \, \sin c(Qr) \, dr \qquad (2.47)$$

The magnitude of the atomic scattering factor is proportional to the atomic number at $2\theta = 0$ and decreases with increasing scattering angle as shown in Figure 2.17.

The total scattering intensity is made up of

- the incoherent scattering from individual atoms, ($i = j$, therefore $r_{ij} = 0$, therefore $I_{incoh} = \sum f_i^2(s)$ only), which is isotropic and contains no information related to the atomic arrangement in the solid

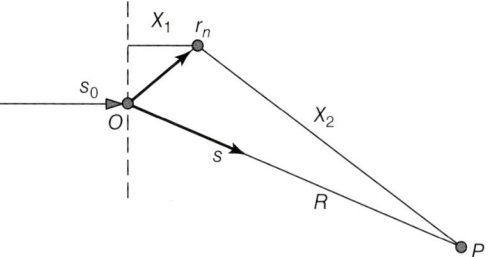

Figure 2.16 The scattering vectors.

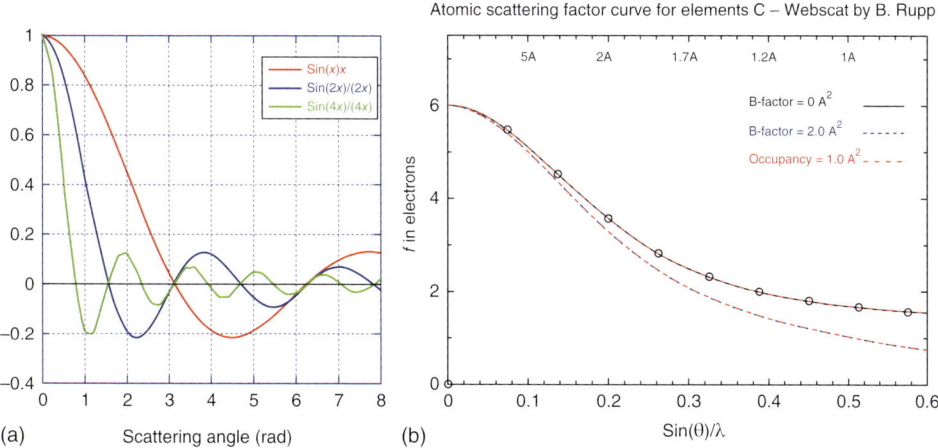

Figure 2.17 (a) The function $\sin x / x$. It has maxima at 0, 7.7, 14.1 and so on and minima at 4.48, 10.9 and 17.3. (b) The atomic scattering function for carbon.

- the coherent scattering, (I_{coh} for $i \neq j$), because of interference between the scattered rays from each i,j pair.

It is clear from Equation 2.46 that the coherent intensity is strongly dependent on the interatomic separation, r_{ij}; it is highest for $r_{ij} = 0$ and decays rapidly for increasing values of r_{ij} (see Figure 2.17). This is of particular importance to amorphous solids that do not possess long range order.

2.2.3.1 Amorphous Solid

Consider a monoatomic homogeneous amorphous solid, so that the atomic scattering factor is $f = f_i = f_j$. Let $\rho(\mathbf{r})$ be the atomic density of the solid at a distance $r = |\mathbf{r}|$ from the atom chosen to be at the origin. Then $\rho(r)$ tends to the average density, ρ_0, when r becomes large. Noting that $4\pi r^2 \rho_r(r) dr$ represents the number of atoms situated in a shell between r and $r + dr$, Equation 2.46 can now be written as

$$I(s) = f\left[1 + \rho_0 \int_0^\infty 4\pi r^2 [\rho_r/\rho_0 - 1] \frac{\sin(2\pi s r)}{2\pi s r} dr\right] \quad (2.48)$$

The ratio, ρ_r/ρ_0, is the radial distribution function or PDF.

As a rule, the positions of atoms in real amorphous solids are not known. It is then a matter to either (i) predicting the radial distribution function (PDF), which defines the spatial configuration of the atoms, that is, the atomic arrangement, or (ii) carrying out the experiment and calculating the atomic arrangement by means of the Fourier transform of the scattered intensity (Equation 2.48).

However, in an ideal monoatomic amorphous solid (IAS) as described here the atomic arrangements are known from Equation (1.65), and furthermore, the corresponding distribution is isotropic. In this case, the vector $\mathbf{r}$ will assume all directions in space with equal probability. The observed diffraction intensity will only be a function of $s = |\mathbf{s}| = (2 \sin \theta)/\lambda$, where θ is the angle between $\mathbf{S}$ and $\mathbf{S}_0$. So the scattered intensity can be represented by a function that is symmetrical around the vector $\mathbf{S}_0$. Examples of predicted X-ray scattering intensity from IAS and Debye equation will be shown in Chapter 3.

2.2.3.2 Ehrenfest Formula

We express the Debye formula in a simplified form as

$$I(Q) = 1 + \frac{1}{N} \sum_i^{n^*} n_i^* \frac{\sin x}{x} \quad (2.49)$$

where

N is the total number of scattering atoms,
n^* is the number of pairs of identical atoms all separated by the same distance, L_i,
$x = QL_i$,
$Q = 4\pi \sin \theta / \lambda$.

The $I(Q)$ curve has a number of maxima that occur at values of Q where the function $dI(Q)/dQ \rightarrow d/dx(\sin x/x) = 0$. Carrying out the differentiation leads to

$$2L \sin \theta = 1.23 \, \lambda, \text{ for the first maximum.} \quad (2.50)$$

This is known as the *Ehrenfest formula*.

2.2.3.3 Polyatomic Solid

For a polyatomic alloy, the Debye equation can be generalized to

$$\frac{I(s)}{N} = I_1 + I_2 - I_3, \quad (2.51)$$

where

$$I_1 = \sum_{i=1}^{N} x_i f_i^2 \quad (2.52)$$

$$I_2 = \sum_{i=1}^{N} \sum_{j=1}^{N} x_i f_i f_j \int_0^\infty 4\pi r^2 \rho_{ij}(r) \frac{\sin(2\pi sr)}{2\pi sr} dr \quad (2.53)$$

$$I_3 = \left[\sum_{i=1}^{N} \right]^2 \int_0^\infty 4\pi r^2 \rho_0 \frac{\sin(2\pi sr)}{2\pi sr} dr \quad (2.54)$$

The equation can be abbreviated to

$$\frac{I(s)}{N} = \langle f^2 \rangle + \langle f^2 \rangle \rho_0 \int_0^\infty 4\pi r^2 [\rho_a(r)\rho_0 - 1] \frac{\sin(2\pi sr)}{2\pi sr} dr, \quad (2.55)$$

with the following substitutions:

$$\langle f \rangle = \sum_{i=1}^{N} x_i f_i \quad (2.56)$$

$$\langle f^2 \rangle = \sum_{i=1}^{N} x_i f_i^2 \quad (2.57)$$

$$w_{ij} = x_i \frac{f_i f_j}{\langle f^2 \rangle} \quad (2.58)$$

$$\rho_a(r) = \sum_{ij} w_{ij} \rho_{ij}(r) \quad (2.59)$$

where i and j refer to the type of atom, x_i is the atomic fraction of ith component and w_{ij} is the weighted atomic scattering factor.

2.2.4
Factors Affecting Integrated Scattered Intensity

2.2.4.1 Integrated Intensity of Powder Pattern Lines from Crystalline Body

$$I_p = |F|^2 \cdot p \cdot \left(\frac{1 + \cos^2 2\theta}{\sin^2 \theta \cos \theta} \right) \tag{2.60}$$

includes the following factors:

1) the structure factor, F,
2) the multiplicity factor, p,
3) the polarization factor,
4) the Lorentz factor.

2.2.4.2 Integrated Scattered Intensity from Monoatomic Body

$$I_N(Q) = I_o \cdot N \cdot f^2(Q) \left(1 + \frac{DW}{N} \sum \frac{\sin(Qr_{mn})}{Qr_{mn}} \right) \tag{2.61}$$

includes the following factors:

1) the structure factor, F,
2) the polarization factor,
3) the Lorentz factor,
4) the Debye–Waller temperature factor, DW.

Experimental Correction Factors

1) Fluorescent radiation emitted by specimen
2) Diffraction of the continuous spectrum
3) Incoherent Compton scattering
4) Coherent scattering
   a. temperature-diffuse scattering
   b. diffuse scattering because of imperfections of atomic arrangements
5) Collimator and beam-stop scattering
6) Air diffuse scattering.

Factors Affecting the Shape of the Diffraction Peak The shape of the diffraction peaks is in general described by the convolution

$$\text{PSF}(\theta) = \Omega(\theta) \otimes \Lambda(\theta) \otimes \Psi(\theta) + b(\theta) \tag{2.62}$$

where $\Omega(\theta)$ is instrumental broadening, $\Lambda(\theta)$ is wavelength dispersion, $\Psi(\theta)$ is specimen broadening and $b(\theta)$ is background function.

2.3
Glass Transition Measured by Calorimetry

Glass-transition temperature is an indirect evidence of random arrangement of atoms in an amorphous solid. At the glass-transition temperature, T_g, there is a change of heat capacity as a result of the change of the liquids elastic, vibrational and thermal properties. It is a thermodynamic property that derives from the liquid to glass transformation during which the random density fluctuations in the liquid become frozen. This property can be measured only during a calorimetric scan over a temperature range that includes the glass-transition temperature, T_g.

An instrument that is used to measure the thermal properties of materials is called the differential scanning calorimeter (DSC). Other calorimetric instruments are a true calorimeter (rarely used for routine applications), differential thermal analyser (DTA), thermogravimetric analyser (TGA) and other related instruments, such as dynamic mechanical analyser. The most essential part of the instrument is an insulated enclosure which in the case of a DSC contains two platinum pans, one for the sample and one for the reference. Each pan has a thermocouple and a heating element built into its base. Small aluminium capsules with lids are used to hold the sample and the reference material. The usual amount of the sample can vary between 5 and 20 mg. An encapsulated sample is placed in one of the platinum pans, which is now called the *sample holder*. The other platinum pan, containing the empty aluminium capsule, is used as a reference, and hence, it is called the *reference holder*. Aluminium is a suitable reference because its thermal properties are known and changes little with temperature in comparison to changes observed in samples in the range from about -100 to $+600\,°C$.

The holders are surrounded by an insulating enclosure so that the temperature of the holders, and their contents, is controlled by the inbuilt heating elements only. The temperatures of the reference holder, T_r, and that of the sample holder, T_s, are measured by the inbuilt sensing elements (thermocouples or temperature-sensitive resistors). There are two types of DSC instruments: a heat flux and power compensation calorimeters. In a power compensation calorimeter, the difference in temperature, $T = T_r - T_s$, is received by the control panel of the DSC, which supplies electric current independently to the heating elements of both holders, and is always trying to keep the temperature of the two holders to be the same. Usually, a program of constant, linear temperature rise, $q = dT/dt\ (K\,s^{-1})$, is selected. Every element of volume of the sample and the reference material must arrive at the programmed rate, as selected. The reference and sample materials absorb power (heat) in proportion to their masses, heat capacities and the selected heating rate:

$$dQ/dt_r = m_r\ C_{pr}\ q, \quad dQ/dt_s = m_s\ C_{ps}\ q \tag{2.63}$$

where $dQ/dt\ (J\,s^{-1})$ is the rate of heat absorption, $C_p\ (J\,(K\,kg)^{-1})$ is the heat capacity at constant pressure, m (g) is mass and the subscripts r and s are used to denote reference and sample, respectively. On the other hand, the power supplied to the

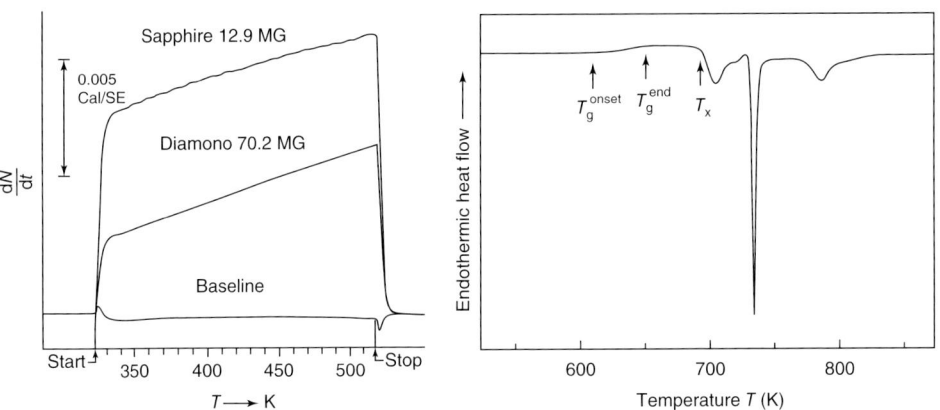

Figure 2.18 DSC recording for crystalline sapphire and diamond and Zr-based amorphous metallic glass. (a) Baseline, calibration and sample analysis and (b) DSC curve of the as-cast $Zr_{41.2}Ti_{13.8}Ni_{10}Cu_{12.5}Be_{22.5}$ BMG.

reference and sample holders is given by

$$dQ/dt_r = k_1 q, \quad dQ/dt_s = k_2 q + k_3 \Delta T \qquad (2.64)$$

where k_1, k_2 and k_3 are calibrated instrumental constants. Notice that when ΔT is not zero, the power supplied to the sample holder is different to that supplied to the reference holder. This differential amount of power becomes the output of the DSC and is recorded as a function of time (together with the rising temperature).

Assuming negligible thermal changes in the reference material, the area under the curve against time (time is proportional to temperature), between any two points on the time axis, is equal to the change in heat content of the sample (enthalpy):

$$\Delta H = k_4 \left(\frac{m}{q}\right) \int \frac{C_p}{dt} dt \qquad (2.65)$$

where k_4 is another instrumental calibration constant.

Examples of DSC recordings for metallic glass and semicrystalline polymer samples are shown in Figure 2.18.

References

Egami, T. and Billinge, S.J.L. (2003) Structural analysis of complex materials, Elsevier Science, Amsterdam.

Finney, J.L. (1970) Random packings and the structure of simple liquids. i. the geometry of random close packing. *Proceedings of the Royal Society of London Series A: Mathematical and Physical Sciences*, **319** (1539), 479–493.

Frank, F.C. and Kasper, J.S. (1959) Complex alloy structures regarded as sphere packings. ii. analysis and classification of representative structures. *Acta Crystallographica*, **12** (7), 483–399.

Hiwatari, Y., Saito, T. and Ueda, A. (1984) Structural characterization of soft-core and hard-core glasses by Delaunay tesselation.

Journal of Chemical Physics, **81** (12), 6044–6050.

Hopkins, A.B., Stillinger, F.H. and Torquato, S. (2010) Densest local sphere-packings diversity. *Physical Review E*, **81**, 041305–041-15.

Mackay, A.L. (1962) A dense non-crystallographic packing of equal spheres. *Acta Crystallogr*, **15** (9), 916–918.

Madras, N. and Slade, G. (1993) *The Self-Avoiding Walk*, Birkhäuser, Boston, MA.

Medvedev, N.N., Geiger, A. and Brostow, W. (1990) Distinguishing liquids from amorphous solids: percolation analysis on the voronoi network. *J. Chem. Phys.*, **93** (11), 8337–8342.

Neder, R.B. and Proffen, Th. (2008) Diffuse scattering and defect structure simulations. Oxford Science Publications. Oxford.

To, L.Th., Daley, D.J. and Stachurski, Z.H. (2006) On the definition of an ideal amorphous solid. *Solid State Sciences*, **8**, 868–879.

Torquato, S. (2002) *Random Heterogenous Materials*, Springer-Verlag, New York.

Welberry, T.R. (2004) *Diffuse X-ray Scattering and Models of Disorder*, Oxford Science Publications.

3
Glassy Materials and Ideal Amorphous Solids

3.1
Introduction

Glassmaking is one of the oldest arts inherited in a continuous line of descent from the ancient civilizations. Like metalworking, it dates back some 5000 years. Many well-preserved examples of ancient glassworker's craft still remain and can be viewed in national museums of countries around the world. Archaeological evidence suggests that the first true glass was made in coastal north Syria, Mesopotamia or Ancient Egypt. The earliest known glass objects may have initially come about as by-products of metalworking (slags) or during the production of Egyptian faience, a vitreous glaze-coated ceramic material made from powdered quartz with small amounts of calcite lime and ash, displaying surface vitrification because of the soda lime and copper pigments to create a bright blue-green lustre. In the modern world, glasses of various kinds are omnipresent materials and entire new technologies have grown up around their manufacture and use.

There are naturally occurring amorphous glasses, such as obsidian mineral (vitrified rock), and solidified organic materials, such as amber (vitrified tree resin). More recent additions to the family of amorphous materials are synthetic polymers (1940s) and metallic glasses (1960s). Jerzy Zarzycki described glass as 'a solid obtained by freezing a liquid without crystallization' (Zarzycki, 1982; Zarzycki, 1991).

The glassy state, which is distinct from either liquid or crystalline state, has relatively recently become the subject of large-scale scientific effort, and our present understanding of it is less complete than our understanding of other forms of matter.

The term *glass* has utilitarian and industrial connotations, such as steel or ceramic or rubber. Its original use for inorganic silica-based glasses has been extended to include organic and metallic glasses and has crept into the language of science of mechanics with such terms as glassy (brittle) behaviour, glassy modulus and so on.

Anglo-Saxon glass has been found across England during archaeological excavations of settlements and cemetery sites, mainly originating from the Roman glass-making centre at Trier, now in modern Germany, and therefore, it is likely that the English word 'glass' originated from the late-Latin/early German term *glesum*.

In French, the word 'verre' comes from the Latin word 'vitrum'.

In Greek, the word for glass 'γυαλί' comes directly from the ancient Greek 'ύαλος' (hyalos, the h in the beginning representing the daseia aspiration always associated with the Greek hypsilon). It is thought that ύαλος itself comes from 'ύδορ (hydor, water) and αλς (hals, sea)'. The former may refer to its transparent appearance, the latter to its origin, presuming that sea sand was used to make silicate glasses.

In Esperanto (according to Wikipedia), 'vitro' estas travidebla, malmola, forta, kemie inerta, kaj biologie nereakciema amorfa solido. Tiel, vitro estas tre utila materialo.

3.1.1
Solidification

From the phenomenological point of view, solidification is the process of converting a liquid into a solid, where solid is defined as a state of a body with memory of its shape of sufficiently long duration (for example, we regard rock a solid even though in geological time rocks can flow, and water a liquid although it stops visibly flowing when it turns into ice). This is broader, but less rigorous than a strictly thermodynamic definition that categorizes substances into gases, liquids and solids, the latter having crystalline forms only. The broader definition allows for glasses to be included in the category of solids.

Liquid is usually transformed into a solid by cooling it. As heat energy is removed from the liquid, at some rate $\dot{q} = \Delta q/\Delta t$, the temperature drops correspondingly at a rate, $\dot{T} = \dot{q}/mc_p$, where m is its mass and c_p its heat capacity. The cooling curve follows a general exponential decay as shown in Figure 3.1. The bifurcation of the curve at point B indicates two possible paths below T_m, one resulting in a polycrystalline solid, and the other in a glassy (amorphous) solid. In either of the two cases, the physical result is a transformation of the liquid into a solid, albeit of different atomic structures. The word *bifurcation* is used here to suggest that a deliberate choice, on our part, of different system parameter values causes a 'qualitative' or topological change in its behaviour.

The evolution of quantum states in the liquid is described *ab initio* by the general Hamiltonian equation:

$$|\phi(t)\rangle = \exp\left(-\frac{iHt}{\hbar}\right)|\phi(0)\rangle \tag{3.1}$$

The physical interpretation of this equation leads to the prediction of interatomic interactions and consequent self-assembly tendencies. In the liquid, these tendencies result in spontaneous and rapid density fluctuations that conform to a Gaussian distribution, as shown in Figure 3.2. Therefore, the density of the liquid at any point, **r**, can be thought of as a time-averaged constant density, ρ_0, plus a

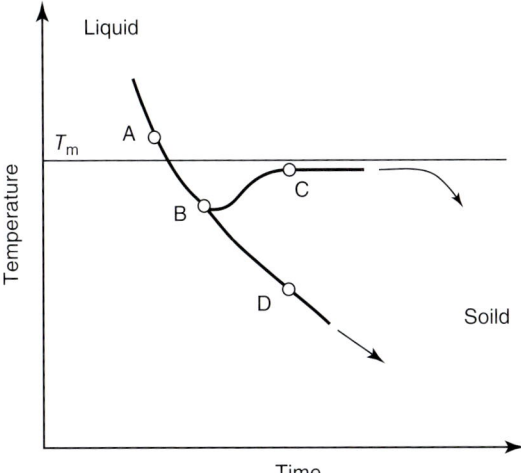

Figure 3.1 A portion of a cooling curve of a system undergoing solidification. T_m is the equilibrium melting temperature. If nucleation and crystal growth occur, the path follows A-B-C; if crystallization does not occur, the path will continue along A-B-D to vitrification.

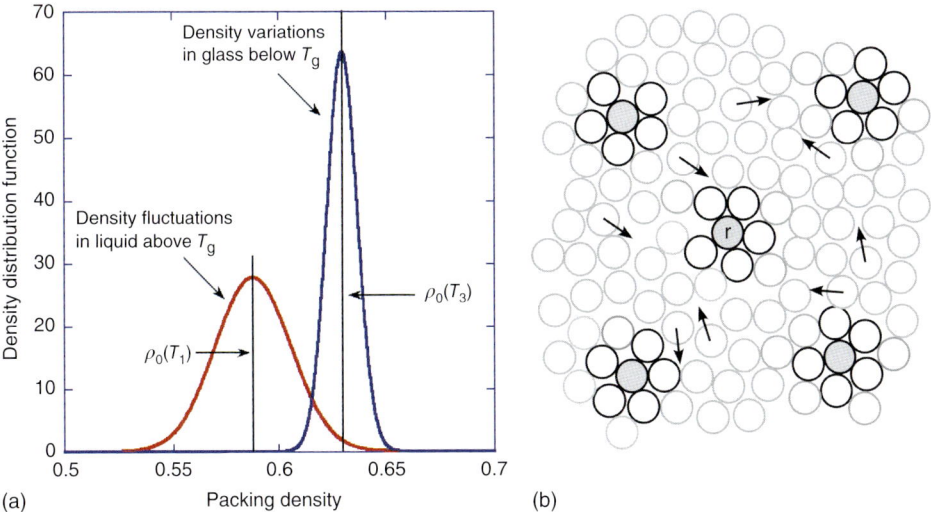

Figure 3.2 (a) A graph showing the extent of density fluctuations in liquid (red) and density variations in glass (blue). (b) Schematic drawing of atomic jumps and density fluctuations in a liquid.

superimposed periodically varying density ρ_p, such that

$$\rho_l(\mathbf{r}, T) = \rho_0(T) + \rho_p(\mathbf{r}, T) \tag{3.2}$$

Approximately two-third of the fluctuations are within ±1 standard deviation. At the high end of density, the fluctuations are manifested in ceaseless formation and solvation of atomic clusters of density close to that of the glass. Depending on certain physical and chemical conditions, these clusters can grow in size to metamorphose into crystalline nuclei or cannot grow (for reasons to be explained later) and remain in random arrangement as local density fluctuations. In the case of continuing cooling, the former process leads to crystallization, and in the latter case it should result in vitrification, that is, glass formation.

3.1.1.1 Solidification by Means of Crystallization

Along the path, A-B-C, as the temperature drops from T_A to T_B below the equilibrium melting temperature, T_m, the Gibbs free energy of a corresponding crystalline solid becomes less than that of the liquid, $G_s < G_l$, so that $\Delta G = \Delta H_f - T_B \Delta S_f \neq 0$ provides the thermodynamic driving force for initiation of the solidification process. The subscript, f, indicates the change in H and in S on fusion.

At the temperature T_B, the high-density clusters grow in size by absorption of atoms onto their surfaces in accordance with appropriate self-assembly rules, thus resulting in near-crystallographic order. This is an exothermic process ($\Delta H_f = H_s - H_l > 0$). A sufficient number of growing nuclei can overcome the heat loss, then the temperature of the system will increase spontaneously, as indicated in Figure 3.1 by the path B-C. At T_C the rate of growth just compensates for the heat loss, and the system reaches near equilibrium condition, so that $\Delta G \cong 0$ and $\Delta S \cong \Delta H_f / T_C$.

These clusters, also referred to as *embryos*, become stable *nuclei* on reaching the critical size

$$r_{crit} \geq 2\gamma_A / g_V^c \tag{3.3}$$

Within the crystal-like cluster of volume V, γ_A is the surface energy per unit area A of the interface between the solid cluster and the surrounding liquid, and g_V^c is equal to the internal energy of interactions per unit volume. In the classical theory of crystallization, it is assumed that g_V^c and γ_A are constant for a given chemically pure system. The arrangement of atoms in these nuclei is assumed to be the same as the arrangement of atoms in the corresponding crystal phase of the solid. In this representation of the solidification process, there are three separately activated steps:

1) formation of clusters of initially highly disordered/random atomic structure, where the rate of formation is given by $\dot{N} = n^* \dot{\phi}$, with n^* as the average number of clusters with critical size given by equation (3.3), and $\dot{\phi} = \sum k_a \phi_a - \phi_d \sum k_d$ is the net arrival of atoms to the clusters with k_a and k_d as rate constants for addition and dissolution of atoms from the surface,
2) rearrangements of atoms towards regular crystalline order, driven by minimization of elastic strain free energy, and controlled by diffusion rate: $J = -D \Delta c$,
3) crystal growth, described by Johnson–Mehl–Avrami kinetic equation with implied functionalities on temperature, time and structure.

Step 1 removes the difficulty with the formation of highly improbable nuclei of perfect-ordered structure. Step 2 allows for the formation of ordered nuclei (required for step three) by annealing and ordering of the imperfect clusters. Step 3 leads to the first-order thermodynamic transformation from liquid to solid at a constant crystallization temperature.

In majority of cases of solidification by crystallization, the final solid is polycrystalline. However, the fundamental theory of the crystalline structure of solids, namely, the theory of crystallography, describes ideal crystals with perfectly ordered arrangements of atoms. None of the imperfections in real materials are considered. Thus, the theory of crystallography serves as the basis for defining the crystalline nature of these solids.

3.1.1.2 Solidification through Vitrification

A system can be made to follow the path, A-B-D, that disallows crystal growth and leads to glass formation under specific physical and/or chemical conditions. These conditions usually imply rapid cooling rates called *quenching*, as represented in Figure 3.1. Evolution of latent heat from this solidification process is insufficient to overcome the heat loss because of the applied high cooling rate, so the temperature of the system falls rapidly. Reducing temperature results in lowering the atomic mobility and diffusion rates in both liquid and solid cluster phases, which in turn slows down atomic rearrangements and density fluctuations. It is well documented that in this process, the liquid viscosity is described by the Vogel–Fulcher–Tammann law (Trachenko and Brazhkin, 2009):

$$\eta \propto \exp \frac{B}{(T - T_V)} \quad (3.4)$$

where B and T_V are constants characteristic of a particular system. It follows from the above equation that the viscosity diverges rapidly towards infinity as temperature of the system approaches T_V. Very high viscosity implies solidity, which brings us back to the definition of a solid.

Another way to understand this solidification process is to consider the liquid viscosity as being influenced by the presence of solid clusters (particles) growing in number with decreasing temperature. Then the viscosity can be described by the Rosco–Einstein relationship, which can be expressed as follows:

$$\eta \propto \exp \frac{H}{(\phi_{crit} - \phi)^\gamma} \quad (3.5)$$

where H, γ and ϕ_{crit} are constants of the system, and $0 \leq \phi \leq \phi_{crit}$. As the volume fraction of high-density solid clusters increases and approaches a critical value, $\phi \to \phi_{crit}$, the viscosity of the supercooled liquid also diverges to infinity (Wang et al., 2007), as follows from Equation 3.5.

With cooling, the average volume of the supercooled liquid decreases (Figure 3.3) and the variance of the density fluctuations decreases (Figure 3.2). This relationship continues below glass-transition temperature, with one fundamental difference. Whilst the density fluctuations in the liquid are stochastic in

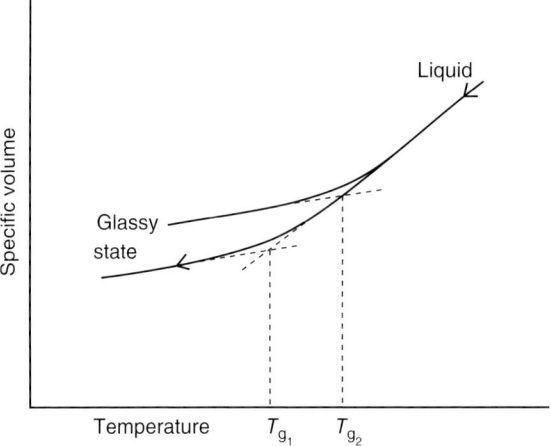

Figure 3.3 Variation of specific volume on cooling of a glass-forming liquid. Different glass states are achieved at different cooling rates.

both time and space, the density variations in the glass become spatially frozen and time independent, resulting in a solid, non-flowing body.

It has been agreed to by convention that the glass-transition temperature, below which the supercooled liquid becomes glass, occurs when its viscosity reaches the value of 10^{12} Pa s. It must be remembered, however, that this is only a convention and that the glass-transition temperature is variable, depending on many factors, such as rate of cooling (or heating), applied pressure, chemical composition, presence of other phases and so on.

There is another process of glass formation based on chemical reaction during which a liquid or a mixture of liquids containing reactive functional groups is converted into a glassy solid by random formation of (usually) covalent intermolecular bonds. It proceeds through a stage called *gelation*, after which the substance becomes elastomeric, and ends with *vitrification* that results in a solid body described by a definition at the start of this section.

In the early theories of Gibbs–diMarcio (Gibbs–diMarcio 1978) and Cohen and Grest, the mean field calculations on Flory–Huggins lattice model predicted a second-order transition in the Ehrenfest sense, usually denoted by T_2, and the dependence of glass-transition temperature on material properties was derived as $dT_g/dP = \Delta\kappa/\Delta\alpha$. Later, the Gibbs–diMarcio argument was amended by introducing an additional parameter for the glass, $z(T,P)$, which links the glass structure to its volume, $V_g(T,P,z)$, and enthalpy, $H_g(T,P,z)$ (Speedy, 1999). Now, the dependence has an additional term that reflects the inherent structure of the glass:

$$\frac{dT_g}{dP} = \frac{\Delta\kappa + \delta V_z/V}{\Delta\alpha} \tag{3.6}$$

Another important equation for glasses relates the glass-transition temperature to the cooling rate and the activation energy, Q_a, of the vitrification processes:

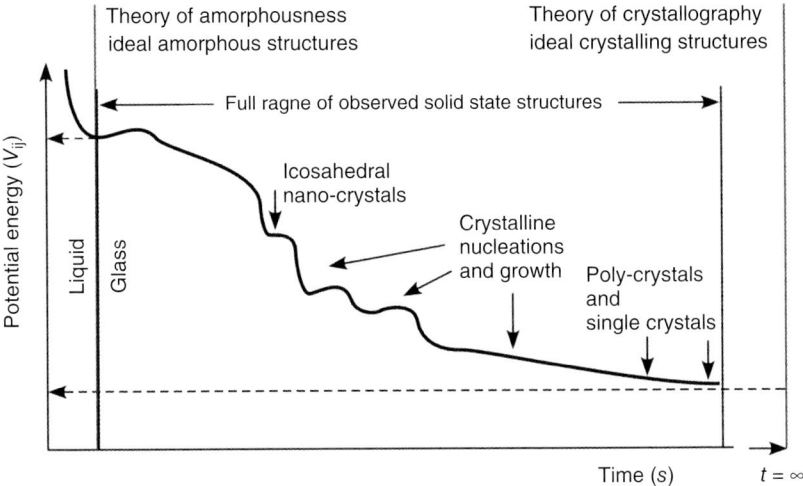

Figure 3.4 Schematic diagram of possible structural evolutions in a hypothetical substance that can form glass as well as grow into a single crystal. The relevance of the theories of amorphousness and crystallography to the world of materials science is indicated.

$$\frac{d\ln |\dot{q}|}{dT_g} = \frac{Q_a}{RT_g^2} \tag{3.7}$$

R is the universal gas constant. This relationship accounts for the variation of glass density with the internal atomic arrangements (inherent structure), which is formed at different cooling rates, shown schematically in Figure 3.3. The formation of random clusters implies a distribution of coordinations that results in local density variations (see Figure 2.7), and the density of glass is varied by formation of imperfections in the random packing of atoms or by reducing and redistributing free volume.

Theory of amorphousness, as outlined in this book, provides the basis for defining the amorphous nature of glassy solids by the full description of the ideal amorphous solid (IAS). In the second part of this chapter, a number of glassy materials are idealized and their atomic arrangements are described in terms of the IAS model defined in Chapter 1. The relationship between the theory of amorphousness and theory of crystallography and the full range of solid-state structures is shown schematically in Figure 3.4.

In considering amorphous structures, we discount solids that are crystalline materials but disordered to such an extent that long-range atomic order becomes undetectable. This corresponds to the case when atomic arrangement is disrupted, say, by heavy particle bombardment leading to the so-called 'amorphous' structure (Johannessen et al., 2007) or by heavy mechanical working (deformation). In principle, these solids can be restored to crystalline order by a suitable annealing. There is an adequate understanding of the observed changes in diffuse X-ray scattering caused by disorder (Welberry, 2004, Nelson, 2002).

3.1.2
Cognate Groups of Amorphous Materials (Glasses)

All amorphous materials (natural and synthetic) can be divided into two distinct broad categories:

Metastable (crystallizable) Glasses. Crystallization is suppressed by thermodynamic factors, such as low value of enthalpy and/or large interfacial energy in combination with sufficiently high cooling rate. However, on annealing (heating) the glasses develop crystallinity.

Inherently Non-crystallizable (Equable) Glasses. In these glasses, there are topological barriers to crystallization in the form or irregular molecular structures, also called *geometrical frustration*.

The first category includes metallic and metalloid glasses, some inorganic glasses and linear macromolecular polymers in which the atomic arrangement is random in the glass, but crystallinity can be developed on annealing or by slow cooling in the first place.

The second category is represented best by materials such as atactic organic polymeric glasses (e.g., atactic polystyrene (a-PS) or atactic poly(methyl methacrylate) (a-PMMA) and by cross-linked polymers (e.g., diglycidyl ether of bisphenol A epoxy with 4,4′-diaminodiphenyl sulphone hardener) the latter forming random three-dimensional network. The randomness of the chemical reaction of copolymerization results in irregular molecular structure that precludes ordered arrangements.

Both characteristics can be combined, as in semi-crystalline polymers or metallic glasses with nanocrystalline particles. However, all glassy solids display common amorphous characteristics; the two most important are:

- X-ray scattering patterns in the form of an amorphous halo
- glass-transition temperature, measured by calorimetric, rheological or other methods.

From the point of view of chemistry, all of the known amorphous materials can be divided into (at least) four cognate classes as listed in the following.

3.1.2.1 Metallic Glasses

The newest group of materials that solidify with amorphous atomic structure are metallic glasses of the metastable category. Examples of commercial metallic alloys are Vitreloy, Metglass and others. There are numerous metallic glasses created and studied in many research laboratories around the world.

In these materials, crystallization is suppressed by alloying a number of elements with selectively chosen atomic radii (geometrical frustration) or by choosing specific composition close to the eutectic. The term *metallic* symbolizes (to a first approximation) non-directional centre-symmetric atomic bonding. It allows a degree of freedom in forming varied atomic cluster arrangements with coordination numbers typically twice as high as compared to the covalent or ionic bonding in organic and inorganic glasses in which coordination between atoms is limited to a small and narrow range. Such man-made metallic glasses are heterogeneous in composition on atomic scale (Torquato, 2002). X-ray or electron scattering patterns show a broad peak (amorphous halo) as attributed to amorphous solids. Most metallic glasses have glass-transition temperature above 600 K and, therefore, are very stable at room temperature. The liquid-to-solid transition is reversible to some degree, although crystallization will occur if annealing continues close to or above the glass-transition temperature (Ma, 2005; Wang et al., 2005). Annealing between glass-transition and recrystallization temperatures usually results in formation of nanocrystals with quasicrystalline structure. With continuing exposure to high temperature, the nanocrystals will undergo phase changes and usually develop into classical stoichiometric intermetallic compounds.

Comprehensive rules for the formation of metallic glasses have been published by Inoue (2000), Ma (2005) and Yavari (2007). A good understanding of the conditions leading to glass formation is gained from the so-called temperature–time–transformation (TTT) diagram, as shown in Figure 3.5. It can be seen from the diagram that the melt of the alloy, initially held at 1100 K, must be quenched to below 600 K in a short interval of 10 s to bypass the onset of crystallization.

Metallic glasses should be distinguished from the so-called *high-entropy alloy*s (HEAs). These are relatively new advanced polycrystalline materials that contain multiple principal elements of near-equimolar ratios. The basic principle and interest in HEAs is that solid–solution phases are stabilized by their significantly high entropy of mixing compared to intermetallic compounds. The configurational entropy is dominant; $\Delta S_{conf} = R\ln n$, where n is the number

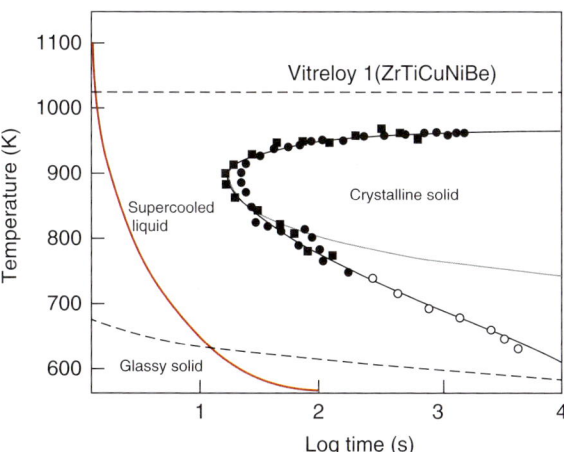

Figure 3.5 Temperature–time–transformation diagram for Vitreloy. Cooling curve—red, melting temperature—horizontal broken line and glass-transition temperature—the lowest broken curve.

of elements in the alloy. For $n = 1$, $\Delta S_{conf} = 0$, for $n = 5$ $\Delta S_{conf} = 1.61\ R$ for $n = 9$ $\Delta S_{conf} = 2.2\ R$. For a review article on HEAs, see Yeh (2013).

3.1.2.2 Inorganic Glasses

Amorphous silica, that is, silicon dioxide SiO_2, can be naturally occurring or synthetic and can be surface hydrated or anhydrous. Synthetic amorphous silica can be broadly divided into two categories of stable materials: (i) vitreous silica or glass that is made by fusing quartz at temperatures greater than 1700 °C and (ii) microamorphous silica, which includes silica sols, gels, powders and porous glasses.

Silicon atom is covalently bonded in a tetrahedral arrangement to four oxygen atoms. Each of the four oxygen atoms is covalently bonded to at least one silicon atom. The atomic structure of these materials is characterized by directional bonds that form three-dimensional spatial networks with an average coordination number in the range between 3 and 5 (the coordination can vary significantly between the liquid and glassy solid). The bond distances and bond angles in amorphous silica are similar to those of cristobalite, Si–O bond distances are 0.16 nm and O–Si–O bond angles are 144°. Surface silanol groups can be isolated from one another, so that intramolecular hydrogen bonding does not occur. Initially formed low-molecular weight species condense to form ring structures so as to maximize siloxane and minimize silanol bonds. A random arrangement of rings leads to the formation of complex structures of generally spherical particles less than 100 nm in diameter.

Studies of radial distribution function (RDF) curves of a vitreous silica have indicated that regions of local atomic order can exist that approximate the cristobalite structure. Whilst amorphous silica is closely related to the cristobalite structure, this local order is believed to be limited to crystalline domains of up to 2 nm in diameter that have completely random orientations and that are thus statistically distributed. Sharp X-ray diffraction patterns (lines) would, therefore, not be obtained (Figure 3.6).

In commercial glasses, the random network is achieved by disrupting the pattern of covalent bonding, for example, by additions of soda and lime to silica, thus preventing the natural tendency of silica to crystallize. At high temperature, large proportions of the bonds dissociate and the three-dimensional network dissolves into a mixture of low-molecular weight fragments, resulting in a relatively low-viscosity liquid. On cooling, the network reforms with viscosity increasing approximately as (critical mass − mass of network)$^{-n}$, where $n \geq 1$. The liquid-to-glass transition is reversible and repeatable. Annealing above the glass-transition temperature may lead to partial crystallization.

Heavy metal fluoride glass is a highly transparent glass developed for use in optical fibers that transmit infrared rays. Chalcogenide glasses consists of elements from the chalcogen group, including selenium, sulphur and tellurium. The glass is transparent to infrared light and is useful as a semiconductor in some electronic devices. Chalcogenide glass fibres are also used in devices used to perform laser surgery.

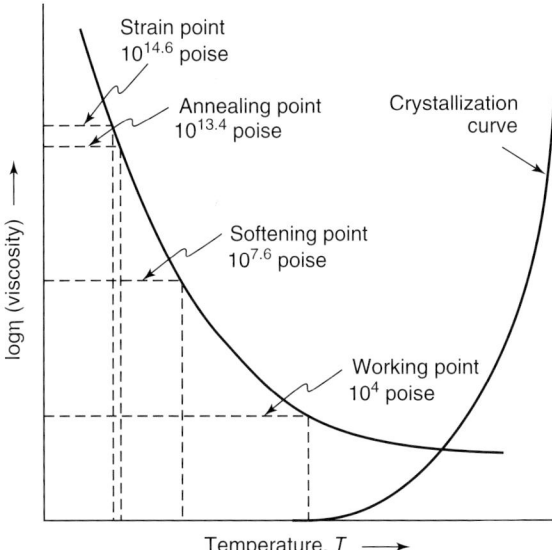

Figure 3.6 Processing window for glass. Temperature must be high enough to make glass flowable, but prolonged exposure to high temperature will start crystallization.

Rules for the formation of these glasses have been established mainly by Zachariasen (1932). Most inorganic glasses of this type form the so-called 'strong glasses', with a narrow distribution in coordination numbers in the glass (Angell, 1985).

Amorphous silicon is a special glassy material. In one form, it contains a significant hydrogen content, creating Si–H bonds. However, a chemically pure a-Si can also be produced (see, for example, Tanaka (1999)).

In this class, we should also include amorphous carbon. The term *amorphous carbon* is restricted to the description of carbon materials with localized π-electrons as described by Anderson. Deviations in the C–C distances greater than 5% are present in such materials, as well as deviations in the bond angles because of the presence of dangling bonds. The above description of amorphous carbon is not applicable to carbon materials with two-dimensional structural elements present in all pyrolysis residues of carbon compounds as polyaromatic layers with a nearly ideal interatomic distance of 0.142 nm and sizes greater than 1 nm.

3.1.2.3 Organic Glasses

These glasses naturally divide into three subclasses:

1) random three-dimensional (cross-linked) networks of copolymers,
2) polymeric glasses based on atactic (non-cross-linked) macromolecules,
3) polymeric glasses based on linear (iso- or syndio-tactic) macromolecules.

The first subclass comprises random networks that form on copolymerization of a wide range of molecules with combined average functionalities greater than

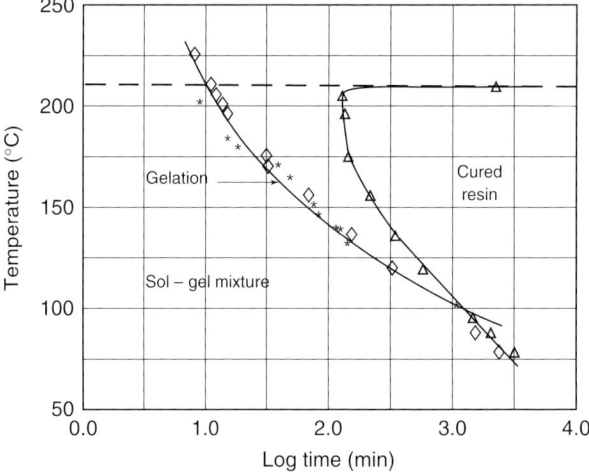

Figure 3.7 Temperature–time–transformation diagram for epoxy-hardener curing system, showing the two stages: gelation and vitrification.

2, for example, epoxy (functionality of 2) and with diaminodiphenylsulfone (DDS) (functionality of 4) as mentioned above (Flory, 1967; Min et al., 1993).

The cross-linking reaction between the components takes place in the mixed liquid state (sol) that allows spatially random bonding to form, resulting in a random, permanent, non-reversible molecular network. The initial liquid transforms into glass through the phenomenon of vitrification by increasing the mass of the cross-linked network (gel) at the expense of the liquid (sol). Such a network can only be disrupted by thermal decomposition or oxidation at sufficiently high temperatures. Heating and annealing below decomposition temperature does not lead to 'melting' or any molecular bond changes. No crystallization can take place. These glasses do not become viscous liquids on heating above glass-transition temperature.

Figure 3.7 shows a TTT diagram for curing (copolymerization) of epoxy with a hardener. In this case, the initial liquid is a mixture of unreacted components, typically held at room temperature. On heating to, say, 150 °C, the chemical reaction starts and proceeds until full cure is achieved, which should be approximately 10^3 min according to the diagram.

The second subclass includes branched polymers with atactic architecture of the macromolecular chain (including random dendrimers). These macromolecular materials are restricted to form glasses on cooling from the molten state because the irregular chain structure prevents self-ordering and registering of the monomer units. Common examples of these glassy polymers are a-PMMA and a-PS. The polymeric materials have permanent covalent bonds, unchanging between liquid and solid state. Annealing above the glass-transition temperature does not lead to crystallization. The liquid-to-glass transformation is, therefore, completely reversible and repeatable (providing oxidation and chain degradation is prevented).

The third subclass includes linear polymers that can be vitrified by quenching, that is, fast cooling. In that sense, their behaviour is similar to that of the metallic glasses so that on annealing, these glasses will develop crystallinity. Common examples of these polymers are poly(ethylene terephthalate) (PET) widely used for drink bottles.

The rules for formation of amorphous polymeric solids were described by Angell (1985) and Flory (1967).

3.1.2.4 Amorphous Thin Films

Materials produced by methods akin to molecular vapour deposition techniques can form amorphous layers. Usually, there is a lack of detailed information about the local and long-range arrangements of atoms in these films; their amorphous nature is usually declared on the basis of absence of any crystalline X-ray diffraction peaks. (It is not generally known if they exhibit glass-transition temperature). On heating, the amorphous structure rearranges and the transformation is non-reversible.

Low-angle X-ray scattering from thin films has been reported, and the structure and properties of amorphous thin films has been described by Gan (2008).

Thin layers between crystalline grains may also exist in amorphous state, as described by Gleiter (1998). In fact, the original concept in physical metallurgy of 'amorphous cement' at the grain boundaries was proposed nearly a century ago by Rosenhein, as recounted in the book by Cahn (2001).

3.2
Summary of Models of Amorphous Solids

There is a large body of published works on dense packing of spheres. The quintessential aspects of such models are the rules established for packing (addition of spheres). Although many methods have been invented to add spheres in a 'random' way, some do not result in ideal random packings because of preferred directionality because of the effects of specific forces, or by implied notions of a lattice with resulting bias towards regular structures. Some of the earliest models with well-defined rules for packing of spheres were published by Finney (1970), Adams and Matheson (1972) and Bennett (1972), and for the creation of a random 3D network by Ordway (1964). In the model of Bennett spheres are added onto sites formed by underlying three-adjacent spheres, a key feature also used for the creation of an IAS, described later herein. However, Bennett's model starts with a regular tetrahedron and, consequently, contains imperfections in the random packing in the form or regular simplexes. A critical evidence for these imperfections is the split in the second peak of the pair distribution function (PDF), described adequately by Clarke and Jónsson (1993) and

Donev et al. (2005). The model of Ordway has been developed further into an elaborate network model by Speedy and Debenedetti (1996).

The literature on the general principles for packing of spheres, with or without lattices, is extensive. For example, by Stoyan et al. (1995), Conway and Sloan (1998), Zong (1999), Torquato (2002) and many other notable contributions. In Chapter 1, we have outlined a unique model of an IAS as described previously elsewhere (To et al., 2006). In order to model solids, one has to consider rigid arrangements of hard spheres, which naturally leads to the so-called closed packing of spheres. Closed packing of spheres refers here specifically to the requirement that all spheres are in fixed positions (Stachurski, 2003) but not necessarily requiring maximum packing density (Torquato et al., 2000; Jiao et al., 2008). Random close packing of spheres should be distinguished from random placing of dispersed spheres in space, of which one example is the Poisson point process (Daley and Vere-Jones, 2008; Quine and Watson, 1984; Torquato, 2002).

The full realization of the amorphous state in solid matter came about as a confluence of several branches of science; the physics and chemistry of colloids and rubber, the technology of silica glass production and the new experimental science of X-ray diffraction. Perhaps the earliest conjecture about the non-crystalline nature of glass came from the work of W.H. Zachariasen on the chemistry of network forming silica glass and the contemporary X-ray experiments on inorganic glasses. In 1934 at the Annual Meeting of the American Ceramic Society in Cincinnati, USA, B.E. Warren put forward a hypothesis that the observed X-ray diffraction patterns from Pyrex-brand resistance glass and another one of vitreous silica were either due to extremely fine crystals of cristobalite, with individual crystals being so small so that a broad diffuse ring is produced or that the X-ray pattern is produced by a non-crystalline random network in which distances between atoms are not fixed to a lattice. He showed that the intensity of scattering is predicted accurately using the Debye equation with the assumption of random distribution of interatomic distances. He concluded that the agreement between the calculated and experimental X-ray scattering curves was sufficiently satisfactory to allow one to consider the random network hypothesis of the glassy state as completely substantiated. Since then many models of amorphous state of solids have been invented, serving different purposes. All models of amorphous structure of solids are either those based on packing of geometrical objects (hard spheres or other solids) or those based on molecular dynamics (MD) and can be loosely gathered under five headings.

3.2.1
Lattice with Atomic Disorder

The earliest models of amorphousness were based on disorder from crystalline lattice. Such models were proposed by Flory (1974) for amorphous structure in organic polymers, and later Gibbs and diMarcio (1978) to calculate and predict physical properties of these substances, such as entropy and glass transition (Figure 3.8).

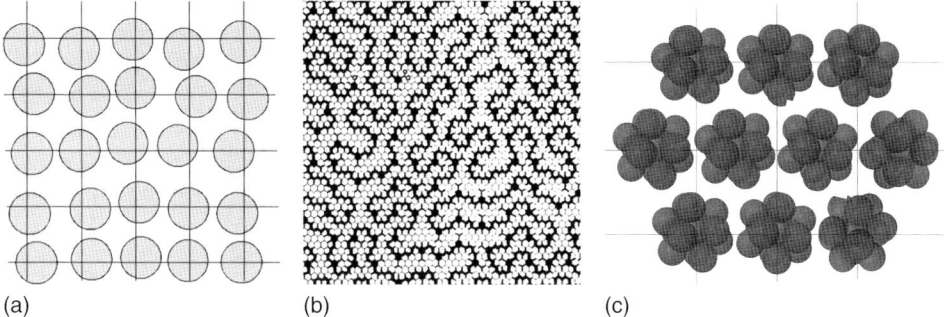

Figure 3.8 Amorphous structure of solids represented as disorder from crystalline lattice or for metallic glasses by disoriented clusters along a lattice.

3.2.2
Disordered Clusters on Lattice

With the discovery of amorphous metallic alloys, it was proposed that their atomic arrangements could be described by clusters containing solute atoms and arranged on a lattice but with random orientation in space. The main proponent of this model is D.B. Miracle (see Nature (2004)).

3.2.3
Geometric Models for Amorphous Networks

A geometrical model for construction of random tetrahedral network was published by Ordway (1964). It was conceived as a structure in which the relation between adjacent units is defined rigidly by the coordination and limitations on bending or rotations of bonds. It involves adding the so-called tetrapods to existing ones following simple topological rules.

3.2.4
Packing of Regular but Incongruent Clusters

In the mid-1900s, Frank and Kasper published papers on the atomic arrangement of intermetallic compounds. When describing the topology of the clusters, they chose clusters with coordination shells that included triangular faces only. In classifying the clusters, they distinguished three cases, depending on the shape of the triangles formed by three adjacent atoms (all atoms of the same size):

- *Case 1.* Clusters in which coordination shell atoms that make equilateral triangles with the center (inner) atom.
- *Case 2.* Clusters in which coordination shell atoms make equilateral triangles; the shell atoms make isosceles triangles with the center.
- *Case 3.* Clusters in which coordination shell atoms make irregular triangles only; the shell atoms make isosceles triangles with the center.

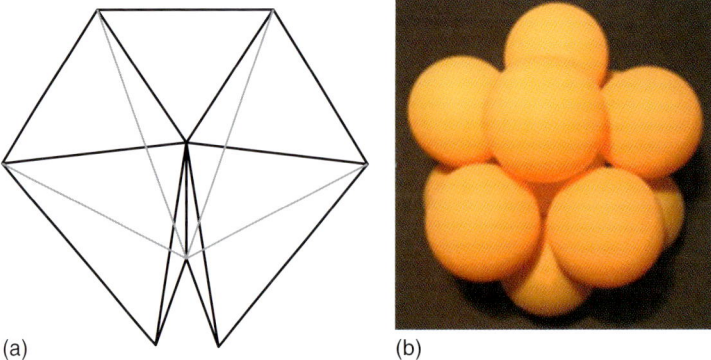

Figure 3.9 Regular but incongruent clusters can be forced to pack with small distortions, such as closing the gap between the five tetrahedra on the left, or packing twenty non-regular (slightly distorted) tetrahedra into an icosahedron on the right.

In metallic glassy alloys, condensation from liquid to solid driven by densification on rapid cooling overrides local preferred interatomic interactions so that contacts with nearest neighbours are made unselectively. The inevitable variance of coordination from cluster to cluster has evoked the idea that atomic arrangements in these alloys could be described in terms of random packing of regular Frank–Kasper (F-K) polyhedra. Such a structure would account seemingly for the disordered nature of these materials (Figure 3.9).

This has given rise to models of amorphous structure in which regular clusters are packed with small distortions into a solid arrangement. Thus, regular icosahedral clusters can be forcefully fitted together to form a solid contiguous packing (Frank and Kasper, 1958), and regular tetrahedral clusters can also be forced into pseudorandom packing (Borodin, 1999).

However, on closer inspection, inconsistencies with this view become apparent. First, the F-K polyhedra possess rotational symmetry, and if packed without distortion, relatively sharp X-ray or electron diffraction spots should occur (instead of diffuse scattering rings). In fact, this is observed when annealing metallic glass leads to the onset of nanocrystallization. Second, if atomic relaxations are allowed to occur so as to fit the polyhedra contiguously, then they are no longer F-K polyhedra satisfying the requirements of case 2 above, and furthermore, the connecting atomic structures will contain polyhedra other than those based on tetrahedral groupings, with quadrangles besides triangles. Strictly speaking, the F-K polyhedra (of case 2) are not an appropriate choice to describe ideal amorphous atomic arrangements, because the atoms in the coordination shells form equilateral triangles.

A geometric model for *pseudorandom* packing of hard spheres has been published by Bennett (1972). It is intended to model a pure non-directionally bonded monoatomic substance. The construction involves placing of spheres, one at a time, at a surface site of an existing cluster. The seed used is an equilateral triangle

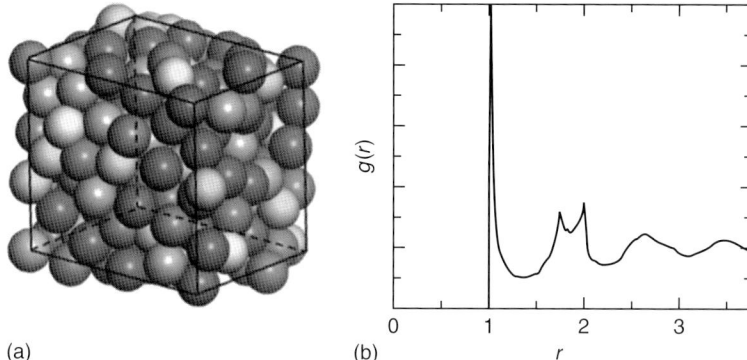

Figure 3.10 (a) A view of a typical simulation cell in molecular dynamics. Atoms are represented as spheres, but in reality, they are centres of forcefields and have no fixed radius. (b) Radial distribution function showing double (split) second peak, characteristic of the presence of regular clusters in random packing.

of three spheres in contact. Addition of one sphere creates the first regular tetrahedron. Further addition of spheres creates a pseudorandom packing of spheres, without long-range order, but with recurring short-range order in the form of close-packed clusters.

The important results from this model are

- pair correlation function shows a major first peak corresponding to the sphere-to-sphere contact distance
- a double (split) second peak, which is indicative of face-centered cubic (fcc)/hexagonal close-packed (hcp) fragments (Figure 3.10).

3.2.5
Irregular Clusters – Random Packing

The third F-K case includes clusters with irregular arrangements of the coordination atoms, comprising non-equilateral triangles as shown in Figure 3.11. Addition of atoms onto three-adjacent-sphere sites of this cluster leads to endless propagation of the irregularity. We regard this case as the basis for the development of a theoretical model of an IAS. The IAS model of a rigid solid requires the spheres (representing atoms) to be impenetrable and in contact with each other. This is a stricter definition of coordination number than that used by Frank and Kasper and others, which makes the topological arguments well defined.

3.2.6
Molecular Dynamics

To obtain amorphous models, crystalline structures are randomized using a set of MD annealing steps that reproduce structural features of actual amorphous solids. This is a completely different type of models from the previous four groups, in which all atoms/molecules are no longer represented by spheres of

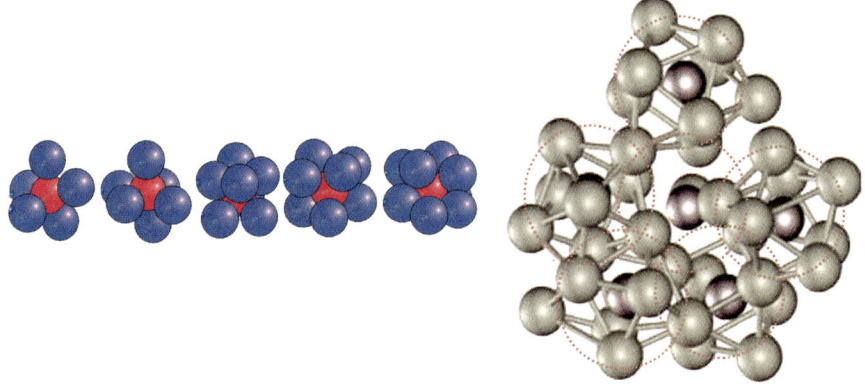

Figure 3.11 Random clusters as building blocks of random packing of spheres.

fixed radius, but instead are centres of interatomic forces with a continuous (differentiable) force–distance relationship (see Introduction to Chapter 1). In these models, randomness of packing is achieved by applying finite displacements (of the order of 0.1 of atomic separation) to each atom in random direction and then solving for a balance of local interatomic forces between neighbouring atoms. This mimics the thermal motions in real solids and, therefore, is becoming the preferred simulation method. Its drawback is the limited capacity to carry out a very large number of displacement cycles for a sufficiently large number of atoms in the simulation cell. This is both time and memory limitation. Present day publications in this field indicate simulations running for 10^6 of cycles, with cell capacity typically 10^5 or fewer atoms. This should be compared with thermal vibrations in solids that are of the order of 10^{14} s^{-1}, and a mole of substance containing 10^{23} atoms.

It will be shown below that the IAS model is a suitable input into an MD simulation, saving on the initial stage of setting up of the cell and its densification.

3.2.7
Monte Carlo Method

Monte Carlo method is a computerized mathematical technique that allows to account for risk in quantitative analysis and decision making. It was invented by Stanislaw Ulam (Figure 3.12). The technique is used widely in disparate fields such as finance, energy, manufacturing, engineering, research, insurance, oil and gas exploration, transportation and the environment.

During a Monte Carlo simulation, values are sampled at random from the input probability distributions. In this way, the simulation provides a much more comprehensive view of what may happen. It gives information about not only what could happen but also how likely it is to happen.

In physics-related problems, Monte Carlo methods are quite useful for simulating systems with many coupled degrees of freedom, such as fluids, disordered

Figure 3.12 Stanislaw Ulam, inventor of the computational Monte Carlo method.

materials, strongly coupled solids and cellular structures. It not a model of random structure, but rather a method of randomizing a structure.

Stanislaw Marcin Ulam was a Polish-American mathematician, born on 3 April 1909 in Lemberg, Austrian Empire (now Lviv, Ukraine).

Ulam studied mathematics with Kazimierz Kuratowski at the Lviv Polytechnic Institute, where he gained his D.Sc. in 1933 under the supervision of S. Banach. In 1935, John von Neumann invited Ulam to come to the Institute for Advanced Study in Princeton, New Jersey. From 1936 to 1939, he spent summers in Poland and academic years at the Harvard University in Cambridge, Massachusetts, where he worked to establish important results regarding ergodic theory. On 20 August 1939, he sailed for America for the last time with his brother Adam Ulam. He became an assistant professor at the University of Wisconsin-Madison in 1940, and a US citizen in 1941.

While at Los Alamos, Ulam developed the 'Monte Carlo method' that searched for solutions to mathematical problems using a statistical sampling method with random numbers.

3.3
IAS Model of a-Argon

Argon is a perfect candidate to form an IAS because it is inert, its interatomic forces are non-directional to a good degree, and it solidifies into a close-packed cubic crystal (Dobbs and Jones 1957). To form glass, liquid argon would have to be quenched to below approximately 55 K at very high cooling rates to prevent crystallization. The glass-transition temperature is assumed to be two-third of its

126 | *3 Glassy Materials and Ideal Amorphous Solids*

known melting temperature of 83.8 K. As an IAS of argon is unlikely to be obtained by experimental means, it is instead simulated by the IAS software. Its characteristics are predicted, recorded and analysed in the following sections.

3.3.1
IAS Parameters

To begin the IAS construction of amorphous argon (a-Ar), we need to select the physical parameters required to simulate the structure. The parameters and some selected physical properties shown in the following have been chosen to simulate amorphous solid a-Ar. The parameters contained in Tables 3.1 and 3.2 are sufficient to simulate a Round Cell of amorphous solid of a-Ar using the IAS software.

The choice of atomic radii is critical; the final properties of the simulated body, such as its density, X-ray scattering pattern and other physical properties, are directly governed by this choice. The size of an atom depends on a number of parameters, such as

- the chemical environment (type of chemical bonding)
- the physical state (gas, liquid or solid, crystalline or amorphous)
- the temperature.

To help choose the parameters, a diagram has been constructed of volume versus temperature from the three publications listed under Table 3.2. The size of the fcc unit cell at 55 K is $a = 0.5453$ nm. Therefore, the calculated size of the corresponding atom radius is $R = 0.193$ nm. The density of the amorphous form of argon (a-Ar) at 55 K is predicted by the packing ratio method to be $(0.63/0.74) \times$

Table 3.1 Atomic parameters for argon at T_g.

Substance	Composition (%)	Atomic radius (nm)	Atomic type/chemistry
Ar	100	0.194	Noble gas at RT

Table 3.2 Physical properties of solid crystalline argon at 83.8 K.

Element		Density (g cm^{-3})	Relative atom mass	Molar volume (cm^3 mol^{-1})	Atom volume (10^{-23} cm^3 atom^{-1})
Ar	fcc	1.62	39.95	29.08	4.83

Data from Dobbs and Jones (1957), MacCrander and Crawford (1977) and Rahman (1964).
In 1 mole of the alloy, there are $N_A = 6.022 \times 10^{23}$ atoms
Molar volume = relative atomic mass/density Atomic volume = molar volume/N_A.

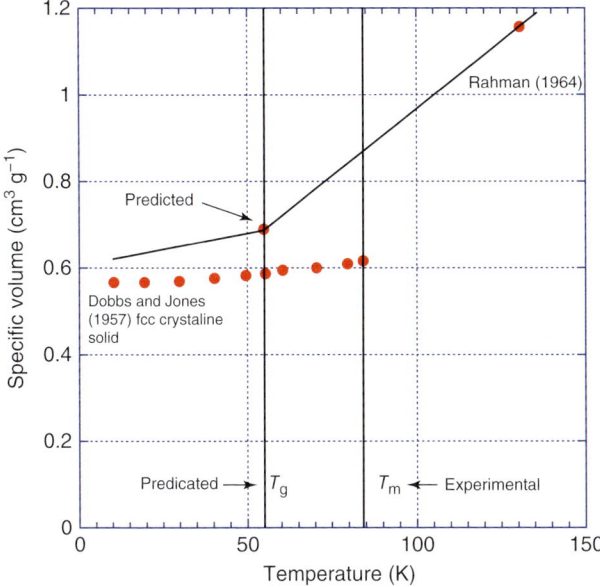

Figure 3.13 Volume–temperature graph for solid and liquid argon.

$1.701 = 1.45\,\text{g cm}^{-3}$, which corresponds to a specific volume of $0.689\,\text{cm}^3\,\text{g}^{-1}$ (Figure 3.13).

3.3.2
Round Cell Simulation and Analysis

The simulated Round Cell of a-Ar (and its cross section) using the IAS software is shown in Figure 3.14. The following statistical characteristics, shown in the following, have been derived from the simulated model.

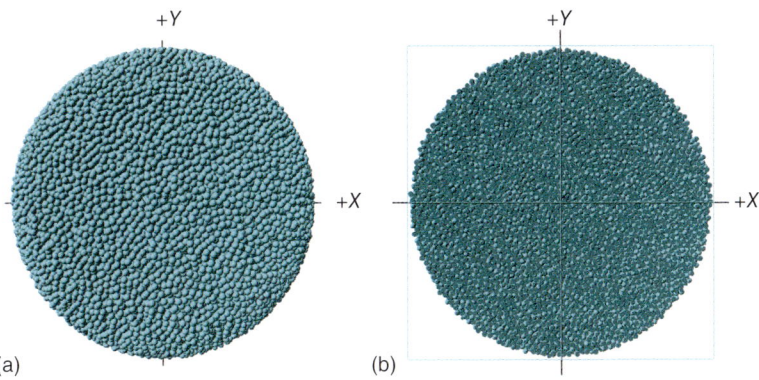

Figure 3.14 A view of the Round Cell and its cross section of solid amorphous argon. 10^5 spheres of the same diameter.

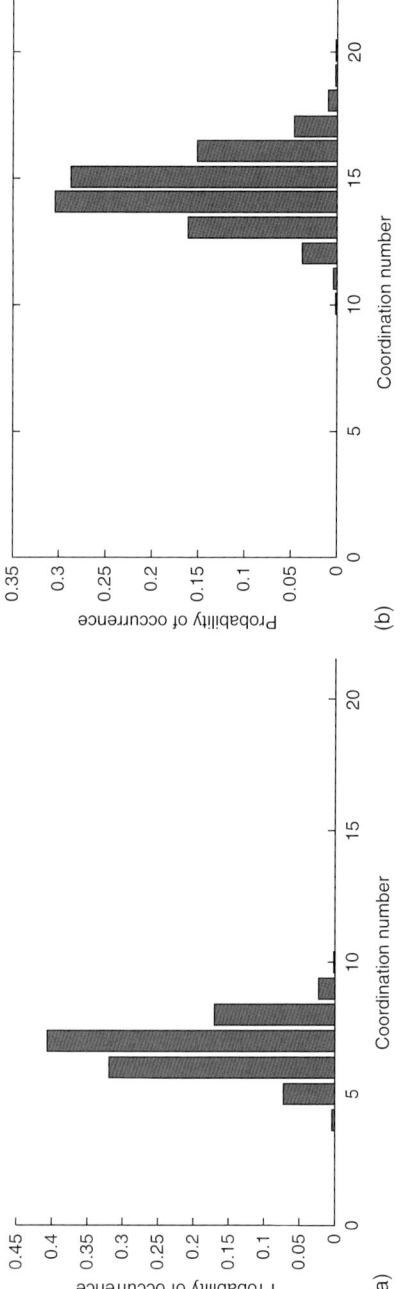

Figure 3.15 Coordination distributions by contact and the Voronoi method for the IAS model of Ar based on 10^5 spheres.

3.3.2.1 Coordination Distribution Function

The irregular primary clusters in the Round Cell have a distribution of coordination numbers as shown in Figure 3.15. The most frequent coordination by direct contact is $k = 7$, whereas it is $k_V = 14 - 15$ by the Voronoi method, also shown in Figure 3.15. The average coordination number by the direct contact method is $\bar{k} = 6.68$ and that by the Voronoi method is $\bar{k}_V = 14.9$.

The distribution for a-Ar shown in Figure 3.15 is similar (but not identical) to that for the cluster distribution obtained by the Fürth model (see Section 1.3.3). This should be expected because the distribution shown in Figure 3.15 is from a Round Cell of 10^5 spheres, whereas that in Section 1.3.3 is predicted by the Fürth formula. However, the similarity of the distributions indicates that the Fürth model is a good approximation for the IAS model.

3.3.2.2 Voronoi Volume and Configuration Distribution Functions

To the first approximation, the average Voronoi volume per atom can be calculated as follows:

$$\overline{V_V} = \frac{1}{p_f}(4\pi R_{Ar}^3/3) = \frac{1}{0.63}(4\pi \cdot 0.185^3/3) = 0.0421 \text{ nm}^3 \qquad (3.8)$$

The mean of the distribution in Figure 3.16 is close to 0.043 nm^3, in reasonable agreement with the calculated value.

The configuration distribution functions, shown in Figure 3.16, resemble theoretical functions derived in Section 1.3.4. The curves for the most abundant k values (5–8) are well defined; there is large scatter because of small number of clusters for $k = 4, 9, 10$.

3.3.2.3 Radial Distribution Function

The RDF, predicted from the simulated Round Cell, is shown in Figure 3.17. Its main peak is positioned at 0.37 nm, twice the assumed radius of the atom. The initial slope is not infinity because the determination of the PDF was obtained using Equation 1.55 with $\Delta r = 0.026$ nm. The area under the first peak is proportional to the average number of Voronoi nearest neighbours, $\bar{k}_V = 14.9$ (see Equation 1.56).

The IAS PDF can be compared to that obtained for liquid argon by MD method, included in the same figure. There is close correspondence between the appearance and positions of the peaks; for the liquid, the first peak occurs at 3.7 Å (0.37 nm), and for the IAS, it occurs at 0.386 nm because of larger atomic radius adopted in the simulation. The second peak in the MD results appears smooth as expected of L-J liquid with wider distribution of second-nearest-neighbour interatomic distances. The IAS results show unexpected shape of the second peak, which suggests some regular clustering although this is not supported by analysis of the packing.

In the MD simulation of liquid argon by (Rahman, 1964), the following parameters were used: density at 94.4 K, $\rho = m/V = (864 \cdot 39.95 \cdot 1.6747 \cdot 10^{-27})/((10.229 \cdot 3.4)^3) = 1.374$ kg m^{-3}, and atomic packing fraction, $ap_f = V_{atoms}/V_{cell} = (864 \cdot (4\pi/3)R^3)/((10.229 \cdot 3.4)^3) = 0.534$ and L-J potential $\epsilon = 1.65 \times 10^{-21}$ J.

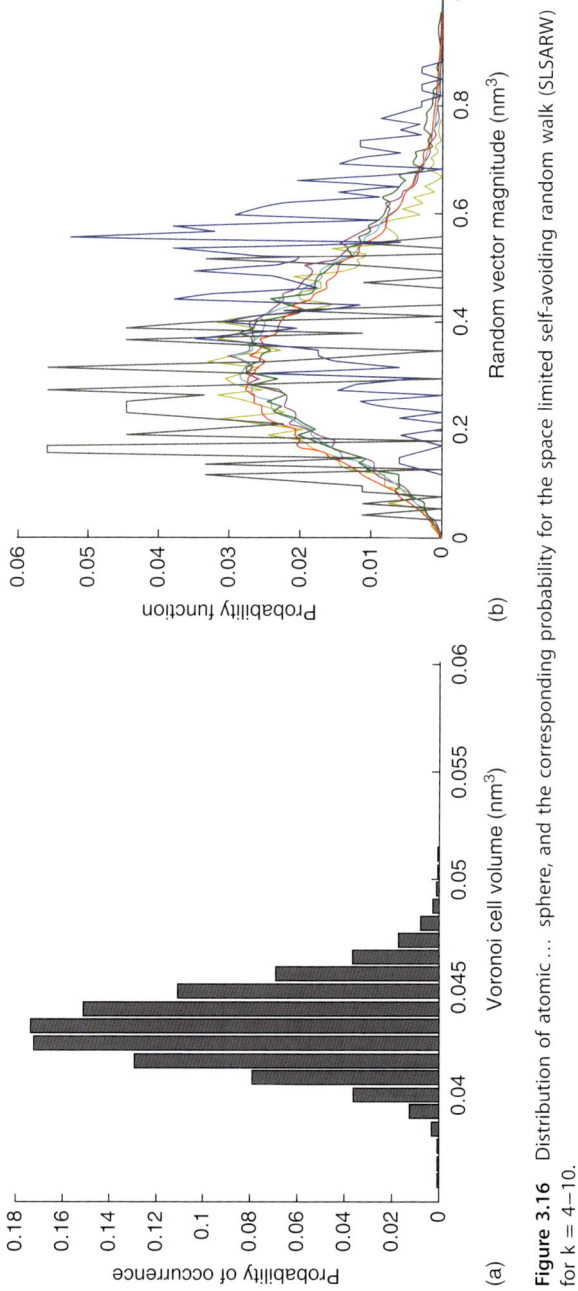

Figure 3.16 Distribution of atomic... sphere, and the corresponding probability for the space limited self-avoiding random walk (SLSARW) for k = 4–10.

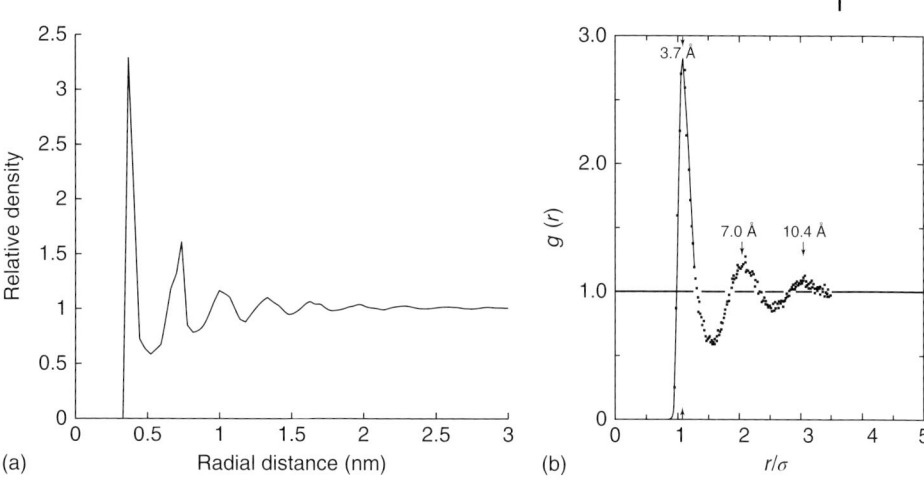

Figure 3.17 (a) Radial distribution function derived from the IAS model of a-Argon with $\Delta r = 0.026$ nm. (b) Radial distribution function for liquid argon at 94.4 K. (*Source*: Reprinted with permission from Rahman (1964), Copyright (1964) by American Physical Society.)

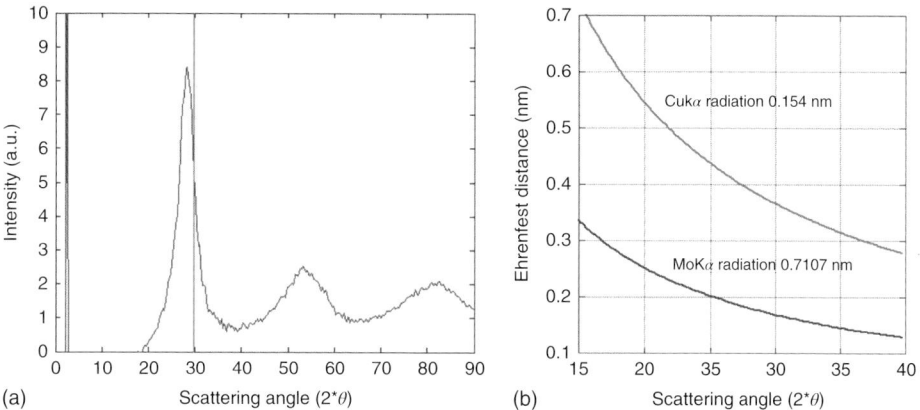

Figure 3.18 (a) Predicted X-ray scattering intensity for the IAS model of a-argon. (b) The relationship between scattering angle and Ehrenfest distance (Equation 2.50).

Notice the relatively low atomic packing fraction for liquid compared to that for the IAS (solid).

3.3.2.4 X-ray Scattering from the IAS Model

The position of the main scattering peak for amorphous argon can be predicted with the Ehrenfest formula (Equation 2.50)

$$\sin\theta = \frac{1.23 \cdot 0.154}{2 \cdot (2 \cdot r_{Ar})} = 0.2441 \rightarrow \theta = 14.1° \tag{3.9}$$

giving the scattering angle for the main peak as 28.3° for CuKα radiation (Figure 3.18).

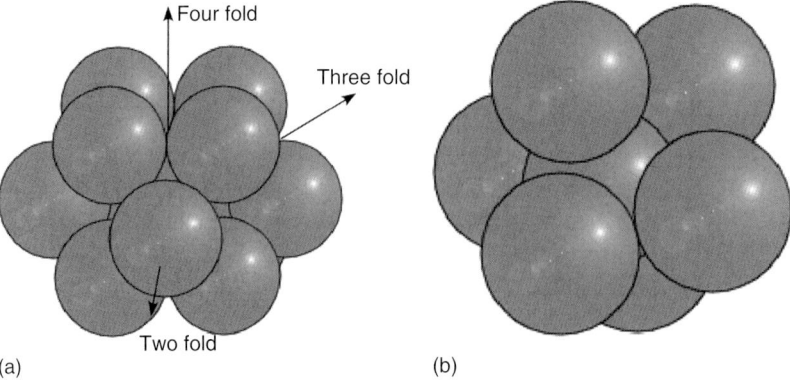

Figure 3.19 (a) A cluster of crystalline argon with twofold, threefold and fourfold axes of rotational symmetry as well as mirror symmetry elements. (b) An irregular primary cluster of amorphous argon without any symmetry elements.

3.3.2.5 Crystalline and Amorphous Cluster

The geometrical appearance of a crystalline and random cluster of solid argon in both states is shown in Figure 3.19. The crystalline cluster, corresponding to the fcc structure, has coordination number of 12; the random cluster, corresponding to the amorphous structure, has the most probable coordination number of 7.

The packing fraction of the crystalline cluster is 0.74, the most probable packing fraction of the random cluster is 0.635 (Figure 2.7).

3.3.3
Summary of a-Ar IAS Structure

Amorphography: monoatomic IAS Class I
Chemical Variety: $m = 1, R_{Ar} = 0.194$ nm
Chemical Composition: Argon 100%
Glass-Transition Temperature: $T_g = 55$ K (predicted)
Crystalline Melting Point: $T_m = 83.8$ K (experimental)
Atomic Positions: Represented by closed packed spheres but randomly distributed in space in the range $[2R; \infty)$.
  The randomness of the atomic positions is evidenced by
  - distribution of Voronoi volumes (see Figure 3.16)
  - PDF (see Figure 3.17)
  - X-ray scattering pattern.
Atomic Correlations: Direct contacts between spheres in the IAS model results in correlations that are evidenced by the PDF (see Figure 3.17)
The total number of contacts is $K = (1/2)N\bar{k}$, where $\bar{k}$ is the average coordination number and N is the total number of spheres.
Free Volume: $v_f = 0$
Packing Fraction: $p_f = 0.63 \pm 0.05$
Predicted Density: $\rho_{amorphous} = 1.45(\pm 0.12) \times 10^3$ kg m^{-3}.
Predicted Density: $\rho_{crystalline} = 1.701 \times 10^3$ kg m^{-3}

3.4 IAS Model of a-NiNb Alloy

3.4.1 Introduction

Primary uses of this alloy include bearing assemblies, casting, step soldering, as components of electric circuits to moderate changes in current, and radiation shielding. NiNb-based amorphous alloys have high absorption capacity, with hydrogen-to-metal ratio of 1.8–2.30 wt.% and excellent embrittlement resistance because of their unconventional structure. The two elements also form an intermetallic compound, Ni_3Nb.

3.4.2 IAS Model of the Alloy

The parameters shown in Table 3.3 were used to create a model of the NiNb alloy of 62:38 composition with the results shown in Figure 3.20. The cross section, in particular, reveals the random nature of the packing. The other characteristics of the packing corroborate with this observation.

3.4.2.1 Coordination Distribution Functions

Analysis of the IAS model of NiNb alloy gives results for the coordination numbers by direct contact and Voronoi tessellation as shown in Figure 3.21. The direct contact distribution is qualitatively similar to that derived by the Fürth method

Table 3.3 Physical properties of the elements (public domain data).

Element		Relative atomic mass	Density (g cm^{-3})	Molar volume (cm^3 mol^{-1})	Atomic volume (10^{-23} cm^3 atom^{-1})
Ni	fcc	58.69	8.91	6.59	1.09
Nb	bcc	92.91	8.57	10.84	1.80

bcc, body-centred cubic.
In 1 mole of the alloy, there are $N_A = 6.022 \times 10^{23}$ atoms.
Molar volume = Atomic mass/Density.
Atomic volume = Molar volume/N_A.

Table 3.4 Atomic parameters (public domain data).

Element	Atomic radius (nm)	Atomic concentration (%)	Atom type/chemistry
Nickel	0.125	62	Transition metal
Niobium	0.146	38	Transition metal

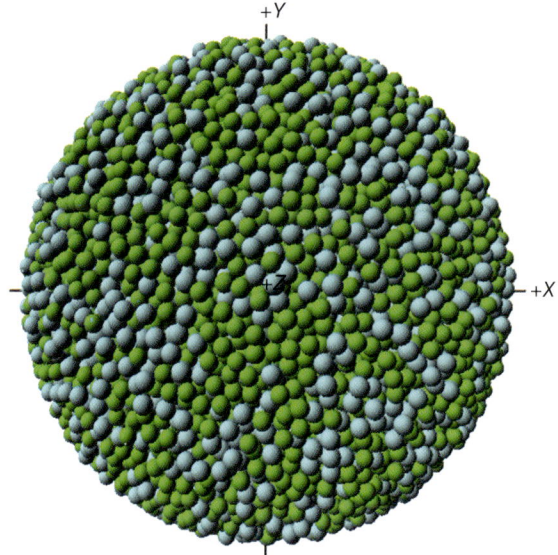

Figure 3.20 IAS Round Cell of the a-NiNb model with 10^4 atoms. Ni is in green, Nb is in blue.

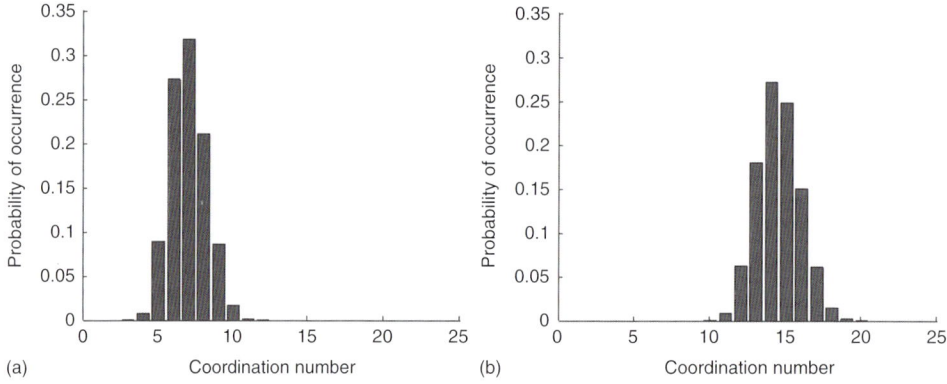

Figure 3.21 Distribution of coordination numbers by direct contact and Voronoi method derived from the IAS model of the NiNb alloy.

in Chapter 1 (see Figure 1.15), but with a small difference. Here we can see a very small probability of clusters with coordination $k = 11$ that is absent in the monoatomic clusters. Noting that the ratio of radii is Ni/Nb ~ 0.85, we find from Table 1.1 that the maximum possible number of contacting outer spheres is 14 for a cluster comprising Nb as the inner sphere and Ni as all of the outer spheres. So $k = 11$ is possible, although the probability of that occurring in the IAS packing is quite small. The same remarks apply to the distribution of neighbours obtained by the Voronoi method.

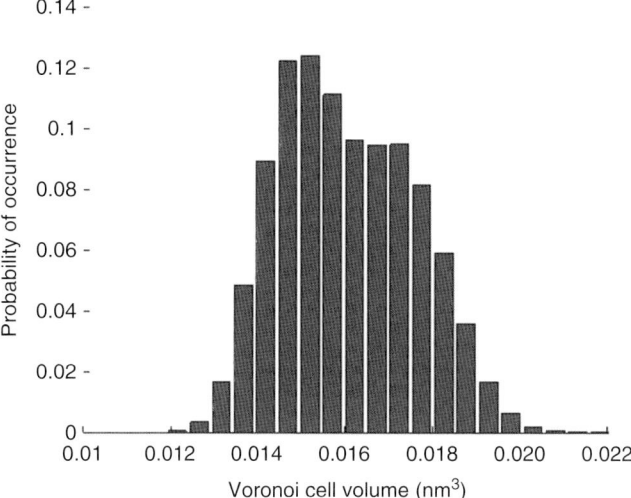

Figure 3.22 Distribution of coordination numbers by direct contact and Voronoi method derived from the IAS model of the NiNb alloy.

3.4.2.2 Voronoi Volume Distribution

A more interesting result is obtained for the Voronoi volume distribution; it appears to be a sum of two distributions, one for Ni atoms and the other for Nb atoms. The expected average Voronoi volume for the Ni atoms alone can be calculated as $(1/0.63) \cdot (4\pi\, R_{Ni}^3/3) = 0.0130$ nm^3, whereas that calculated for the Nb atoms alone is $(1/0.63) \cdot (4\pi\, R_{Ni}^3/3) = 0.0207$ nm^3. These values are at the extreme ends of the distribution in Figure 3.22, and therefore, it is not a simple sum of two separate distributions, but rather a convolution of the two, with the result appearing in the graph. The composition of the alloy, which is approximately two-third of Ni to one-third of Nb, is reflected in the asymmetry of the graph.

3.4.2.3 Pair Distribution Function

The PDF for the IAS model, spanning the radial distance from 0 to 2 nm, is shown in Figure 3.23. The graph of the high-resolution PDF, spanning the radial distance from 0.24 to 0.295 nm, is also shown in the same figure. The first peak appears at 0.25 nm, which correspond to the interatomic contact distance of Ni pair, and the last peak appears at 0.292 nm, which corresponds to the interatomic distance of Nb pair. The middle peak represents the Ni–Nb pairs. All three peaks combine to give the first peak of the main PDF graph.

3.4.2.4 Probability of Contacts

According to Equation 1.63, the probability of contacts between the two types of atoms, Ni and Nb, can be calculated as shown in Table 3.5.

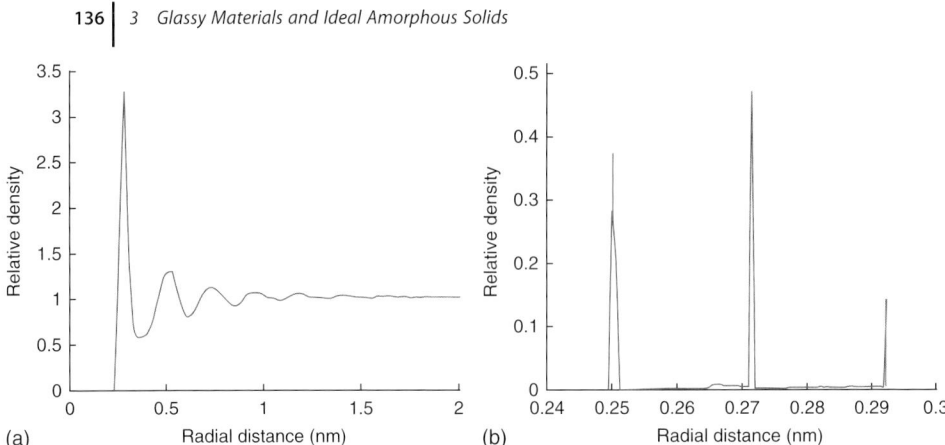

Figure 3.23 Radial distribution function derived from the IAS model of a-NiNb alloy.

Table 3.5 Probability of contacts from Equation 1.63 and from Figure 3.23.

	Ni	Nb	Ni	Nb
Ni	0.384	0.471	0.374	0.476
Nb	—	0.145	—	0.174

The frequency of contacts between the elements can be also measured from Figure 3.23 as the height of the peaks. Then a direct comparison between these values and those calculated in Table 3.5 can be made as shown in Figure 3.24. It can be noted that to a first approximation, a linear relationship is followed. However, there are deviations from the straight line, and it is conjectured that these are due to the relatively small size of the Round Cell (10^6 atoms), insufficient for more accurate statistics. The deviations should become smaller with increasing size of the Round Cell.

3.4.3
X-ray Scattering from a-NiNb Alloy

3.4.3.1 Experimental Results

Figure 3.25 shows synchrotron experimental data for this amorphous alloy (2012). The scattering intensity is plotted against the wave vector, $q = \sin\theta/\lambda$. The first main peak appears at $q = 2.95$ nm^{-1}. From the relationship $\sin\theta = q \cdot \lambda$ and appropriate transformation to CuKα radiation, we obtain for the first peak $\theta = 21.4°$, which gives the scattering angle $2\theta = 42.8°$.

There are two overlapping peaks appearing between $q = 4$ and 6. The crystalline phases of this system include Ni_3Nb, Ni_6Nb_7 and Ni_8Nb, each with different crystal structure and unit cell parameters. Ni_3Nb has the highest enthalpy of formation (31 kJ mol^{-1}), compared to Ni_8Nb (15 kJ mol^{-1}) Joubert et al., (2004).

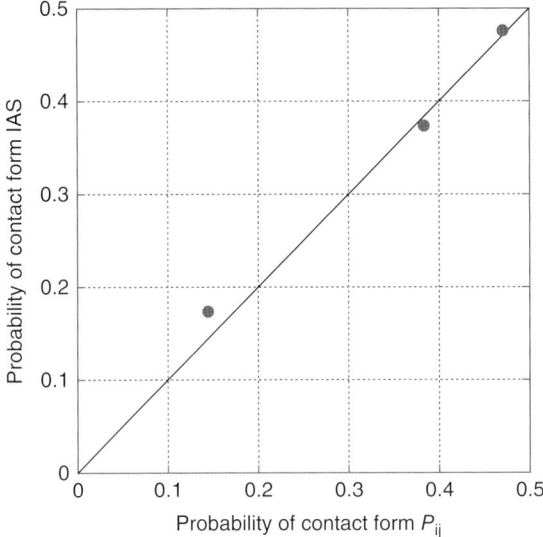

Figure 3.24 Comparison of contact probability calculated by Equation 1.63 with that measured from Figure 3.23.

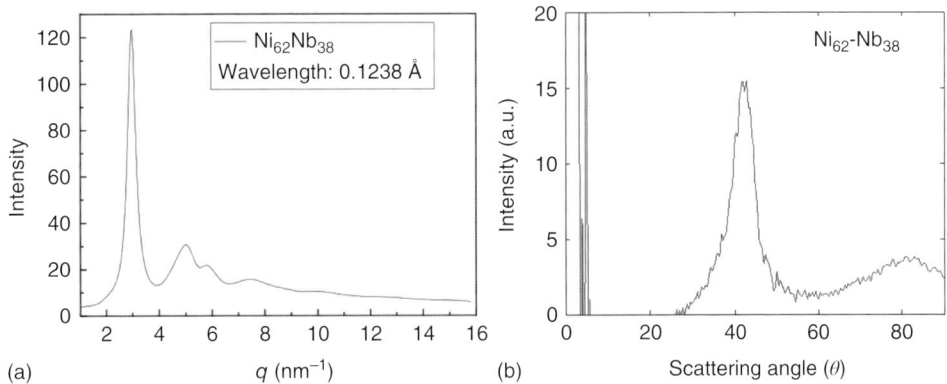

Figure 3.25 (a) Neutron scattering profile for the NiNb alloy. (Reprinted with permission from Wang et al. (2012), Copyright (2012) by American Physical Society.) (b) x-ray scattering profile for the same alloy predicted from the IAS model using the Debye equation.

On this basis, let us assume that the amorphous alloy contains fragments of the Ni_3Nb crystallites (nanocrystals) with orthorhombic unit cell of dimensions: $a = 0.5116, b = 0.4565, c = 0.4258$ (Nash and A., 1986). For the orthorhombic system, the basic equation governing the Bragg angle is $\sin^2(\theta) = Ah^2 + Bk^2 + Cl^2$, with A, B, C as unknown constants to be determined by analysis. No doubt that these nanocrystals would contribute to the intensity of the main peak, but such a contribution would be subsumed under the main Debye scattering. It is conjectured that a secondary effect is the appearance of the two peaks between $q = 4$ and $q = 6$ in the experimental result.

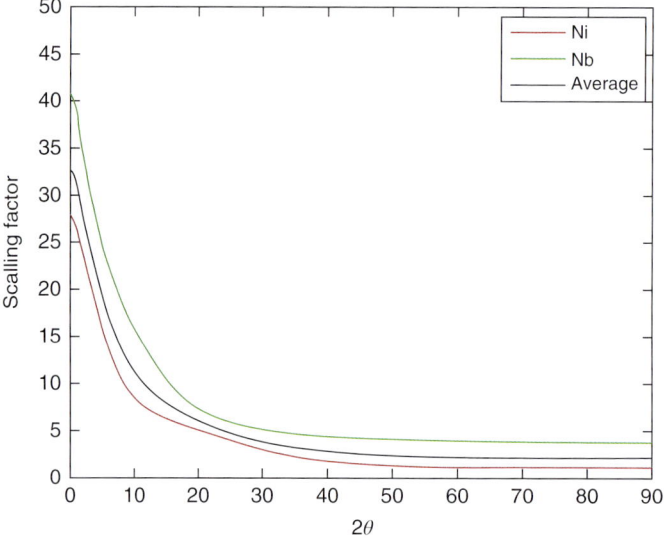

Figure 3.26 Atomic scattering factors for nickel and niobium.

3.4.3.2 Theoretical Results

The computations of X-ray scattering of the a-NiNb alloy using the Debye equation require the atomic scattering factors and the coordinates of the atomic positions, obtained from the IAS model. The atomic scattering factors are shown for Ni and Nb in Figure 3.26 together with their average calculated as $f_{NiNb}(2\theta) = (62/100) \cdot f_{Ni}(2\theta) + (38/100) \cdot f_{Nb}(2\theta)$.

The final result for the intensity versus scattering angle, calculated by the Debye equation, is shown in Figure 3.25 in comparison with the experimental curve. Its main peak occurs at 42.6°, very close to that derived from the experimental data. The second peak, which is a single peak, appears at a scattering angle of 82°. Taking this angle as input, the corresponding value for the scattering vector is calculated as $q = \sin\theta/\lambda = 5.3$ nm^{-1}, which is very close to the first of the two overlapping peaks.

3.4.4
Density of a-Ni62-Nb38 Alloy

3.4.4.1 Crystalline Alloy

Density of the crystalline alloy calculated from molar quantities:

$$\rho_{molar} = \left(\sum \text{at.frac} \times \text{at.wt.}\right) / \left(\sum \text{mol.vol.} \times \text{at.wt.}\right)$$
$$= 71.7/8.20 = 8.74 \text{g cm}^{-3} \text{(data from Tables 3.3 and 3.6)} \quad (3.10)$$

3.4 IAS Model of a-NiNb Alloy

Table 3.6 Calculation of volume fraction of the alloy elements.

Element	Atomic fraction	Number of atoms (10^{23})	Volume of atoms (cm^3)	Volume fraction of element
Ni	0.62	3.734	4.070	0.497
Nb	0.38	2.288	4.118	0.503
Total	1.00	6.022	8.188	1.000

Table 3.7 Calculation of weight fraction of the alloy elements.

Element	Atomic fraction	Mass of atoms (g)	Weight fraction of element
Ni	0.62	36.39	0.507
Nb	0.38	35.31	0.493
Total	1.00	71.70	1.000

Density of the crystalline alloy calculated by *atomic fraction*:

$$\rho_{at.frac.} = [0.62 \times 8.91 + 0.38 \times 8.57] = 8.78 \text{g cm}^{-3} \quad (3.11)$$

Density of the crystalline alloy calculated by *volume fraction*:

$$\rho_{vol.frac.} = [0.497 \times 8.91 + 0.503 \times 8.57] = 8.74 \text{g cm}^{-3} \quad (3.12)$$

Density of the crystalline alloy calculated by *weight fraction*:

$$\rho_{wt.frac.} = [0.507 \times 8.91 + 0.493 \times 8.57] = 8.74 \text{g cm}^{-3} \quad (3.13)$$

Note the agreement of densities calculated by all four methods. The above calculations are shown in full to demonstrate the methods used. It will be shown later, for the next two amorphous solids (a-MgCuGd and a-ZrTiCuNiBe), that the only correct methods are by molar quantities and volume fraction (Table 3.7).

3.4.4.2 Amorphous Alloy

According to the IAS model, the packing fraction of the amorphous alloy is $pf_a = 0.63 \pm 0.005$.

Nickel in the crystalline state has packing fraction, $pf_{Ni-c} = 0.740$, and Niobium packing fraction is $p_{Nb-c} = 0.68$. Then the density of an amorphous alloy is predicted by the packing ratio method as follows:

$$\rho_{am.alloy} = \left(\frac{pf_a}{pf_c}\right)_{Ni} (\text{vol.fr.} \times \rho)_{Ni} + \left(\frac{pf_a}{pf_c}\right)_{Nb} (\text{vol.fr.} \times \rho)_{Nb} \quad (3.14)$$

$$\rho_{(0.63)} = (0.851)(0.497 \times 8.91) + (0.926)(0.503 \times 8.57) = 7.76 \text{g cm}^{-3}$$

Expression 3.14 is based on the assumptions that the volumes occupied by atoms are the same in the amorphous alloy as in the crystalline alloy. This is correct for the IAS model of hard sphere packing but may not be true in real solids.

3.4.5
Summary of a-NiNb IAS Structure

Amorphography: biatomic IAS Class I
Chemical Variety: $m = 2 : R_{Ni} = 0.125, R_{Nb} = 0.146$ nm
Chemical Composition: Ni 62%, Nb 38%
Glass-Transition Temperature: $T_g = 1066$ K
Crystalline Melting Point ($Ni_6 Nb_7$): $T_m \sim 1563$ K
Atomic Positions: Represented by closed packed spheres randomly distributed in space in the range $[2R; \infty)$. The randomness of the atomic positions is evidenced by

- distribution of Voronoi volumes (see Figure 3.16).
- pair distribution function and contact probability (see Figures 3.23 and 3.24).
- X-ray scattering pattern (see Figure 3.26).

Atomic Correlations: Direct contacts between spheres in the IAS model results in correlations that are evidenced by the fine PDF (see Figure 3.17).
The total number of contacts is $K = (1/2)N\bar{k}$, where $\bar{k}$ is the average coordination number and N is the total number of spheres.
Free Volume: $v_f = 0$
Packing Fraction: $p_f = 0.63 \pm 0.05$
Predicted Density: $\rho_{amorphous} = 7.76(\pm 0.62) \times 10^3$ kg cm^{-3}
Measured Density ($Ni_6 Nb_7$): $\rho_{crystalline} = 8.74 \times 10^3$ kg m^{-3}

3.5
IAS Model of a-MgCuGd Alloy

3.5.1
Physical Properties of the Elements

Published tables of chemical elements (Internet, textbooks and other publications) can be used as a source of atomic dimensions. Tables (3.8 and 3.9) show parameters collated for the elements of the amorphous alloy of Mg65-Cu25-Gd10 (a-MgCuGd).

3.5.2
IAS Simulation of a-MgCuGd Alloy

For the IAS simulation, the choice of atomic radii is critical; the final properties of the simulated alloy, such as its density, X-ray scattering pattern and other physical

Table 3.8 Atomic parameters (public domain data).

Element	Atomic radius (nm)	Atomic concentration (%)	Atom type/chemistry
Magnesium (Mg)	0.146	65	Alkaline earth metal
Copper (Cu)	0.128	25	Transition metal
Gadolinium (Gd)	0.176	10	Rare earth element

Table 3.9 Physical properties of the elements (public domain data).

Element		Relative atomic mass	Density (g cm^{-3})	Molar volume (cm^3 mol^{-1})	Atomic volume (10^{-23} cm^3 atom^{-1}]
Mg	hcp	24.3	1.738	14.0	2.35
Cu	fcc	63.54	8.92	7.12	1.18
Gd	hcp	157.25	7.90	19.9	3.31

In 1 mole of the alloy, there are $N_A = 6.022 \times 10^{23}$ atoms.
Molar volume = Atomic mass/Density.
Atomic volume = Molar volume/N_A.

Table 3.10 Atomic parameters used for the IAS simulation.

Element	Atomic radius (nm)	Atomic concentration (%)	Atomic volume (nm^3)
Magnesium (Mg)	0.150	65	0.0141
Copper (Cu)	0.135	25	0.0103
Gadolinium (Gd)	0.180	10	0.0244

properties, are directly governed by this choice. We recall the property of scale invariance: if $S_a + X$ is a packing for an amorphous solid, then so is $S_{ca} + cX$ for every positive c. Therefore, it is only the relative size of atoms that matters for the modelling purposes. A relevant information is isobtained from extended X-ray absorption fine structure (EXAFS) measurements. In consideration of the EXAFS results, the values of atomic radii have been adjusted to those shown in Table 3.10.

The average atomic volume for the alloy is calculated as 0.65 at.vol.$_{Mg}$+ 0.25 at.vol.$_{Cu}$+0.10 at.vol.$_{Gd}$ = 0.0142 nm^3. These values differ significantly from those in Table 3.9, which were calculated in a different way for respective crystalline elements.

The data from Table 3.10 are now entered into the IAS graphical interface as shown in Figure 3.27 to begin the simulation. The result is a Round Cell as shown in Figure 3.28, which also includes an MD simulation cell for comparison.

142 | *3 Glassy Materials and Ideal Amorphous Solids*

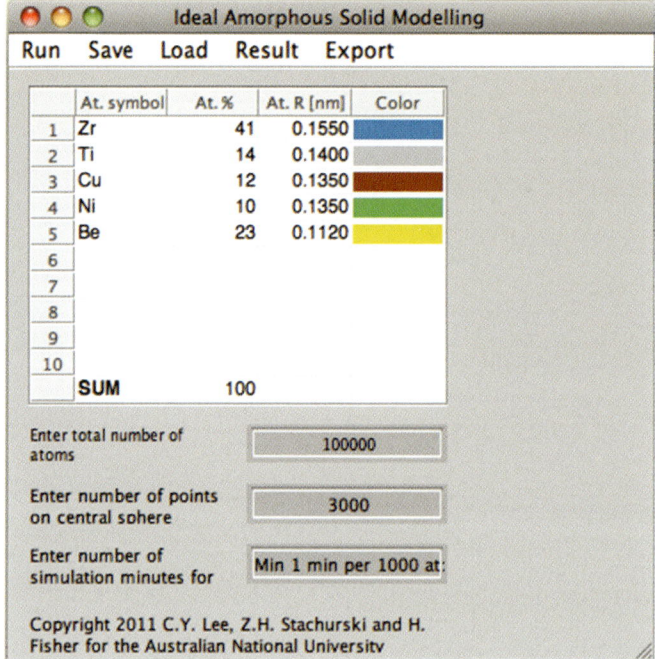

Figure 3.27 The graphical user interface for the IAS software used for simulation of a-MgCuGd.

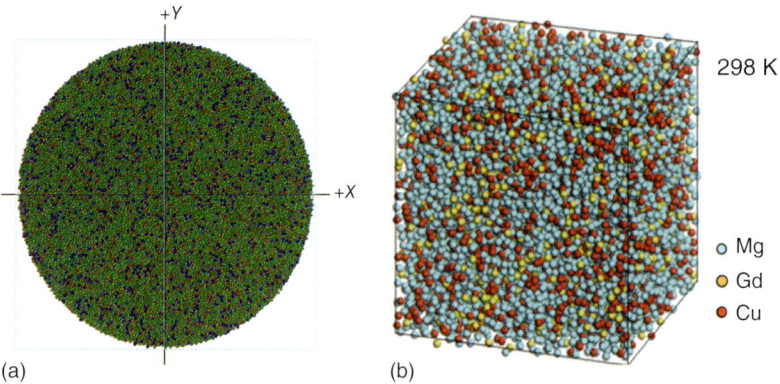

(a) (b)

Figure 3.28 (a) IAS model comprising 10^6 atoms of Mg65-Cu25-Gd10 amorphous alloy: Mg is in green, Cu is in orange, Gd is in purple. (b) MD model of a similar alloy of Mg61-Cu28-Gd11 comprising 2×10^4 atoms. (Reprinted with permission from Wang *et al.* (2011), Copyright (2011) by American Physical Society.)

Figure 3.28 shows typical appearance of simulation cells: Figure 3.28a for the IAS model and Figure 3.28b from MD method for comparison. The temperature of simulation is indicated for the MD result but not for the IAS result. This is because temperature is not relevant for the IAS model as it is based on the packing of hard spheres. However, the temperature is taken into account when selecting the diameters of the spheres. The interatomic distances are dependent on temperatures as well as the resulting density of the simulated model. In the MD simulation, the interatomic distances and their distributions will vary because of the soft-sphere condition (e.g., L-J potential).

3.5.2.1 Coordination Distribution Functions

The coordination distribution functions, shown in Figure 3.29, are based on the touching sphere method with maximum frequency of clusters occurring at $k = 7$, and for Voronoi neighbour method, for which the maximum occurring at $k = 14$. This is typical of hard packing of spheres and similar to the previous cases for a-Ar and a-FeNi described previously.

3.5.2.2 Configuration Distribution Function

The configuration distribution functions for the IAS model of the alloy are shown in Figure 3.30. The appearance of the functions is as expected, although interestingly, they have similar peak positions regardless of the value of k.

3.5.2.3 Radial Distribution Function

RDF is a sum over all pairs of atoms and, therefore, represents an envelope over all distances with individual distributions, reflected by the width of the peaks shown in Figure 3.31. In the figure, one can see a comparison of the RDFs for this alloy obtained from both the IAS simulation and from MD simulation. The agreement is excellent. The PDF obtained using the IAS method shows slightly greater detail because of the fact that it was computed with 10^5 atoms, whereas the MD was obtained from a cell containing 2×10^4 atoms.

Note that the main peak occurs at the interatomic distance corresponding to compositional average: $2\overline{R} = 0.65 \times 0.292 + 0.25 \times 0.256 + 0.10 \times 0.352) = 0.289$ nm. In the IAS PDF, the outline of the peak is made up from six individual underlying peaks of the possible contacts (see Figure 3.32). In the MD-derived PDF, the shape of the peak also involves a distribution of the individual nearest-neighbour contacts, as shown in Figure 3.32. It may be observed that the shapes of the peaks are very similar, which indicates that the atomic arrangements in the two models are sufficiently similar, and also that the PDF (as usually obtained from X-ray measurements) is not a very sensitive measure of the arrangement in amorphous solids. Extremely accurate scattering data are needed to very high scattering angles, and the PDF is averaged over a large sample; consequently, it will not reveal fluctuations in the structure. Nevertheless, PDF is a good starting point for structural analysis and any model must agree with observed scattering patterns.

The advantage of the IAS model is the ability to increase the resolution when calculating the PDF, as explained in Chapter 1. The result of such a high-resolution

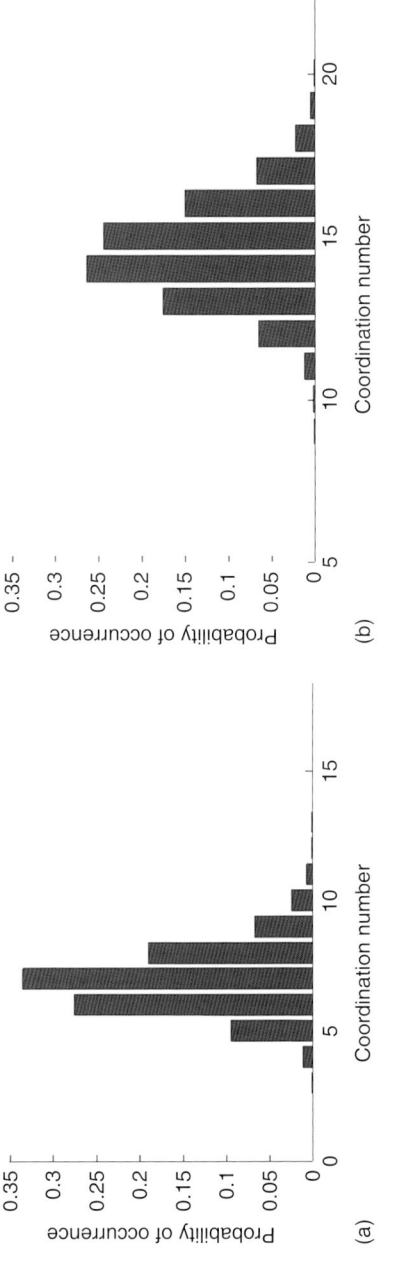

Figure 3.29 Coordination distribution functions for the IAS model of a-MgCuGd. A graph based on (a) touching contacts and (b) Voronoi tessellation.

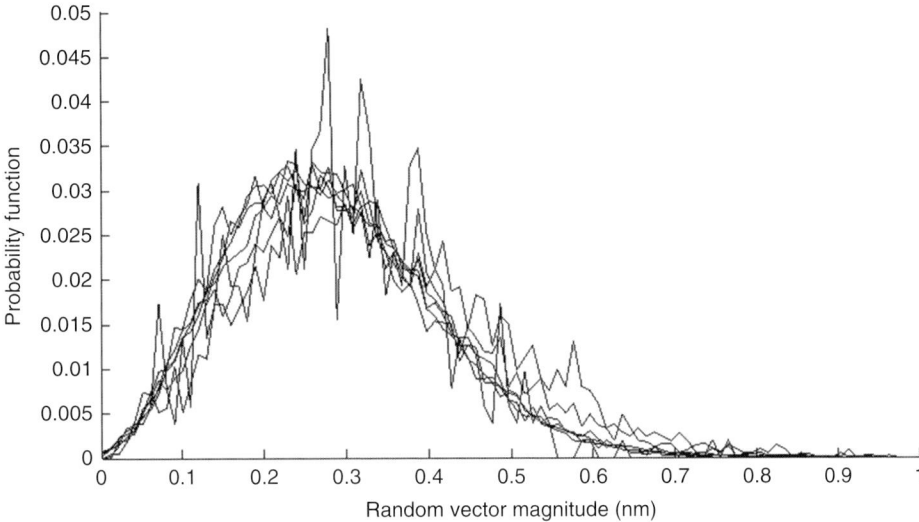

Figure 3.30 The configuration distribution functions, $P_k(\zeta)$, for the IAS model of the a-MgCuGd alloy.

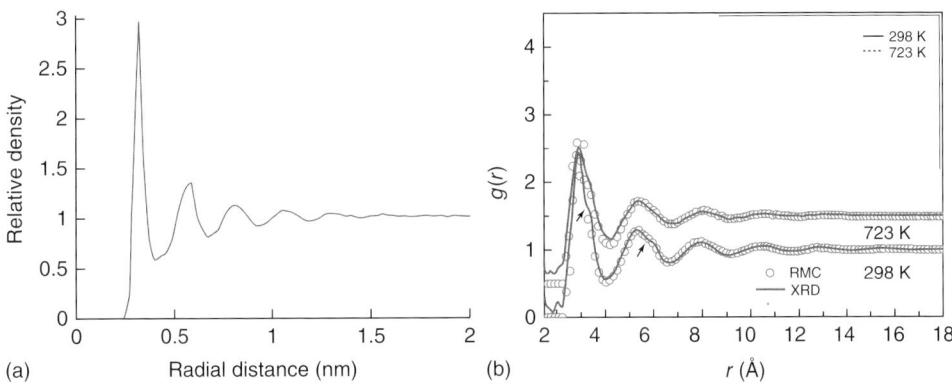

Figure 3.31 (a) PDF derived from the IAS model of the a-MgCuGd alloy. (b) PDF for the MD model Mg61-Cu28-Gd11 amorphous alloy. (Reprinted with permission from Wang et al. (2011), Copyright (2011) by American Physical Society.)

PDF for the a-MgCuGd model is shown in Figure 3.32, limited to the range, $0.26 \leq r \leq 0.38$ nm. The peaks show the occurrence of the centre-to-centre distances for specific pairs of atoms against radial distance. The distances correspond exactly to those shown in Table 3.11. The heights of the peaks are proportional to the frequency of occurrence of the specific pairs, which can be compared favourably to predicted probability of contacts shown in Table 3.12.

The temperature is not relevant for the IAS model because it is based on packing of hard spheres. However, the interatomic distances are dependent on temperature, as is the resultant density of the simulated model. In the IAS model,

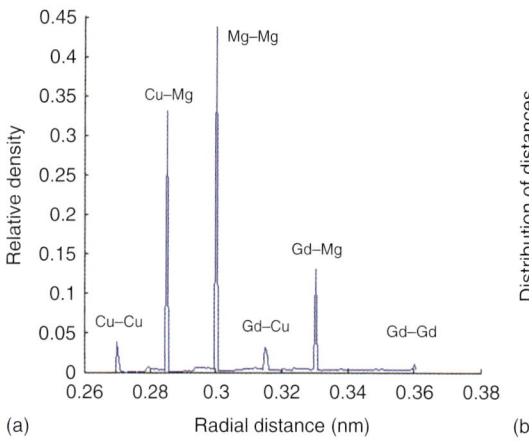
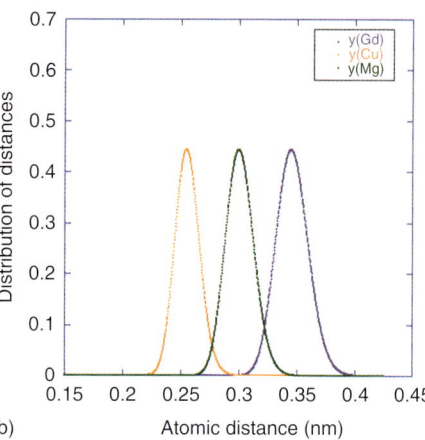

Figure 3.32 (a) PDF obtained for the IAS model of the Mg65-Cu25-Gd10 amorphous alloy with increased resolution ($\Delta r = 0.001$ nm). The height of the peaks indicates the relative frequency of contacts, respectively. (b) Distribution of interatomic distances for the Mg–Mg, Cu–Cu and Gd–Gd pairs obtained from the MD model. (Reprinted with permission from Wang et al. (2011), Copyright (2011) by American Physical Society.)

Table 3.11 Interatomic distances from IAS and MD simulations.

Atomic distance	Cu–Cu	Mg–Mg	Gd–Gd	Cu–Mg	Cu–Gd	Mg–Gd
IAS	0.256	0.292	0.352	0.274	0.304	0.322
Wang et al.	0.255	0.300	0.345	0.271	0.286	0.315

Table 3.12 Probability of contacts between atoms using Equation 1.63.

	Mg	Cu	Gd
Mg	0.4225	0.08125	0.0650
Cu	—	0.0625	0.0250
Gd	—	—	0.010

this is represented by the appropriate choice of the atomic diameters before simulation.

The interatomic distances, their distributions and coordination numbers in an MD model will vary because of the soft sphere condition (L-J potential) as shown in Figure 3.32.

The interatomic distances between the various possible pairs of atoms is shown in Table 3.11. For the IAS model, they are simple sums of individual radii as given in Table 3.8. For the MD model, they are the average values derived from simulation.

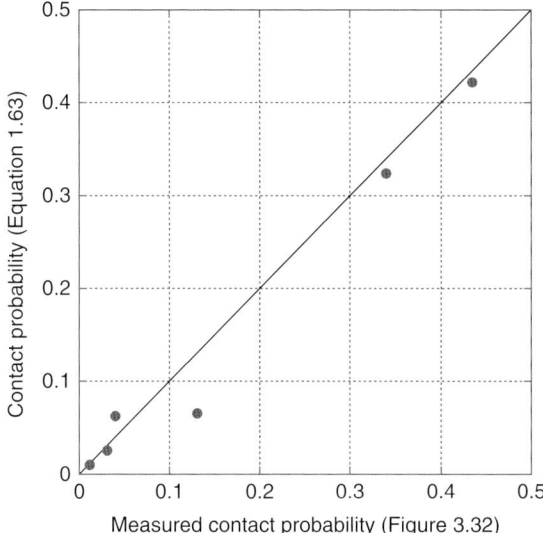

Figure 3.33 Comparison of contact probability calculated by Equation 1.63 with that measured from Figure 3.32.

3.5.2.4 Probability of Contacts

According to Equation 1.63, the probability of contacts between the three types of atoms, Mg, Cu and Gd, can be calculated as shown in Table 3.12.

The frequency of contacts between the elements can be also measured from Figure 3.32 as the height of the peaks. Then a direct comparison between these values and those calculated in Table 3.12 can be made as shown in Figure 3.33. It can be noted that to a first approximation, a linear relationship is followed. However, there are deviations from the straight line, and it is conjectured that these are due to the relatively small size of the Round Cell (10^6 atoms), insufficient for more accurate statistics. The deviations should become smaller with increasing size of the Round Cell.

3.5.2.5 Cluster Composition According to IAS

Given the probability of contacts, the composition of the alloy and the information derived from Figures 3.29 and 3.32, one can infer the structure of the most probable clusters in the IAS packing model as shown in Figure 3.34 and as described in the accompanying table.

Mg-cluster + 5Mg + 2Cu + 1Gd;	CN = 8
Cu-cluster + 5Mg + 2Cu;	CN = 7
Gd-cluster + 7Mg + 2Cu + 1Gd;	CN = 10

Figure 3.34 Predicted composition of selected primary clusters in the IAS model of the Mg65-Cu25-Gd10 amorphous alloy (Mg is in green, Cu is in brown and Gd is in purple).

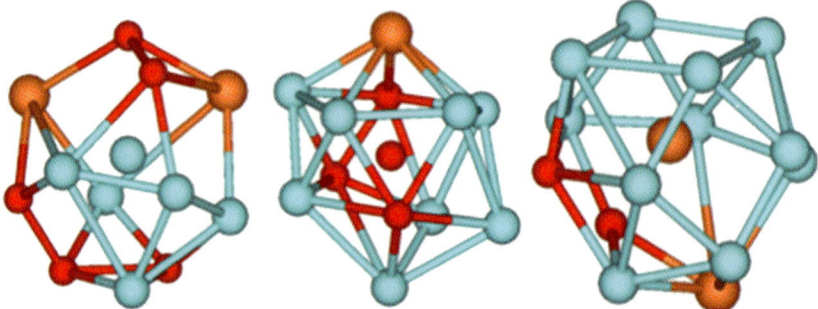

Figure 3.35 Primary clusters for the MD model of the Mg61-Cu28-Gd11 amorphous alloy as depicted by Wang et al., (2011) (Mg is in blue, Cu is in red and Gd is in orange).

In the IAS model, there is a large variety of clusters with distributions of compositions and coordination numbers as evidenced by the graphs in Figure 3.29. The three hypothetical clusters shown in Figure 3.34 account closely for the total composition of the alloy and represent the most probable average primary clusters.

3.5.2.6 Cluster Composition According to MD

This is to be compared to the clusters proposed for the MD model shown in Figure 3.35 and the accompanying table.

Mg-cluster + 5Mg + 5Cu + 2Gd; CN = 12
Cu-cluster + 3Cu + 8Mg + 1Gd; CN = 12
Gd-cluster + 11Mg + 2cu + 1Gd; CN = 14

The coordination number of clusters with varied sizes of atoms/spheres can be predicted with the help of the method described in Chapter 1. This is shown in Table 3.13, in which the maximum number and the most probable coordination number are given for the hypothetical clusters comprising copper atoms around

Table 3.13 The coordination number for clusters of two atom types.

Inner/outer atom	Cu/Gd	Mg/Gd	Cu/Mg	Mg/Mg
Atomic radius ratio	0.727	0.829	0.877	1.00
Maximum coordination number[a]	17	14	13	12
Most probable coordination number[b]	10	8	7	7

a) From Table 1.1
b) From Figure 1.15.

gadolinium, magnesium atoms around gadolinium and copper atoms around magnesium.

According to the IAS model, the most probable cluster coordination number is 14 or 15. However, the MD model simulation gives coordination between 12 and 14 for the most probable clusters. Although this is inconsistent with the theory, it must be remembered that the actual model contains a distribution of clusters of a large variety of compositions and coordinations, which will account for the overall properties of the alloy. It can be concluded, with some confidence, that the cluster compositions derived from the two independent methods (IAS and MD) are close enough for accepting as representative of the actual amorphous alloy. Further confirmation will come from a study of X-ray scattering.

3.5.3
X-ray Scattering from a-Mg65-Cu25-Gd10 Alloy

3.5.3.1 Flat Plate X-ray Scattering Pattern

Figure 3.36 shows the recorded pattern of X-ray scattering from a ribbon of amorphous MgCuGd alloy. It was taken on an X-ray generator equipped with an accurate Marresearch mar345 detector. The experimental conditions were Siemens 40 kV, 30 mA, slit size 0.4×0.4 mm, beam-stop 4.95 mm, exposure time 7200 s, sample-detector distance 200 mm and wavelength 0.07107 nm. In order to be able to make quantitative analysis of the pattern, the linear dimensions in the plane of the figure need be calibrated. This can be done against a known standard. For this purpose, a diffraction pattern of crystalline powder of silicon has been taken on the same apparatus as shown in Figure 3.37. The diameters of the rings can be measured and related to linear dimensions as follows.

3.5.3.2 Calibration based on Si Powder Pattern

Accurate Bragg diffraction angles for pure silicon can be obtained from published X-ray diffraction files. It will be found that with CuKα radiation ($\lambda = 0.154$ nm), the Bragg peak positions for crystalline Si are as shown in Table 3.14.

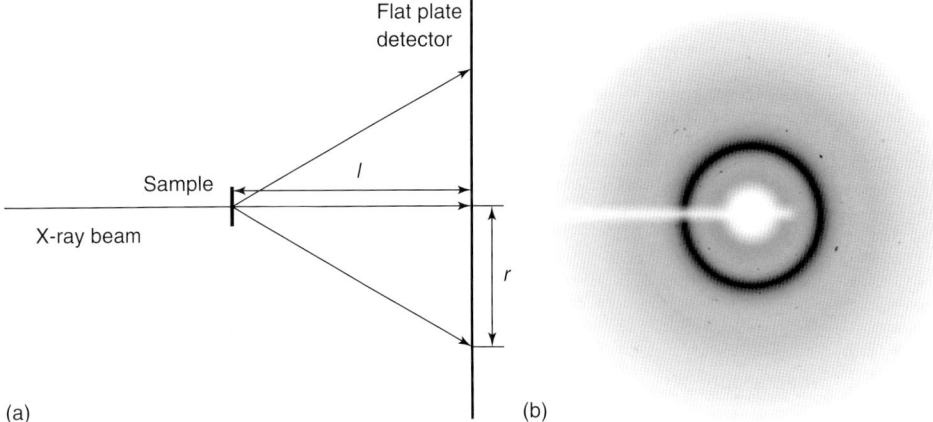

Figure 3.36 (a) The geometry of flat plate X-ray scattering. The pattern on (b) is from amorphous MgCuGd ribbon.

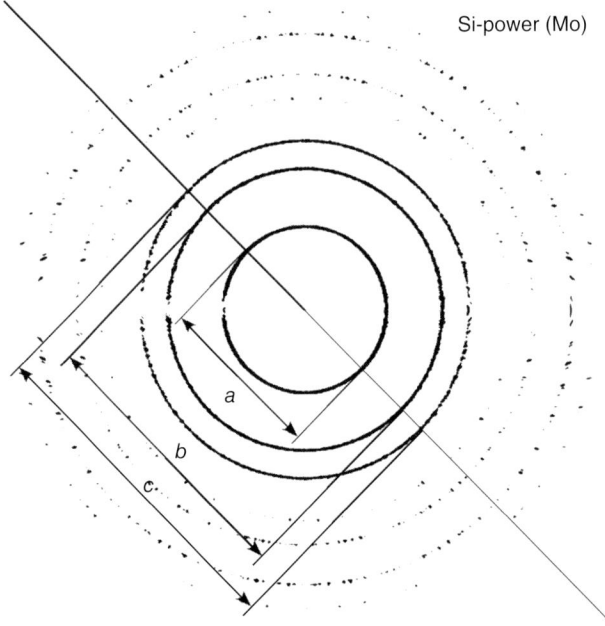

Figure 3.37 Calibration of the X-ray scattering pattern using crystalline Si powder, and measurement of the amorphous MgCuGd pattern.

The Bragg angles based on CuKα wavelength are transformed into Bragg angles based on MoKα wavelength using the equation

$$\sin\theta_{Mo} = \frac{\lambda_{Mo}}{\lambda_{Cu}} \sin\theta_{Cu} \tag{3.15}$$

Table 3.14 Bragg diffraction data for pure crystalline Si with CuKα radiation.

Crystal plane (hkl)	111	220	311	331	422	333
Corresponding Bragg angle, 2θ	28.4	47.5	56.2	69.0	76.5	88.3

Table 3.15 Transformation of Si Bragg angles from CuKα to MoKα wavelength.

Published $2\theta_{Cu}$	$\sin\theta_{Cu}$	$\sin\theta_{Mo}$	$2\theta_{Mo}$	$100 \times \tan 2\theta_{Mo}$
28.4	0.245	0.113	12.97	20.3
47.5	0.402	0.186	21.43	39.2
56.2	0.471	0.217	25.06	46.7
69.0	0.566	0.261	30.26	58.3
76.5	0.619	0.285	33.12	65.2
88.3	0.696	0.321	37.45	76.6

With $\lambda_{Mo}/\lambda_{Cu} = 0.461$, the transformation is shown in Table 3.15. The last column in the Table expresses the distance on the flat plate of the scattering pattern. The linear dimensions, a, b, c an so on, drawn in Figure 3.37 for the Si pattern will be proportional to the number shown in the last column of Table 3.14. Consequently, this information can be used to analyse the scattering pattern for the MgCuGd ribbon. Again, the diameters of the rings can be measured and transformed into corresponding scattering angles. The initial profile of the pattern, shown in Figure 3.38, is plotted against linear dimension. The linear dimension is then translated into scattering angle, 2θ, as also shown in the figure.

The X-ray scattering patterns contain a large background component because of X-ray fluorescence. The a-MgCuGd ribbon specimen contained large surface irregularities and nanocrystals that contributed to the background. The applied correction is of the form $I_{corrected} = I_0 - A\exp(-2\theta/660)$, with $I_0 = 140$ and $A = 1.0$.

Using Equation 3.15, the peak positions from Figure 3.38 can be transformed into peak positions for CuKα radiation wavelength to correspond with the computed results shown in Figure 3.39.

The X-ray scattering intensity for an amorphous solid of MgCuGd can be predicted by the Debye equation 3.16, with atomic positions **r** obtained from the IAS simulation and atomic scattering factors derived from published literature.

$$I(\mathbf{s}) = 1 + f_{average} \sum [g(\mathbf{r}) - 1]\exp(-2\pi i \mathbf{s} \cdot \mathbf{r}), \qquad (3.16)$$

where $f_{average} = [f_{Mg} + f_{Cu} + f_{Gd}/3]$ was used in the calculations with the results shown in Figure 3.39. This not only simplifies the calculations but also introduces an error. It is deemed that the error is small on the basis of comparison of experimental and predicted scattering peaks (Figures 3.38 and 3.39 and Table 3.16). The

152 | *3 Glassy Materials and Ideal Amorphous Solids*

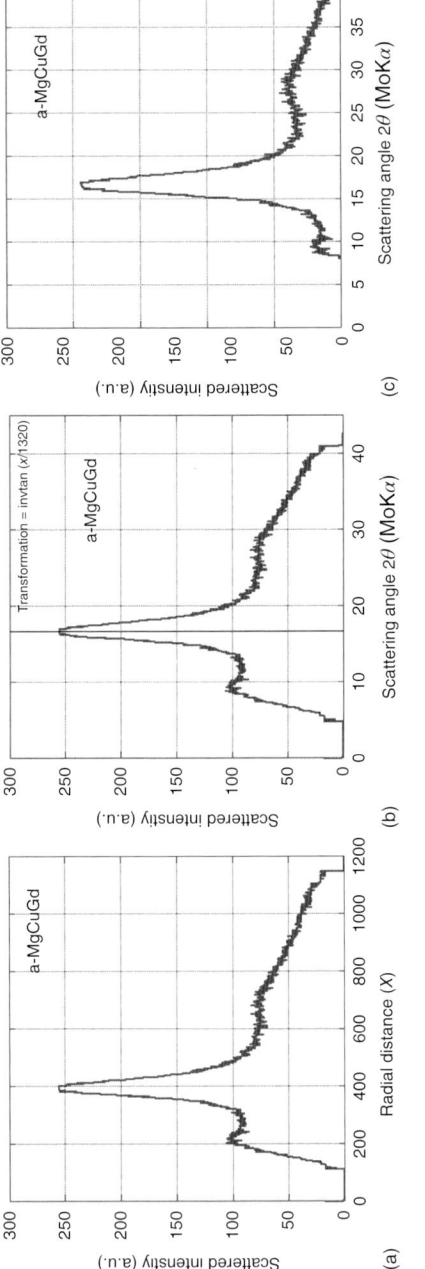

Figure 3.38 Profile of the MgCuGd X-ray scattering from Figure 3.37. (a) Raw data, (b) after transformation of the x-axis into corresponding scattering angle. (c) Profile of the MgCuGd X-ray scattering corrected for background radiation because of fluorescent radiation emitted by the specimen.

3.5 IAS Model of a-MgCuGd Alloy

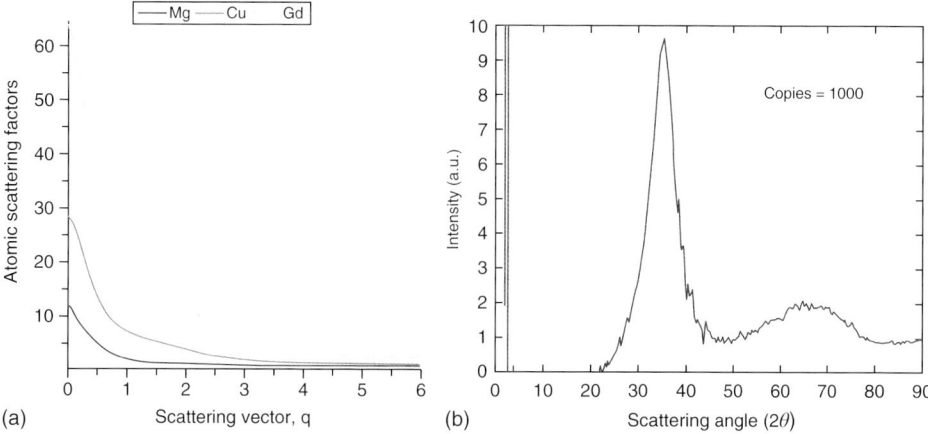

Figure 3.39 (a) A plot of atomic scattering factors for the three elements: Mg, Cu and Gd, obtained from http://www.ruppweb.org/Xray/comp.scatfac.htm.
(b) Predicted X-ray scattering for the MgCuGd amorphous alloy using the IAS packing data and the atomic scattering factors.

Table 3.16 Peak positions for a-MgCuGd.

Peak	Peak position	Peak position
Main peak	$2\theta_{Mo} = 16.5°$	$2\theta_{Cu} = 36°$
Second peak	$2\theta_{Mo} = 29.9°$	$2\theta_{Cu} = 68°$

agreement between predicted and measured X-ray scattering intensities for the amorphous alloy is satisfactory.

3.5.3.3 Uncertainties and Corrections

- The shape of the peak with its position can be fitted by an empirical function (pseudo-Voigt),

$$I_p(x) = A[\gamma L(x) + (1-\gamma)G(x)] \quad (3.17)$$

where A and $0 \leq \gamma \leq 1$ are adjustable parameters and
- the Lorentz function: $L(x) = I_{max}/[1 + (\Delta x)^2]$ with $x = 2\theta_p$
- the Gauss function: $G(x) = I_{max} \exp(-(\Delta x)^2)$ with $x = 2\theta_p$.

$2\theta_p$ is the scattering angle of the peak at its maximum, and therefore, Δx is the angular deviation. A comparison of the two functions is shown in Figure 3.40. For further information, the reader is referred to JCPDS-International Centre for Diffraction Data 2011 ISSN 1097-0002. An example of the pseudo-Voigt fit for the main peak of MgCuGd is also shown in Figure 3.40.

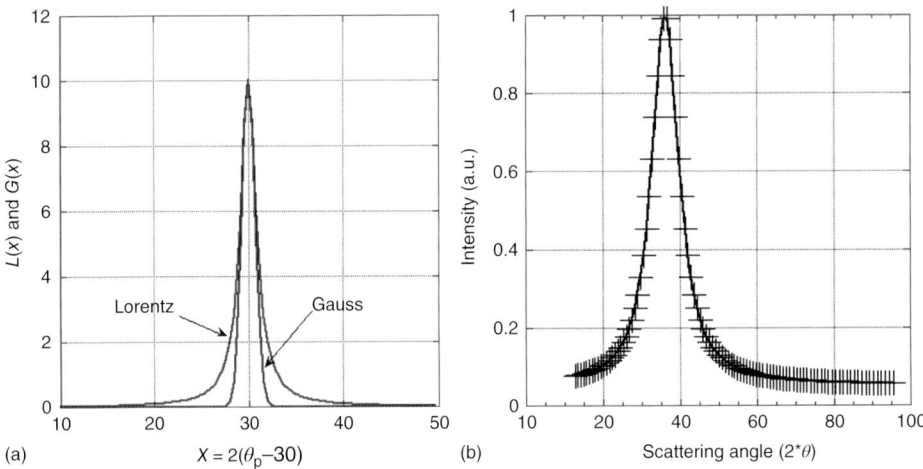

Figure 3.40 Comparison of the Lorentz and Gaussian functions, and fitting of the Lorentz–Gaussian function to the main scattering peak.

- The Gaussian function represents the instrumental broadening, and the Lorentz function represents the specimen broadening.
- The main uncertainty in the IAS method is the diameter of the atoms used. The success of the IAS method relies entirely on having accurate interatomic distances as input into the software program. This is not simple as the distances depend not only on the individual pair under consideration but also on the surrounding chemical elements. An experimental method that can measure these affects is the EXAFS, which uses X-rays to probe the physical and chemical structure of matter at an atomic scale.
- Thermal expansion/contraction will alter interatomic distances, and therefore, temperature effects must be considered. For the MgCuGd alloy, the coefficient of thermal expansion is known, $\alpha = \frac{1}{3} \times 4.2 \times 10^{-5}$ Bednarcik (2011). Thermal strain because of $\Delta T = 300$ K is $\Delta \epsilon_T = \alpha \times \Delta T =\sim 0.4\%$ Expected peak position change, $[I_{max}(300) - I_{max}(0)]/I_{max}(300) \sim 0.5\%$, giving a change of $1.8°$ for the first peak at $2\theta = 36°$ (CuKα).
- There are correction factors, such as incoherent scattering, background, sample and instrument. For further information, the reader is referred to Blessing et al. (1998)

It is assumed that a model for the atomic-scale arrangements exists. Then, the main objective is to verify the proposed IAS model by a trial-and-error method (Hukins, 1981). A crucial test is to compare the predicted scattering trace with experimentally derived data. There are three possible outcomes:

1) the agreement is good in which case the model is accepted;
2) the agreement is not good, in which case it is rejected;
3) the model is modified to improve the agreement until case (i) is reached.

Table 3.17 Calculation of volume fraction of the alloy elements.

Element	Atomic fraction	Number of atoms (10^{23})	Volume of atoms (cm³)	Volume fraction of element
Mg	0.65	3.914	9.18	0.709
Cu	0.25	1.506	1.78	0.137
Gd	0.10	0.6022	1.99	0.154
Total	1.00	6.022	12.95	1.000

Table 3.18 Calculation of weight fraction of the alloy elements.

Element	Atomic fraction	Mass of atoms (g)	Weight fraction of element
Mg	0.65	15.80	0.334
Cu	0.25	15.88	0.334
Gd	0.10	15.72	0.332
Total	1.00	47.30	1.000

Excellent agreement can be seen between the measured intensity (Figure 3.38) and the predicted intensity (Figure 3.39), as evidenced in Table 3.16. Therefore, on the basis of this and previous considerations, it can be concluded that the IAS method provides a good model for the atomic arrangement in the amorphous Mg-Cu-Gd alloy.

The final acceptance of the atomic arrangement may require corroboration with other complementary methods.

3.5.4
Density of Mg65-Cu25-Gd10 Alloy

3.5.4.1 Crystalline Alloy

Density of the crystalline alloy calculated from molar quantities:

$$\rho_{molar} = \left(\sum \text{at.frac} \times \text{at.wt.}\right) / \left(\sum \text{mol.vol.} \times \text{at.wt.}\right)$$
$$= 47.30/12.95 = 3.65 \text{g cm}^{-3} \text{(data from Tables 3.17 and 3.18)} \quad (3.18)$$

Density of the crystalline alloy calculated by *atomic fraction*:

$$\rho_{at.frac.} = [0.65 \times 1.738 + 0.25 \times 8.92 + 0.10 \times 7.90] = 4.15 \text{g cm}^{-3} \quad (3.19)$$

Density of the crystalline alloy calculated by *volume fraction*:

$$\rho_{vol.frac.} = [0.71 \times 1.738 + 0.137 \times 8.92 + 0.154 \times 7.90] = 3.67 \text{g cm}^{-3} \quad (3.20)$$

Density of the crystalline alloy calculated by *weight fraction*:

$$\rho_{\text{wt.frac.}} = [0.33 \times 1.738 + 0.34 \times 8.92 + 0.33 \times 7.90] = 6.12 \text{g cm}^{-3} \quad (3.21)$$

Note the agreement of densities calculated from Equations 3.18 and 3.20 and a disagreement by the other two methods. The above calculations are shown in full to demonstrate the correct molar quantities and volume fractions methods and the incorrect atomic fraction and weight fraction methods. These calculations predict the density of this alloy in its (phase separated) crystalline form as 3.66 ± 0.001 g cm^{-3}.

3.5.4.2 Amorphous Alloy

According to the IAS model, the packing fraction of the amorphous alloy is $pf_a = 0.63 \pm 0.005$.

Each element in the crystalline state has packing fraction, $pf_c = 0.740$ (applicable to hcp and fcc). Then the density of an amorphous alloy is predicted by the packing ratio method as follows:

$$\rho_{\text{am.alloy}} = \left(\frac{pf_a}{pf_c}\right)[(\text{vol.fr.} \times \rho)_{\text{Mg}} + (\text{vol.fr.} \times \rho)_{\text{Cu}} + (\text{vol.fr.} \times \rho)_{\text{Gd}}] \quad (3.22)$$

$$\rho_{(0.63)} = 0.851 \, [0.71 \times 1.74 + 0.137 \times 8.92 + 0.154 \times 7.90] = 3.12 \text{g cm}^{-3}$$

The reported density of such an alloy is close to 4.0 g cm^{-3} (Wang et al., 2011). This is significantly higher than that predicted above for both crystalline and amorphous alloys. Both copper and gadolinium are heavy elements, occupying less than one-third of the alloy volume, however, magnesium is very light occupying approximately two-third of the volume. The large discrepancy in density can be partially accounted for by nanocrystallization and the presence of other defects in the amorphous alloy, and partially by the difficulties in measuring densities of small samples. However, this has not been corroborated by independent measurements and should be further investigated.

Expression 3.22 is based on the assumptions that the volumes occupied by atoms are the same in the amorphous alloy as in the crystalline alloy. This is correct for IAS model of hard sphere packing, but is unlikely to be true in real solids, and this may be the major contribution to the much higher reported density.

3.5.5
Summary of a-MgCuGd IAS Structure

Amorphography: triatomic IAS Class I
Chemical Variety: $m = 3$: $R_{\text{Mg}} = 0.150, R_{\text{Cu}} = 0.135, R_{\text{Gd}} = 0.180$ nm
Chemical Composition: Mg 65%, Cu 25%, Gd 10%
Glass-Transition Temperature: $T_g = 430$ K
Crystalline Melting Point: $T_m = 723$ K

Atomic Positions: Represented by closed packed spheres randomly distributed in space in the range $[2R_{Cu}; \infty)$. The randomness of the atomic positions is evidenced by

- distribution of the Voronoi volumes (Figure 3.16)
- the reasonable agreement between contact probability (Figure 3.33)
- X-ray scattering pattern (see Figure 3.36) main peak at $2\theta_{Mo} = 16.5°, 2\theta_{Cu} = 36°$

Atomic Correlations: Direct contacts between spheres in the IAS model results in correlations that are evidenced by the PDF (Figure 3.17).

The total number of contacts between the spheres is $K = \int \Psi(k) \, dk = (1/2)N\bar{k}$, where $\bar{k}$ is the average coordination number and N is the total number of spheres.
Free Volume: $v_f = 0$
Packing Fraction: $p_f = 0.63 \pm 0.05$
Predicted Density: $\rho_{amorphous} = 3.12(\pm 0.25) \times 10^3$ kg m^{-3}
Predicted Density: $\rho_{crystalline} = 3.66 \times 10^3$ kg m6^{-3}

3.6
IAS Model of a-ZrTiCuNiBe Alloy

The bulk metallic glass (BMG) of composition $Zr_{41.5}$-$Ti_{13.75}$-$Cu_{12.5}$-Ni_{10}-$Be_{22.5}$ is an alloy known as *Vitreloy 1*. It was developed by the California Institute of Technology research team (Caltech, USA) and commercialized by Liquidmetal Technologies in 2003. One particular property of the alloy is its low free volume, which results in low shrinkage during cooling. For all of these reasons, Vitreloy can be formed into complex shapes using processes similar to thermoplastics. Vitreloy has been used for the casing of late-model SanDisk 'Cruzer Titanium' USB flash drives as well as their Sansa line of flash-based MP3 player, and casings of some mobile phones, for making the SIM ejector tool of some iPhone 3Gs made by Apple Inc. in the US. They retain a scratch-free surface longer than competing materials, while still being made in complex shapes. The same qualities lend it to be used as protective coatings for industrial machinery, including petroleum drill pipes and power plant boiler tube. Many other zirconium-based bulk metallic glasses have been developed since. Some properties of the alloy in comparison to aluminium alloy and spring steel are shown in Table 3.19.

For experimental purposes, ingots of the alloy with nominal composition are prepared by arc melting a mixture of pure metal elements in a titanium-gettered argon atmosphere, followed by suction casting into copper moulds to form samples. The physical properties of the alloy elements are shown in Table 3.20.

3.6.1
Transmission Electron Microscopy

High-resolution transmission electron micrographs (TEMs), shown in Figure 3.41, reveal the amorphous structure of the alloy in comparison to a

3 Glassy Materials and Ideal Amorphous Solids

Table 3.19 Selected mechanical properties of Vitreloy 1 and other alloys.

Property	Units	Vitreloy	Aluminium	Spring steel
σ_y	(MN m^{-2})	1900	278	1590
E	(GN m^{-2})	96	70	210
ρ	(g cm^{-3})	6.00	2.70	7.85

Table 3.20 Physical properties of the alloy elements (published data).

Element		Atomic mass	Density (g cm^{-3})	Molar volume (cm^3 mol^{-1})	Atomic volume (10^{-23} cm^3 atom^{-1})
Zr	hcp	91.2	6.51	14.0	2.35
Ti	hcp	47.9	4.51	10.6	1.76
Cu	fcc	63.5	8.92	7.12	1.18
Ni	fcc	58.7	8.90	6.60	1.09
Be	hcp	9.01	1.85	4.87	0.81

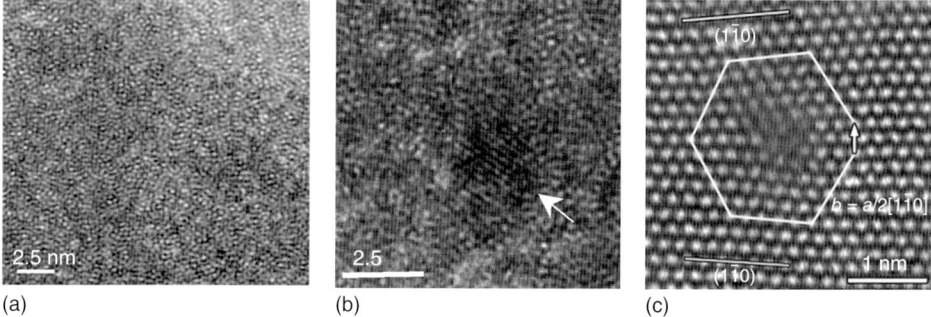

Figure 3.41 TEM (Philips EM430 at 300 kV) of a-ZrTCuNiBe amorphous alloy, as quenched and after annealing at 383 °C for 7000 s with appearance of a nanocrystal, and that of crystalline germanium.

crystalline one shown in Figure 3.41c. The crystalline micrograph is of single crystal germanium containing a dislocation as indicated by the Burgers vector. The ordered arrangement of the atoms is clearly seen.

Figure 3.41a shows the amorphous structure of quenched alloy with random structure. Note the similarity of the pattern to that shown in Figure 1.1. This is a two-dimensional projection of the three-dimensional structure; nevertheless, the randomness of the atomic arrangement is clearly seen. Density fluctuations are a natural phenomenon occurring in liquids and predicted by statistical mechanics. On freezing to below glass-transition temperature, it is expected that the density variations remain because of the rapidly increasing viscosity. Also, on cooling below a critical temperature, spinodal decomposition takes place, leading to compositional fluctuations. Again it is expected that on rapid cooling to the glassy state, the compositional fluctuations also remain frozen-in. Figure 3.41b shows the original sample developing nanocrystallinity after annealing. The three micrographs in Figure 3.41 show the gradual change of atomic arrangements from completely random (amorphous) to fully ordered (crystalline).

3.6.2
IAS Simulation of Amorphous a-ZrTiCuNiBe Alloy

The atomic parameters used for the simulation are shown in Table 3.21. Many IAS models of this alloy have been simulated with up to 10^6 atoms (spheres). An example of the graphical user interface (GUI) is shown in Figure 3.42.

The typical electronic structure of transition metal atoms can be written as $[ns^2(n-1)d^m]$, where the inner d orbital has more energy than the valence-shell s orbital. In physical terms, this means that there is no preference for directional bonding. The range of elements have been carefully chosen to present as much of *geometrical frustration* as possible, thus hindering crystallization and allowing glass formation.

3.6.2.1 Coordination Distribution Function

On analysis of the IAS model with 10^5 atoms (spheres), the coordination distributions, measured by direct contact and Voronoi method, have been plotted and are shown in Figure 3.43. From the plots, one can derive the average coordination number for this alloy; it is $\bar{k} = 7.38$ for the neighbours by contact and $\bar{k}_V = 15.5$ for the nearest Voronoi neighbours. These numbers are not significantly larger than those for monoatomic IAS packing because of the self-compensating effect. Larger coordinations for the larger atoms are counterbalanced by smaller coordinations by smaller atoms. However, these distributions are somewhat broader (compare Figure 3.43 with Figure 3.15).

Evidence for the dependence of coordination on the relative sizes of the atoms is shown in Figure 3.43, where the specific data were extracted from the IAS simulation for clusters with Zr as the inner spheres surrounded by other Zr atoms, and clusters with Zr as the inner sphere surrounded by Be atoms.

Table 3.21 Atomic parameters of Vitreloy 1 components.

Element	Atomic radius (nm)	Atomic volume (nm³)	Atomic concentration (%)	Atom type/chemistry
Zr	0.1550	0.0156	41.2	Transition metal
Ti	0.1400	0.0115	13.8	Transition metal
Cu	0.1350	0.0103	12.5	Transition metal
Ni	0.1350	0.0103	10.0	Transition metal
Be	0.1120	0.00588	22.5	Alkaline earth metal

160 | *3 Glassy Materials and Ideal Amorphous Solids*

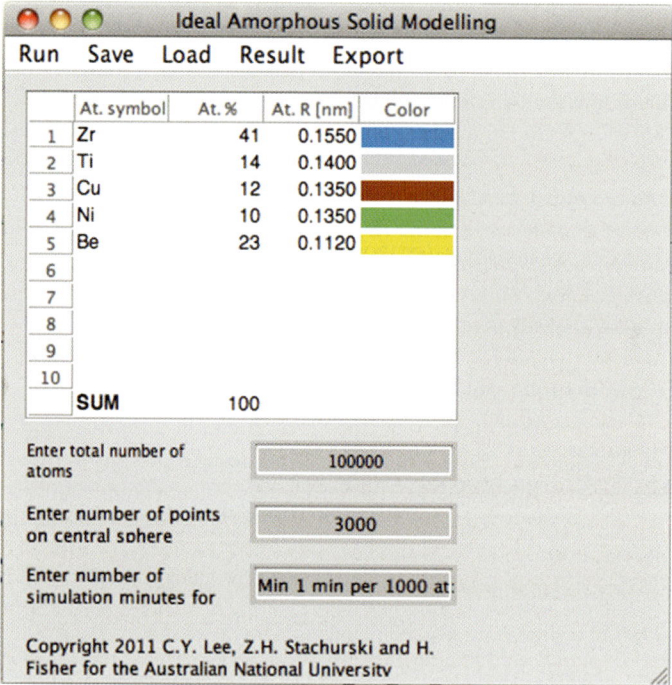

Figure 3.42 Graphical user interface for an IAS simulation of the a-ZrTiCuNiBe alloy.

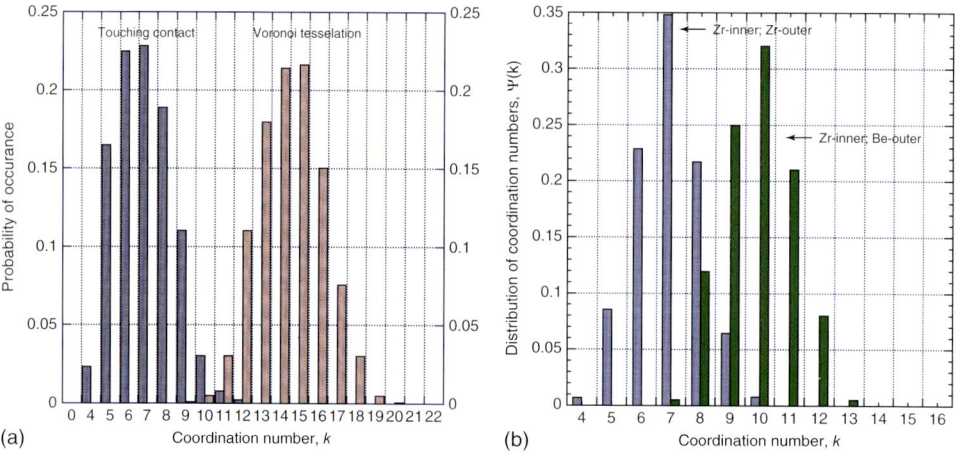

Figure 3.43 Coordination distribution functions for the a-ZrTCuNiBe amorphous alloy.

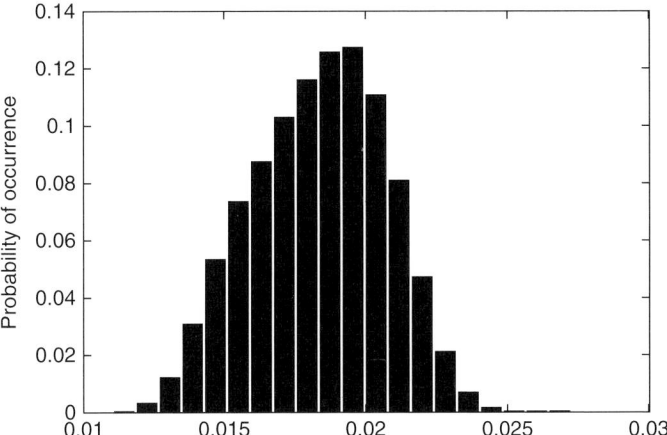

Figure 3.44 Voronoi volume distribution for the a-ZrTCuNiBe amorphous alloy.

3.6.2.2 Voronoi Volume Distribution

The overall Voronoi volume distribution for the alloy is shown in Figure 3.44. The average volume calculated from the data in the figure is 0.0184 nm^3.

The volume occupied by an average atom (sphere) can be calculated with the help of Table 3.21 as $0.412(0.0156) + 0.138(0.0115) + 0.125(0.0103) + 0.1(0.0103) + 0.225(0.00588) = 0.01165$ nm^3. If the packing fraction of the IAS model is 0.633, then the average Voronoi volume, calculated by this simple method, should be approximately 0.01849 nm^3. There is a good agreement between the two values.

3.6.2.3 Radial Distribution Function

Two RDFs have been obtained for the IAS model, one of relatively low resolution ($\Delta r = 0.026$ nm) and the other with higher resolution ($\Delta r = 0.001$ nm). Both are shown in Figure 3.45.

The low resolution has the effect of enveloping the peaks, thus concealing detailed information. The high-resolution individual peaks appearing in Figure 3.45a are merged into one broad peak between 0.22 and 0.32 nm, which includes contributions from all atomic pairs. The first peak of the split (at approximately 0.27 nm) corresponds to ZrBe and TiCu pairs, and the second (at approximately 0.32 nm) to the other ZrX pairs. The remaining much less abundant pairs with distances between 0.22 and 0.32 nm also contribute to the body of the peak. The second peak in the pair distribution function (at approximately 0.55 nm) has no split, which indicates the absence of any regular fcc or hcp cluster arrangements in the IAS.

The Round Cell of the a-ZrTiCuNiBe model (i.e., the position of all atoms) was used as input into classical MD program with L-J 12-6 potential and appropriate interatomic potentials as shown in Table 3.22. The cell was subjected to 10 000

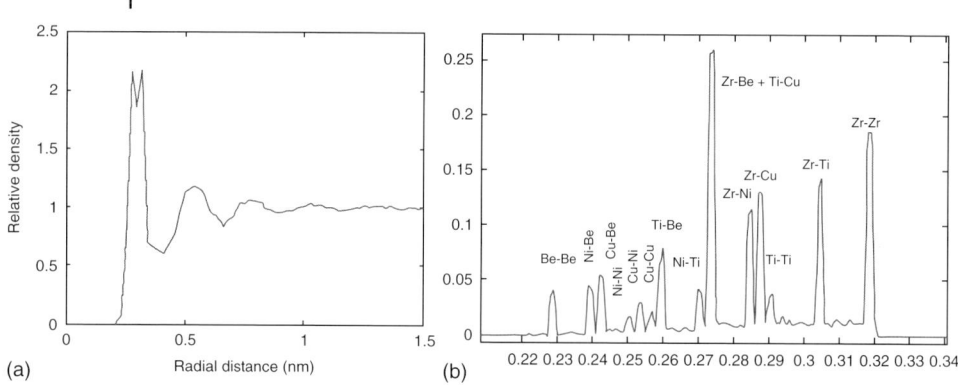

Figure 3.45 Radial distribution functions (RDFs) for the a-ZrTCuNiBe amorphous alloy at low and high resolution.

Table 3.22 Enthalpy of mixing, ΔH (kJ mol^{-1}).

Element	Zr	Ti	Cu	Ni	Be
Zr	14.0	1	23	49	43
Ti		14.2	9	35	30
Cu			13.3	−4	10
Ni				17.5	4
Be					11.7

cycles and the resulting structure was analysed. In particular, the radial distribution function was determined. It is shown here in Figure 3.46 in comparison to RDF determined by Wang et al. (2011) for the same alloy. To compare the two functions, one must take the results obtained at 300 K. It is evident that the match is almost perfect, validating the point that the IAS model is a good representation of the baseline atomic arrangement for amorphous metallic alloys.

3.6.3
Atomic Probe of the a-ZrTiCuNiBe Alloy

A strong evidence for the random and homogeneous distribution of atoms in the metallic alloy comes from the atomic probe measurement, shown in Figure 3.47. This result comes from a sample made by the flat ribbon technique with quenching rates of the order of 10^4 K s^{-1}, and therefore, a good retention is expected of the random atomic dispersion in the liquid, even though interatomic forces are far from equal (Table 3.22).

3.6.3.1 Probability of Contacts

Table 3.23 and Figures 3.45 and 3.48 combine to describe the probability of contacts between the different types of spheres in the IAS model of this alloy. The

3.6 IAS Model of a-ZrTiCuNiBe Alloy

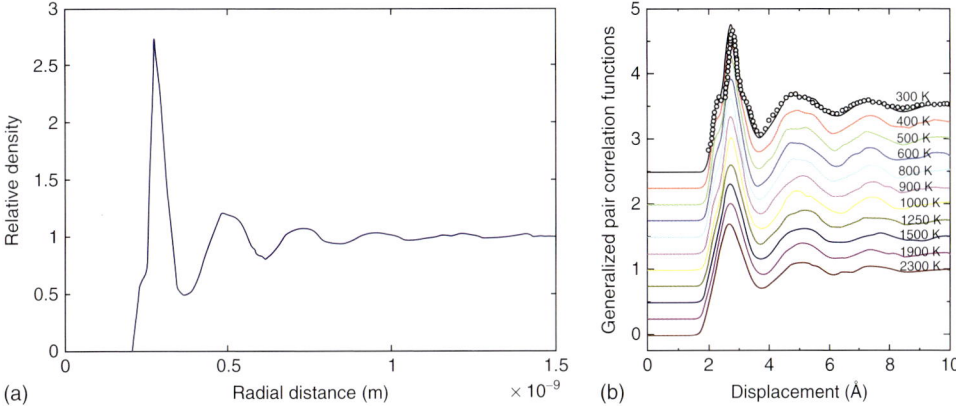

Figure 3.46 Radial distribution functions (RDFs) for the a-ZrTCuNiBe amorphous alloy: (a) From IAS subjected to molecular dynamics run. (b) From Hui et al. (2009), Copyright (2009) reproduced with permission from American Physical Society.

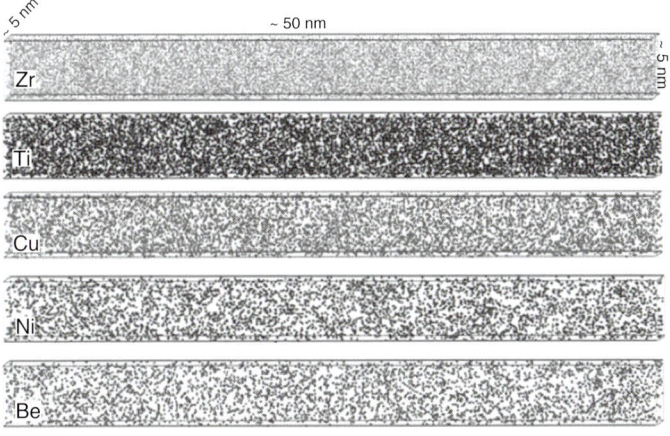

Figure 3.47 Atomic probe for the a-ZrTCuNiBe amorphous alloy. (Reproduced from Martin et al. (2004).

Table 3.23 Probability of contacts between the elements calculated using Equation (1.63)

Element	Zr	Ti	Cu	Ni	Be
Zr	0.1697	0.114	0.103	0.082	0.185
Ti		0.0190	0.035	0.0276	0.0506
Cu			0.0156	0.025	0.056
Ni				0.020	0.045
Be					0.0506

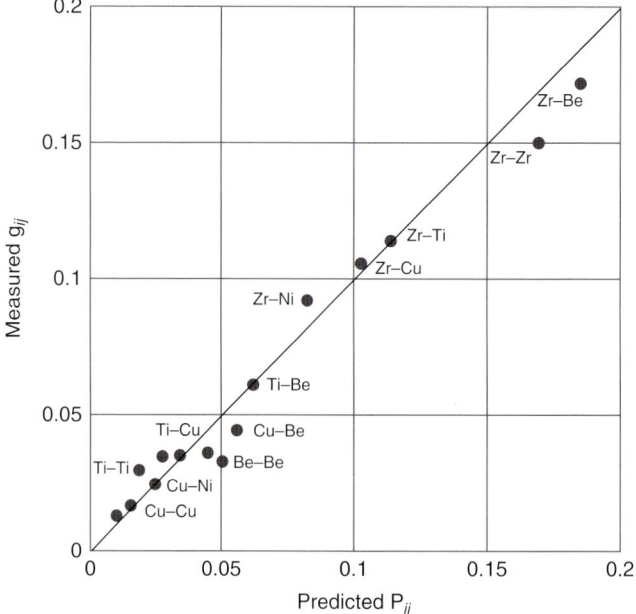

Figure 3.48 Comparison of contact probability for the elements in the IAS model.

expected linear relationship in Figure 3.48 is reasonably well obeyed. The deviations should be studied with a more detailed analysis of the IAS model software, and the probabilistic Equation (1.63).

Evidence from the atomic probe (Figure 3.47) and contact probability (Figure 3.48) confirm the random packing and homogeneous dispersion of the elements despite large variations in the enthalpies of mixing (Figure 3.49; Table 3.22).

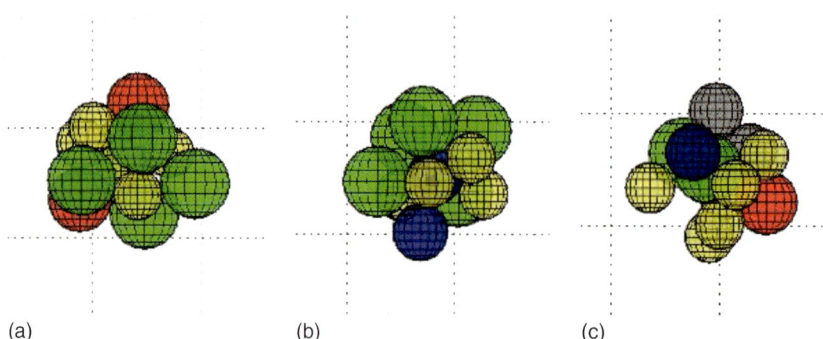

Figure 3.49 Examples of clusters (Voronoi method) in the IAS model of a-ZrTCuNiBe alloy. Be is in yellow, Ni is in grey, Cu is in blue, Ti is in red and Zr is in green.

3.6.4
Selected Clusters from the a-ZrTiCuNiBe Alloy

The IAS Round cell containing atomic positions of the Zr-based amorphous alloy was subjected cluster analysis. The examples of clusters shown in Figure 3.49 represent a selection of the many combinations possible with five elements of different concentrations and sizes.

3.6.5
X-ray Scattering from the a-ZrTiCuNiBe Alloy

The flat plate recording of X-ray scattering from a sample of the a-ZrTiCuNiBe alloy is shown in Figure 3.50, together with its profile derived by measuring the intensity along a radial direction from the centre outwards. The radial distance is transformed into the scattering angle, 2θ, and the corresponding graph is also shown in the same figure.

An IAS Round Cell of this alloy with 10^6 atoms was computed, and the positions of the atoms (spheres) were used to predict X-ray scattering using the Debye Equation (2.46). The result of this computations is shown in Figure 3.51 together with the measured curve for comparison.

The agreement is satisfactory (Table 3.24). However, the agreement is better for the main peak but less so for the secondary peak. In both the cases, the predicted peak position was larger than those measured. This suggests that the chosen radii of the elements were slightly too small. Adjusting the sizes of the radii would amount to fitting the prediction to the experimental result—not a useful scientific activity. Instead, an appropriate study should be carried out to discover the atomic sizes or rather the interatomic distances in the real alloy.

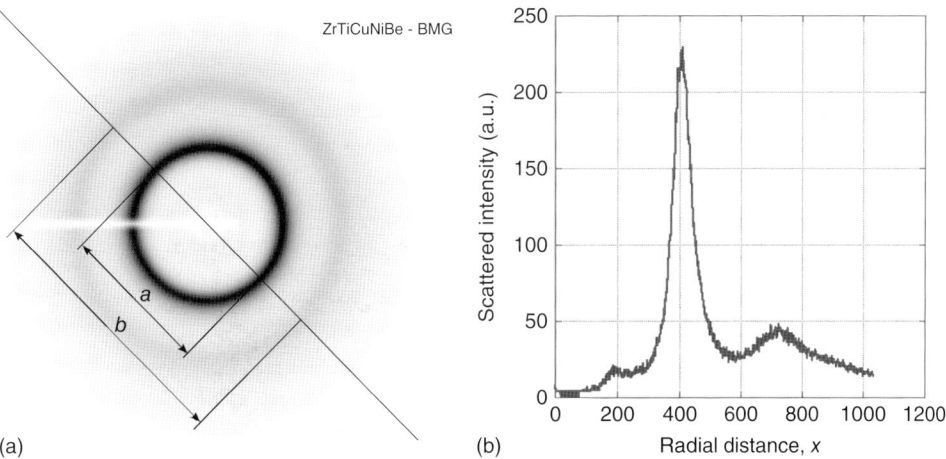

Figure 3.50 Measured X-ray scattering for the a-ZrTCuNiBe alloy and its profile.

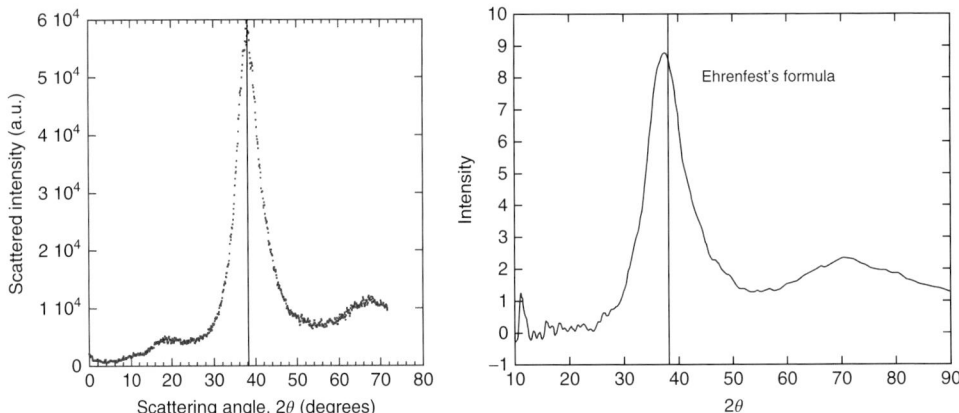

Figure 3.51 Comparison of the intensity profile curves for the ZrTiCuNiBe metallic glass: (a) experimental scattering curve and (b) predicted scattering curve. CuKα radiation.

Table 3.24 Comparison of peak positions for a-ZrTiCuNiBe.

Main peak	2θ MoKα	2θ CuKα	Predicted [CuKα]
	17.4	38.2	38.4
Second peak	2θ MoKα	2θ CuKα	Predicted [CuKα]
	28.5	68	71.2

The sensitivity of the IAS model to the presence of imperfections in the structure is evident in Figure 3.52, where the two graphs show the effects of vacancies in the IAS model achieved by

- removing at random 10% of atoms, thus creating 10% of vacancies but retaining the same overall composition and the same average atom–atom distance
- removing all Cu atoms, thus creating 12.5% vacancies, consequently changing the composition of the alloy and increasing the average pair contact distance from an initial value of 0.1412–0.1431 nm using the number-weighted average values.

It can be seen that in the graph in Figure 3.52a, the intensity of the first peak has decreased correspondingly, in agreement with the Debye equation. At the same time, the background intensity has increased with the content of vacancies as predicted by the Laue formula.

Removing the Cu atoms not only lowers the intensity of the peak but also causes a shift in the position of the peak in accordance with the Ehrenfest formula 2.50. According to the Ehrenfest formula, this should reduce the scattering angle by approximately 1.4°, quite close to the observed shift in the graph.

An important observation has been made by Yavari et al. (2005) that free volume in glasses can be measured directly by X-ray diffraction methods without the

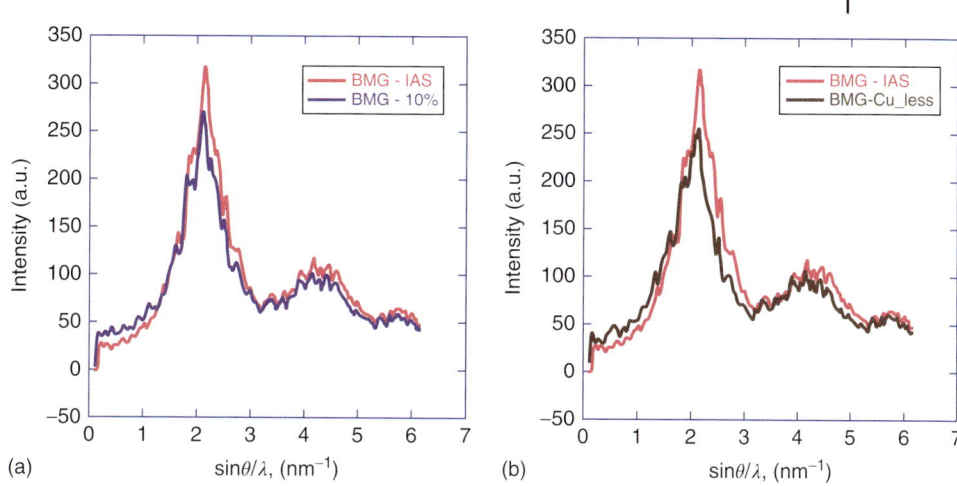

Figure 3.52 X-ray scattering for the a-ZrTCuNiBe amorphous alloy affected by (a) vacancies and (b) removal of Cu atoms. (Reprinted with permission from Lee *et al.* (2010), Copyright (2010) by American Physical Society.)

need for dilatometric measurements. Creation of vacancies in crystalline materials leads to volume changes, but no changes in the interatomic distance (to a first approximation). This was corroborated by both X-ray diffraction and dilatometric measurements that distinguish between volumetric dilation and interatomic distance. When applied to glassy materials, X-ray and dilatation methods show that the average interatomic distance changes in the same way as the volume. This reaffirms the view that free volume in glasses redistributes to disperse throughout the volume, and atomic-sized vacancies are unstable and, therefore, non-existent.

3.6.6
Density of ZrTiCuNiBe Alloy

3.6.6.1 Crystalline Alloy
Density of the alloy calculated from *molar quantities*:

$$\rho_{molar} = 59.95/9.924 = 6.04 \text{g cm}^{-3} \text{ (using data from Tables 3.25 and 3.26)} \quad (3.23)$$

Density of the crystalline composite calculated by *atomic fraction*:

$$\rho_{at.frac.} = [0.41 \times 6.51 + 0.14 \times 4.51 + 0.125 \times 8.92]$$
$$+0.1 \times 8.90 + 0.225 \times 1.85 = 5.72 \text{g cm}^{-3} \quad (3.24)$$

Density of the crystalline composite calculated by volume fraction:

$$\rho_{vol.frac.} = [0.584 \times 6.51 + 0.149 \times 4.51 + 0.089 \times 8.92]$$
$$+0.066 \times 8.9 + 0.111 \times 1.85 = 6.06 \text{g cm}^{-3} \quad (3.25)$$

Table 3.25 Calculation of volume fraction of the alloy elements.

Element	Atomic fraction	Number of atoms (10^{23})	Volume of atoms (cm³)	Element volume fraction
Zr	0.41	2.47	5.80	0.584
Ti	0.14	0.843	1.48	0.149
Cu	0.125	0.753	0.888	0.089
Ni	0.10	0.6022	0.656	0.066
Be	0.225	1.35	1.10	0.111
Total	1.00	6.022	9.924	1.000

Table 3.26 Calculation of weight fraction of the alloy elements.

Element	Atomic fraction	Weight of atoms (g)	Weight fraction of element
Zr	0.41	37.4	0.624
Ti	0.14	6.71	0.112
Cu	0.125	7.94	0.132
Ni	0.10	5.87	0.098
Be	0.225	2.03	0.034
Total	1.00	59.95	1.000

Density of the crystalline composite calculated by weight fraction:

$$\rho_{\text{wt.frac.}} = [0.624 \times 6.51 + 0.112 \times 4.51 + 0.132 \times 8.92]$$
$$+ 0.098 \times 8.90 + 0.034 \times 1.85 = 6.68 \text{g cm}^{-3} \quad (3.26)$$

Note the agreement of calculated densities from Equations (3.23) and (3.25), and a disagreement with the other two methods.

3.6.6.2 Amorphous Alloy

Assume that the amorphous alloy has packing fraction, $pf_a = 0.630$.

In the crystalline state, each element has packing fraction, $pf_c = 0.740$ (hcp or fcc).

Then the predicted density of an amorphous alloy is calculated as

$$\rho_{\text{alloy}} = \left(\frac{pf_a}{pf_c}\right)[(\text{vol.fr.} \times \rho)_{\text{Zr}} + (\text{vol.fr.} \times \rho)_{\text{Ti}} + (\text{vol.fr.} \times \rho)_{\text{Cu}}]$$
$$+ (\text{vol.fr.} \times \rho)_{\text{Ni}} + (\text{vol.fr.} \times \rho)_{\text{Be}} \quad (3.27)$$

$$\rho_{\text{alloy}} = 5.16 \text{g cm}^{-3}$$

The measured density of amorphous Zr-based alloy is published as 6.00 g cm^{-3} (Lu et al., 2003). This is close to the crystalline density calculated by the volume fraction method (Equation 3.25) and that calculated by the molar quantities (Equation 3.23).

Therefore, there is a large discrepancy between this and the predicted value for the amorphous ally of 5.12 g cm^{-3}. This difference may be accounted for in terms of the presence of nanocrystallinity in the alloy (flaws in the amorphous packing).

However, if this were the case, then from the rule of mixtures the volume fraction of the heavier crystalline phase would have to be

$$V_{cryst} = (\rho_{total} - \rho_a)/(\rho_{cryst} - \rho_a) = (6.0 - 5.12)/(6.06 - 5.12) = 0.93$$

3.6.6.3 Vitreloy Alloys

An early alloy, Vitreloy 1
Zr: 41.2 Be: 22.5 Ti: 13.8 Cu: 12.5 Ni: 10

A variant, Vitreloy 4 (Vit4)
Zr: 46.75 Be: 27.5 Ti: 8.25 Cu: 7.5 Ni: 10

New Vitreloy 105 (Vit105)
Zr: 52.5 Ti: 5 Cu: 17.9 Ni: 14.6 Al:10

A more recent development (Vitreloy 106a), which forms glass under less rapid cooling, is
Zr: 58.5 Cu: 15.6 Ni: 12.8 Al: 10.3 Nb: 2.8

3.6.7
Summary of a-ZrTiCuNiBe IAS Structure

Amorphography: pentaatomic IAS Class I
Chemical Variety: $m = 5$:
$R_{Zr} = 0.155, R_{Ti} = 0.140, R_{Cu} = 0.135, R_{Ni} = 0.135, R_{Be} = 0.112$ nm
Chemical Composition: Zr 41.2%, Ti 13.8%, Cu 12.5%, Ni 10%, Be 22.5%
Glass-Transition Temperature: $T_g = 633$ K
Crystallization Temperature: $T_m = 729, 741$ K
Atomic Positions: Closed packed spheres representing randomly distributed atoms in space in the range $[2R_{Be}; \infty)$. The randomness of the atomic positions is evidenced by
- distribution of Voronoi volumes (see Figure 3.44)
- radial distribution function (see Figure 3.46)
- atomic probe (see Figure 3.47)
- X-ray scattering pattern (see Figure 3.50) main peak at $2\theta_{Mo} = 17.4°$ or $2\theta_{Cu} = 38.2°$

Atomic Correlations Direct contacts between spheres in the IAS model results in correlations that are evidenced by the PDF (see Figure 3.45). The total number of contacts between the spheres is $K = \int \Psi(k) \, dk = (1/2)N\overline{k}$, where $\overline{k}$ is the average coordination number and N is the total number of spheres (Table 3.3).

Free Volume: $v_f = 0$
Packing Fraction: $p_f = 0.63 \pm 0.05$
Predicted Density: $\rho_{amorphous} = 5.16(\pm 0.4) \times 10^3$ kg m^{-3}.
Predicted Density: $\rho_{crystalline} = 6.06 \times 10^3$ kg m$^{(-3)}$

3.7
IAS Model of a-Polyethylene (a-PE)

Polyethylene (PE) is one of the most useful synthetic polymers. It comes in a great variety of grades, and its uses include a whole spectrum of applications: from food packaging, to ultra-high-molecular weight material used as liners in hip replacement, to high-modulus fibres for light-weight high-strength industrial ropes, body and vehicle armour (e.g., Spectra fibre is known as one of the worlds strongest and lightest man-made fibres).

Linear pure PE (also called *polymethylene*) is a synthetic polymeric material comprising macromolecular chains of the general chemistry $-[CH_2]_n-$. Under special polymerization conditions, it can be produced with high molecular weight, that is, $n \approx 10^5$, and very low dispersion, that is, all chains approximately the same length. It is a crystalline solid at room temperature, with equilibrium melting point, $T_m = 415$ K. The crystal structure of pure PE is pseudo-body-centred orthorhombic as shown in Figure 3.53. Its density can be estimated accurately from X-ray measurements of the unit cell.

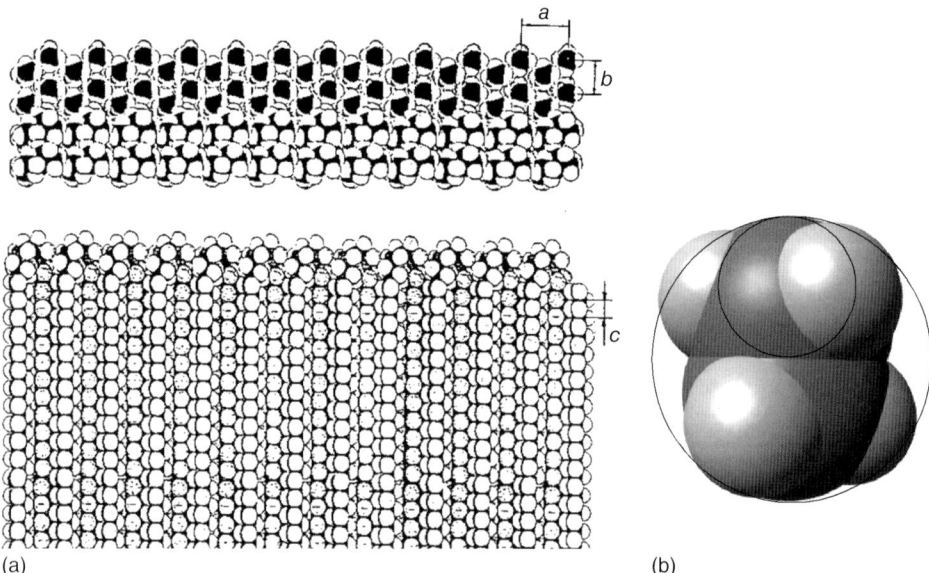

Figure 3.53 Crystallographic drawing of a segment of PE single crystal lamella, and a drawing of CH_2 monomer units, showing that one sphere representation of CH_2 is inaccurate; a larger sphere for C_2H_4 is a better approximation.

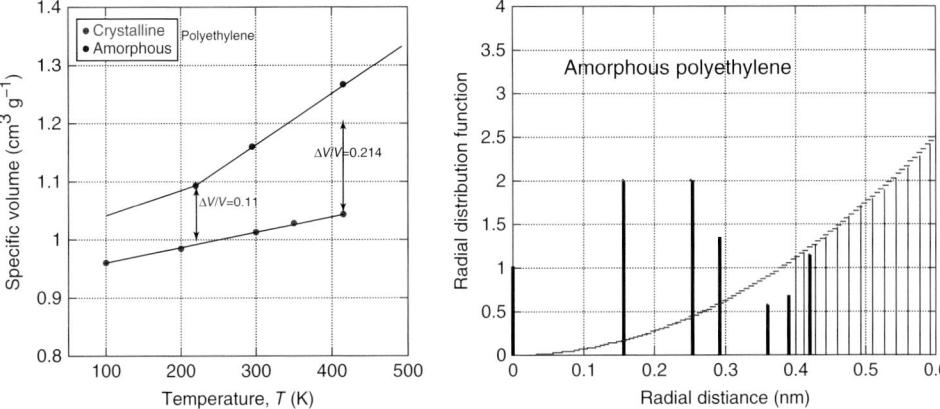

Figure 3.54 Volume–temperature relationship and assumed pair distribution function for the amorphous polymethylene.

The variation of the specific volume of PE as a function of temperature is shown in Figure 3.54. The specific volume of molten PE at the melting temperature is 1.266 cm^3 g^{-1} (density = 0.790 g cm^{-3}). Amorphous PE can be made by extremely fast quenching of the melt, from well above T_m (500 K) to below 150 K, and holding it at that temperature below the glass-transition temperature, T_g. As soon as it is taken out of the cold container, it recrystallizes rapidly.

The specific volume of the melt at 415 K was obtained by physical measurements. The specific volume of the crystal at the same temperature was predicted from measurements of the crystal unit cell at several temperatures and extrapolation to the melting point.

3.7.1
Radial Distribution Function

As a solid consisting of linear macromolecular chains, a-PE can be considered as belonging to IAS Class II (see Section 1.8). Accordingly, it can be modelled as a packing of spheres, each sphere representing either CH_2 or $2(CH_2)$ groups. However, the representation of the molecular groups by spheres shown in Figure 3.53 indicates the difficulty in deciding the diameter of the spheres and can only be resolved by fitting to some experimental data—not a useful approach. A better understanding of the amorphous structure of a-PE can be arrived at by analysis of its molecular chain architecture, and its most probable configurations. A computer-generated cell of amorphous PE containing 10^4 atoms in the so-called 'ball-and-stick' representation, and a segment of a molecular chain in random coil configuration, is shown in Figure 3.55.

The van der Waals radius for the carbon atom is 0.170 nm, and that for the hydrogen atom is 0.117 nm. In a repeat unit of CH_2, the C–H distance is only

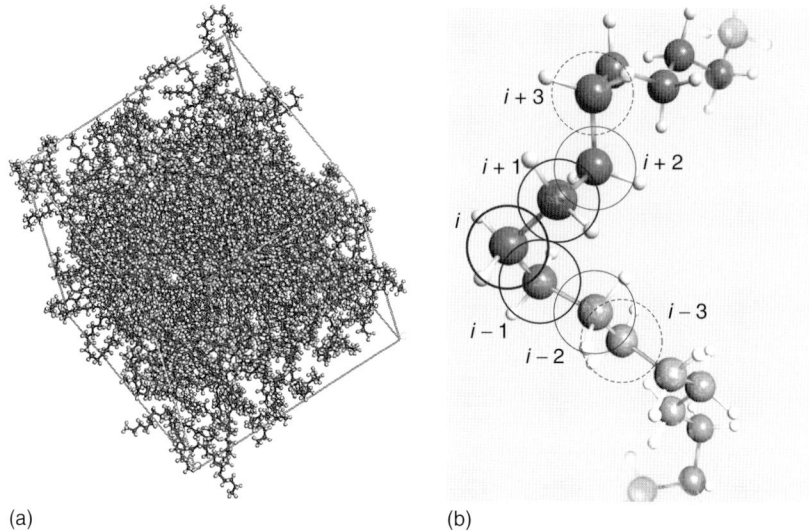

(a) (b)

Figure 3.55 A computer-generated cell of amorphous PE comprising 9000 atoms, and a segment of the macromolecular chain in random coil configuration.

0.109 nm because of the strong covalent binding forces between these two atoms. The CH_2 group is the smallest building block of the polymethylene chain.

The C–C covalent bond along the molecular chain fixes the separation of the repeat units at a distance of 0.154 nm and adjacent bonds form an angle of 109.5°. The H–C–H angle on each repeat unit is about 107° and a plane on which these three atoms lie bisects the C–C–C angle. The molecular parameters listed so far vary little with temperature and, henceforth, will be considered as constant.

Rotation around each C–C bond axis is allowed into three distinct positions between the jointed CH_2 units:

- trans: the hydrogens are on opposite side (0° reference position)
- gauche+: the hydrogens are interlaced as in Figure 3.55 (120° rotation relative to reference position)
- gauche−: the hydrogens are interlaced the other way (−120° relative rotation).

These rotations create random coil configuration of the macrochain if successive bonds assume random positions. However, the first nearest neighbours to any i CH_2 group in the chain are always two covalently bonded CH_2 groups ($i + 1$ and $i − 1$ in Figure 3.55) at a distance of 0.154 nm from the unit at the origin, and they may be in any of the three orientations. The second nearest neighbours are again two CH_2 groups ($i + 2$ and $i − 2$), covalently bonded to their nearest neighbours and found always at a distance of 0.254 nm from the origin, as can be easily verified. Whilst the distance is fixed, their precise location in space depends on the trans or gauche configurations of the first covalent bond.

The third nearest covalently bonded CH_2 groups along the chain are no longer at a fixed distance from the origin. Their positions depend on combinations of rotations of the first and second C–C bond between the units, that is, trans, gauche+ or gauche–. If it is trans, then the distance from the origin is the same as in the crystalline material (~ 0.39 nm). If it is gauche (+ or −), then the distance is close to 0.292 nm. Disregarding the possible differences in local potential free energy, the occupation of the former sites will be $1/3 \cdot 2 = 0.67$ and the occupation of the latter sites will be $2/3 \cdot 2 = 1.33$, as shown by the relative bar heights in Figure 3.54b. The fourth nearest covalently bonded CH_2 groups will be found at three different distances from the origin, characterized by the bond sequences: (i) ttt; (ii) tg+t = tg−t = ttg+ = ttg− and (iii) tg+g = tgg (the tg+g− = tg−g+ sequences are forbidden because of molecular overlap). The occupation/distance will be $(1/7 \cdot 2)/0.508$ nm, $(4/7 \cdot 2)/0.421$ nm and $(2/7 \cdot 2)/0.359$ nm for the three cases, respectively. The occupancy of these sites, as a function of radial distance, is shown in the diagram in Figure 3.54. This simplified evaluation of the occupancy of sites disregards the forbidden *pentene* effect and the difference in potential energy between the trans and gauche conformations Flory (1974). However, it will suffice as a first approximation to the random structure of polymethylene.

As the radial distance from the origin increases beyond approximately 0.43 nm, other CH_2 non-bonded groups become enclosed within. These interact with the groups i to $i \pm 3$ via van der Waals forces, and their distances can vary by much finer increments compared to the covalently bonded groups because of much higher compliance of the bonds. Consequently, their presence can be shown as a smooth distribution (see Figure 3.54). Limiting the radius of consideration (somewhat arbitrarily) to 0.43 nm, it will be found that the number of bonded CH_2 groups is 8.71 and the number of non-bonded CH_2 groups is 4.29 (13 all together within that radius). This amounts to a density of 0.91×10^3 kg m^{-3}, corresponding closely to that indicated in Figure 3.54 at 210 K. At 400 K, the density falls to 0.79×10^3 kg m^{-3}; thermal expansion occurs by increasing the non-bonded distances between the CH_2 groups so that the radius of consideration increases to 0.43 nm with the total number of CH_2 groups remaining the same. The overall variation of density with radial distance is indicated by the broken curve representing the relationship, mass/distance $\propto 4\pi r^2 \rho_{amorphous}$. This summarizes the description of the molecular structure of amorphous polymethylene.

3.7.2
X-ray Scattering

The X-ray scattering pairs in PE are C–C, C–H, H–C and H–H. Equation 2.46 can be separated into the incoherent scattering intensity (for $j = i$), written as

$$I'(s) = \sum_j [Z_C \bar{f}(s)]^2 + \sum_j [Z_H \bar{f}(s)]^2 = [\bar{f}(s)]^2 [\Sigma Z_C^2 + \Sigma Z_H^2] \quad (3.28)$$

and the coherent part (for $j \neq i$), written as

$$I''(s) = [\bar{f}(s)]^2 [\Sigma\Sigma\, Z_C^2\, s_{ij} + 2\Sigma\Sigma\, Z_C Z_H\, s_{ij} + \Sigma\Sigma\, Z_H^2\, s_{ij}] \quad (3.29)$$

where $\bar{f}(s)$ is a normalized atomic scattering function and $Z_C = 6$, $Z_H = 1$.

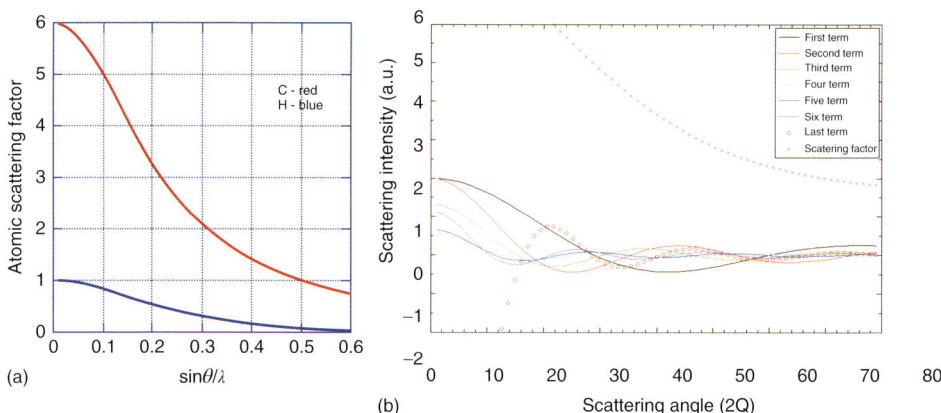

Figure 3.56 (a) Atomic scattering factor for carbon (C) and hydrogen (H). (b) Calculated scattering intensity for the terms in Equation 3.31.

The atomic scattering factors for C and H atoms are shown in Figure 3.56. Each CH_2 group is assumed to be a spherical scattering unit comprising one C atom at the centre and two H atoms spread out uniformly over the radius, C–H = 0.77 nm. The scattering factor for this spherical unit can be approximated as follows:

$$f_{CH_2} = f_C + 2f_H \frac{\sin(0.77 \cdot s)}{0.77 \cdot s} \quad (3.30)$$

Now, the scattered intensity from amorphous PE can be predicted by expanding the leading terms in the Debye Equation 2.46

$$I(s) = Nf_{CH_2}^2 [1 + 2\frac{\sin(0.154\ s)}{0.154\ s} + 2\frac{\sin(0.254\ s)}{0.254\ s} + \frac{4}{3}\frac{\sin(0.292\ s)}{0.292\ s}$$
$$+ \frac{4}{7}\frac{\sin(0.359\ s)}{0.359\ s} + \frac{2}{3}\frac{\sin(0.390\ s)}{0.390\ s} + \frac{8}{7}\frac{\sin(0.421\ s)}{0.421\ s} + \Psi(R,s)] \quad (3.31)$$

The first seven terms correspond to the discrete scattering units shown in Figure 3.54. The last term corresponds to the continuum density distribution function, and therefore, it is in an integral form. The integration can be carried out with the following result:

$$\Psi(R,s) = \int_R^\infty 4\pi r^2 \rho_a \frac{\sin(rs)}{rs} dr = -\frac{4R}{s^2}\pi \rho_a \left[\frac{\sin Rs}{Rs} - \cos Rs\right] \quad (3.32)$$

The radius, $R = 0.43$ nm, is obtained from $\sqrt[3]{3m/4\pi\rho_a}$, where m is the mass of atoms, and the amorphous density $\rho_a = 0.79$ g cm^{-3} at 400 K. Each of the eight terms in Equation (3.31) can now be evaluated. The last term is very small compared to 1 and, therefore, can be neglected. The result of calculations of the first seven terms is shown in Figure 3.56. The graph shows the variation of the calculated intensity.

3.7.2.1 Short-Range Order

It is stressed in Section 1.5 that no order, at any level, exists in the IAS. However, in this Section, it is shown that short-range order (SRO) exists in the amorphous structure of polymethylene. It consists of ordered and predictable positions of adjacent, covalently bonded CH_2 groups, each present at a distance of 0.154 nm from the central group. Randomness of packing increases with radial distance, becoming almost complete beyond approximately 0.5 nm. For this reason, we refer to this solid as IAS Class II to distinguish it from the random packing of non-bonded spheres (IAS Class I).

3.7.3
Summary of a-PE IAS Structure

Amorphography: biatomic IAS Class II
Chemical Variety: $m = 2 : R_C = 0.170, R_H = 0.117, R_C + R_H = 0.109$ nm
Chemical Composition: Carbon (1/3), Hydrogen (2/3)
Glass-Transition Temperature: $T_g = 210$ K
Crystalline Melting Point: $T_m = 413$ K
Atomic Positions: SRO for adjacent covalently bonded CH_2 units, the order reducing with distance along the molecular chain (Figure 3.54). Randomly distributed in space in the range $[r \sim 0.5$ nm; $\infty)$. The randomness of the atomic positions is evidenced by

- RDF (see Figure 3.54)
- Gaussian random coil configuration of macromolecular chains
- X-ray scattering pattern (see Figures 3.57 and 3.58) main peak at $2\theta_{Cu} = 18.7°$

Atomic Correlations: Direct contacts between spheres in the IAS model results in correlations that are evidenced by the PDF (see Figure 3.54).

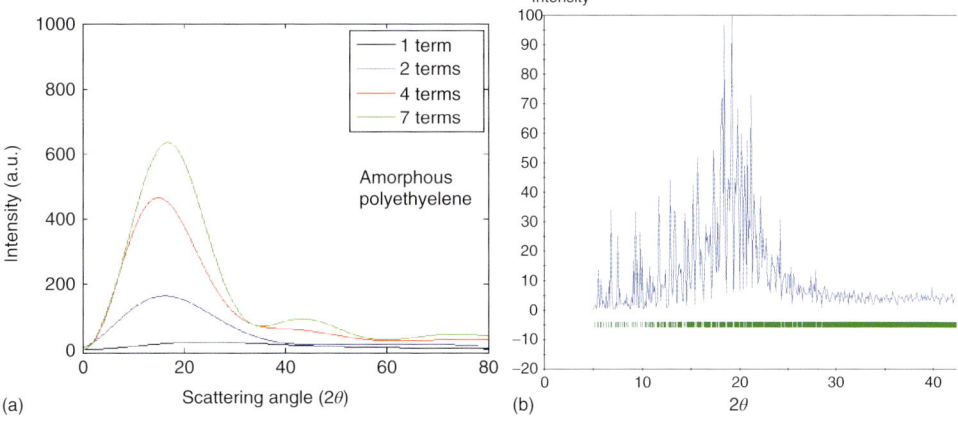

Figure 3.57 Predicted X-ray scattering profile for the IAS model of PE and a model of amorphous PE from Figure 3.55 obtained using Materials Studio. CuKα radiation.

3 Glassy Materials and Ideal Amorphous Solids

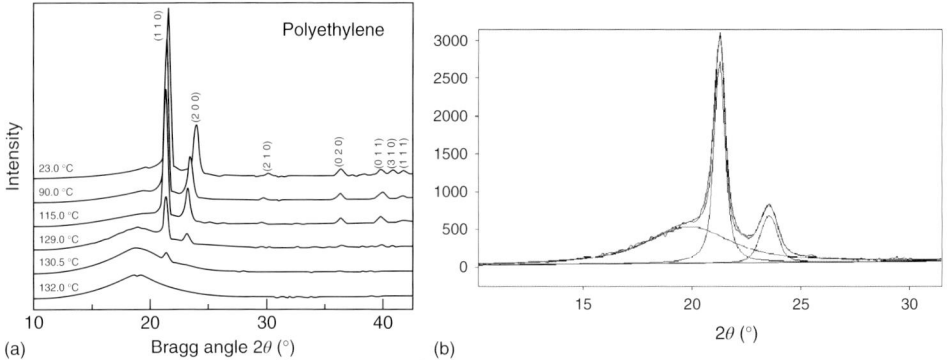

Figure 3.58 Experimental X-ray scattering profile for PE, and the peaks resolved into amorphous and crystalline. CuKα radiation.

3.8
IAS Model of a-Silica (a-SiO$_2$)

Although the ancient artefacts of glass are based on silica, high-transparency silica glass is a product of modern technology. The silica in most glass products is modified with additives that alter the molecular network and consequently the properties of the glass, especially glass-transition temperature and viscosity (flowability). Examples of modified glasses are given in Table 3.27.

When the pure molten silicon dioxide SiO$_2$ is rapidly cooled, it does not crystallize but solidifies as a glass. The geometry of the silicon and oxygen centres in glass is similar to that in quartz and most other crystalline forms of the same composition.

When molecular silicon monoxide, SiO, is condensed in an argon matrix cooled with helium along with oxygen atoms generated by microwave discharge, molecular SiO$_2$ is produced with a linear structure. This form of silica will not be modelled here.

3.8.1
Molecular Parameters for SiO$_2$

Si atom has tetrahedral coordination, with four oxygen atoms surrounding a central Si atom. It is different from polymethylene where each C-atom also has

Table 3.27 Composition (%) of common glasses based on silica.

Type	SiO$_2$	Na$_2$O	MgO	CaO	Al$_2$O$_3$	K$_2$O	B$_2$O$_3$
Vitreous silica	100						
Window glass	72	14.2	2.5	10	0.6		
Pyrex	81	4.5			2.5		12
Fibreglass	57		7	10	16	6	4

tetrahedral coordination but it is connected to 2C and 2O atoms. The van der Waals radius of the Si atom is 0.210 nm, whereas the atomic radius is 0.110 nm. The van der Waals radius of O is 0.152 nm, the covalent radius is 0.66 nm. When bonded together, the length of the Si–O covalent bond is 0.16 nm, and the Si–O–Si angle is 144°. Therefore, the distance between the centres of two oxygens, or two silicons, is $2 \cdot 0.16 \cdot \sin 144°/2 = 0.304$ nm. A segment of the silica molecular chain is shown in Figure 3.59.

3.8.2
IAS and United Atom Models for SiO$_2$

Figure 3.59 shows a Round Cell of SiO$_2$ simulated by the IAS software as random packing of unconnected spheres representing Si and O atoms in the correct proportions. This means that each atoms has multiple contacts, with a distribution similar to that shown in Figure 3.29. Therefore, the IAS model is not a SiO$_2$ network structure as illustrated by the molecular segment in Figure 3.59.

Also, another model, the so-called united atom, was simulated using the IAS software (not shown). In this case, each SiO$_2$ group was modelled by single spheres of diameter 0.304 nm, each with scattering power given by

$$f_{SiO_2} = f_{Si} + 2f_O \frac{\sin 0.16\, s}{0.16\, s} \quad (3.33)$$

The corresponding X-ray scattering profiles calculated by the Debye equation for the two cases (IAS and united atom) are shown in Figure 3.60. Neither of these results is in agreement with the experimentally obtained scattering curve, as shown in Figure 3.61. Both models must be rejected.

A better method of predicting X-ray scattering by inorganic glasses was proposed by Warren (1934), similar to that shown already for amorphous polymethylene. Warren approximated the pair radial distribution function for SiO$_2$ to be as

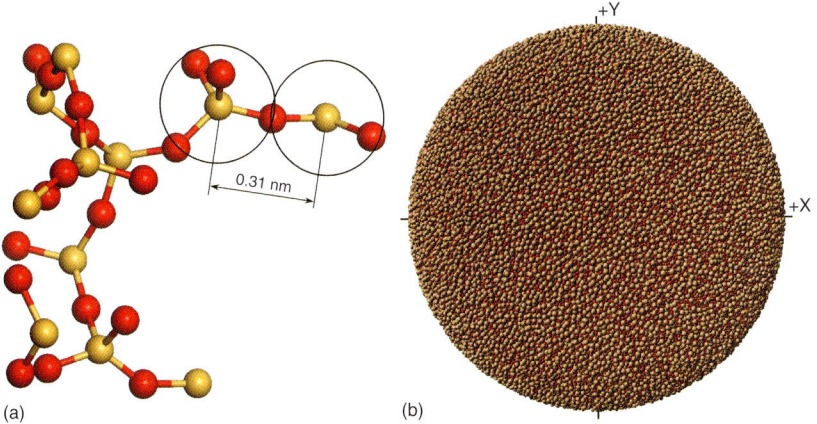

Figure 3.59 A segment of the SiO$_2$ molecular chain, and an IAS simulated Round Cell of silica containing 10^6 atoms.

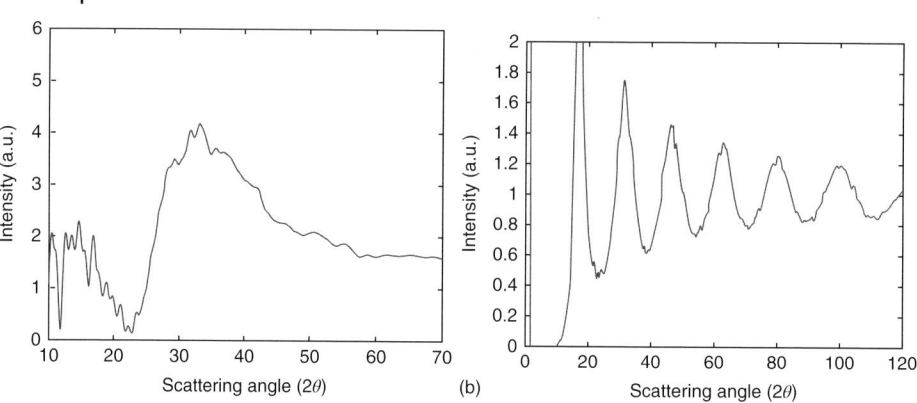

Figure 3.60 (a) X-ray scattering profile for the IAS model of SiO$_2$ (calculated using the Debye equation). (b) X-ray scattering profile for the united atom model of SiO$_2$, also calculated using the Debye equation. (CuKα radiation).

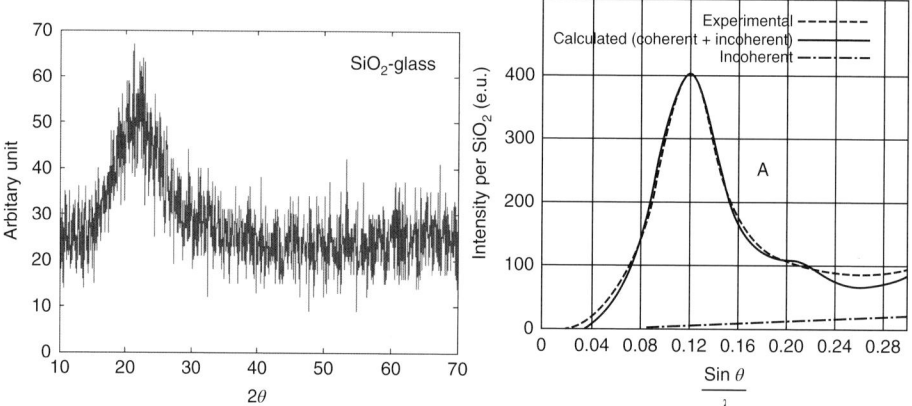

Figure 3.61 X-ray scattering profiles for SiO$_2$. (a) From experiment and (b) a prediction according to Equation 3.34.

shown in Figure 3.62. Then the intensity of scattering for amorphous silica can be expressed as follows:

$$I(s) = Nf_{Si}[f_{Si} + 4f_O \frac{\sin(0.16\ s)}{0.16\ s} + 4f_{Si} \frac{\sin(0.32\ s)}{0.32\ s} + 6f_O \frac{\sin(0.40\ s)}{0.40\ s}$$
$$+ 12f_{SiO_2} \frac{\sin(0.52\ s)}{0.52\ s} - 17\Phi(s)(\ 0.605\ s)] + 2f_O[f_O + 2f_{Si} \frac{\sin(0.16\ s)}{0.16\ s}$$
$$+ 3f_O \frac{\sin(0.262\ s)}{0.262\ s} + 6f_{SiO_2} \frac{\sin(0.4\ s)}{0.4\ s} - 8f_{SiO_2}\Phi(s)(0.455\ s)] \quad (3.34)$$

The function $\Phi(s)$ is evaluated in a similar way to that in Equation 3.32

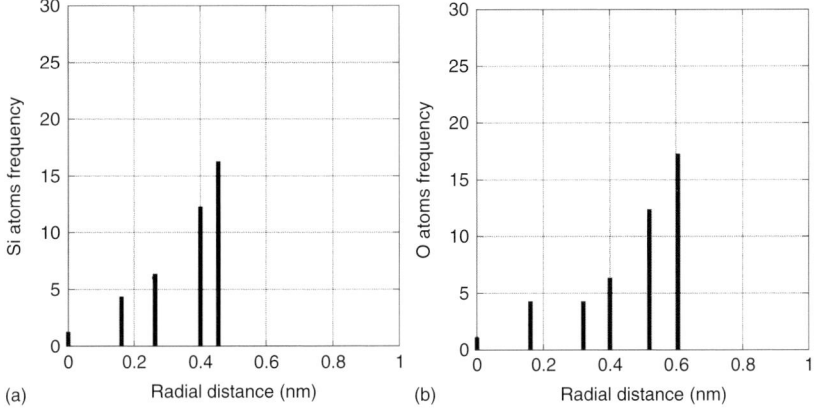

Figure 3.62 Pair distribution functions for distinct Si and O atoms according to Warren.

The success of the Warren method is in the assumption that the scattering unit, that is, SiO_2 or CH_2, is represented as a spherical scattering object that satisfies the assumption of spherical symmetry. The spherical simplification of the scattering groups is reasonable, and the groups have random orientation in space. However, appropriate choice of the spatial distribution of the units is necessary.

These glasses provide good examples of amorphous solids with SRO that is repeated throughout the whole structure. However, the SRO groups are randomly distributed and randomly oriented in space. The above interatomic correlations are assumed to be ideal, as if there were no thermal fluctuations of distances or bond angles. The fluctuations had been estimated to be of the order of 0.5 and 3%, respectively, increasing at higher temperatures. The total effect is small enough to be neglected at this level of computations.

3.8.3
Summary of a-SiO$_2$ IAS Structure

Amorphography: biatomic IAS Class II
Chemical Variety: $m = 2$: $R_{Si} = 0.150, R_O = 0.135$, Si-O $= 0.160$ nm
Chemical Composition: Silicon (1/3), Oxygen (2/3)
Molar Mass: 60.08 g/mol
Glass-Transition Temperature: $T_g = 1475$ K (pure silica)
Crystalline Melting Point: $T_m = 1998$ K
Atomic Positions: SRO for SiO_2 adjacent covalently bonded units, reducing with distance along the molecular chain. Randomly distributed in space in the range $[\sim 4R_{Si-O}; \infty)$. The randomness of the atomic positions is evidenced by

- X-ray scattering pattern (see Figure 3.36). main peak at $2\theta_{Mo} = 16.5°$, $2\theta_{Cu} = 36°$
- the presence of a glass-transition temperature

Free Volume of IAS: $v_f = 0$.

3.9
Chalcogenide Glasses

Network glasses based on chalcogens (chalcogenide vitreous semiconductors (ChVS)) are unique solids widely used in IR optoelectronics and photonics as high-efficiency optical amplifiers in optical communication networks, memory or switching devices, high-resolution inorganic photoresists, media for submicrometer lithography process, antireflection coatings and so on.

The discovery of ternary chalcogenide glasses allowed phase change memory to be used successfully in the rewriteable (DVD+RW) optical disk. High-amplitude laser pulse causes the material to reach melting point, and on rapid quench, the material changes from crystalline to amorphous phase (this is called *RESET current*). A relatively longer, low-amplitude pulse heats up the material to the crystallization temperature only, allowing phase change from amorphous to crystalline (SET current). During writing, a focused laser beam selectively heats areas of the phase-change material above the melting temperature (500–700 °C), into the liquid state. Then, cooled quickly, the random liquid state is frozen-in and the amorphous state is obtained. If the phase-change layer is heated below the melting temperature but above the crystallization temperature (200 ^{circ}C) for a sufficient time, the atoms revert back to the crystalline state. The amorphous and crystalline states have different refractive indexes and can be optically distinguished. In the DVD+RW system, the amorphous state has a lower reflectance than the crystalline state and, during read-out, this produces a signal identical to that of a regular dual layer DVD-ROM disc, making it possible to read DVD+RW discs on DVD-ROM drives and DVD Video players.

Optical glasses of As–Ge–Se composition because of excellent transmission in IR spectral range, high refractive index values and wide spectra of optical phenomena that can be used for the fabrication of various surface relief are perspective for the use in IR optics, optoelectronics. The atomic arrangements in a chalcogenide glass of composition As12–Ge33–Se55 is described in this section.

The chalcogenide glasses show greater complexity in their atomic arrangements than the previous two glasses because of the varying character of interatomic bonds, which leads to partial coordinations and formation of local SRO. This also means that interatomic distances vary depending on composition.

3.9.1
As12–Ge33–Se55 Chalcogenide Glass

From Table 3.28, the interatomic distances are calculated as

$r_{Ge-Ge} = 0.2459$ nm,
$r_{As-As} = 0.2433$ nm,
$r_{Se-Se} = 0.2380$ nm.

It is known from experimental measurements (Figure 3.63) that interatomic distances for mixed pairs are

3.9 Chalcogenide Glasses

Table 3.28 IAS model composition and radii of atoms in AsGeSe glass.

Atoms	Concentration (%)	Radius (nm)	Structure	Character
Ge	33	0.123	Diamond cubic	Metalloid
As	12	0.1216	Simple trigonal	Metalloid
Se	55	0.1183	Hexagonal	Nonmetal

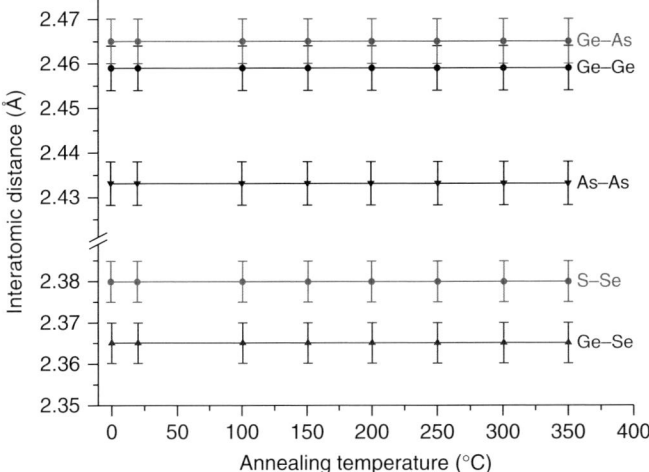

Figure 3.63 Measured interatomic separations of elements in As12–Ge33–Se55 glass by EXAFS. (*Source:* Data from Research School of Physics and Engineering, ANU.)

$$r_{Ge-As} = \tfrac{1}{2}(r_{Ge} + r_{As}) = 0.245 \text{ nm},$$
$$r_{Ge-Se} = 0.2366 \text{ nm} < \tfrac{1}{2}(r_{Ge} + r_{Se}),$$
$$r_{As-Se} = 0.2430 \text{ nm} > \tfrac{1}{2}(r_{As} + r_{Se}).$$

Also, it is conjectured on the basis of EXAFS measurements of the glass that As bonds to As only (forming As-atomic chains), and Se bonds with Ge and with itself (also forming multi-atom molecules). Indirect evidence for this comes from experimental results shown in Figure 3.64. These suppositions are corroborated by other published data (Sen and Aitken, 2002).

3.9.2
Measured Coordination Distribution

The enthalpy of bond formation of the atomic pairs is shown in Table 3.29. In the absence of other effects, the higher the enthalpy the higher the preference for bond formation. The highest affinity is for Ge–Se bonding.

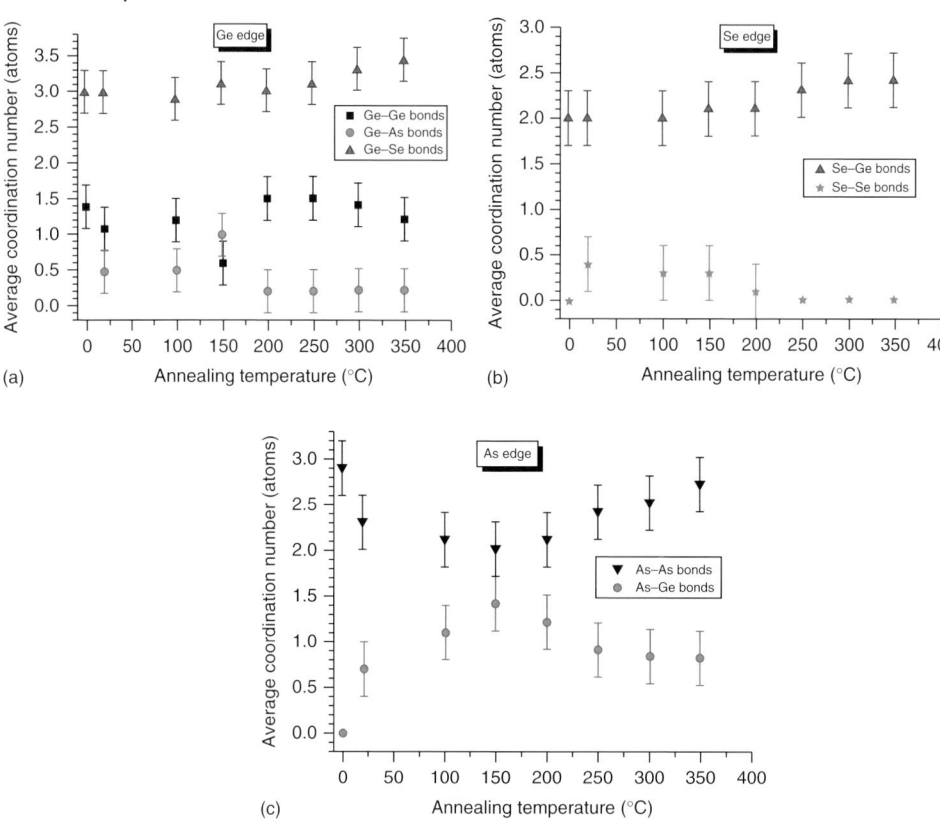

Figure 3.64 Experimentally derived coordination distribution for selected pairs in AsGeSe glass. (*Source:* Data from Research School of Physics and Engineering, ANU.)

Experimental information about the coordination number in the first coordination shell of Ge, As and Se has been studied by fine structure spectroscopic studies. It can be summarized as shown in the graphs in Figure 3.64.

The constrained local coordination environments of Ge, As and Se allows one to conclude the presence of preferred Se–Se and As–As bonds within medium-range topological effects. The atomic arrangements are likely to consist of three-dimensional network of corner sharing tetrahedra and trigonal pyramids in view of the Ge atoms being four coordinated and As atoms being three coordinated.

Table 3.29 Enthalpy of bond formation (kJ mol^{-1}).

Element	As	Ge	Se
As	201.0	196.3	227.2
Ge		186.0	234.0
Se			227.0

3.9.3
Measured X-ray Scattering

The experimental X-ray scattering pattern for the AsGeSe glass is presented in Figure 3.65. The graph shows three peaks positioned close to $2\theta = 14°, 29°$ and $52°$ with CuKα radiation. The first peak at $2\theta \sim 14°$ (not observed if diffractometer is set to start from $20°$) is referred to as the *first sharp diffraction peak* (*FSDP*), or pre-peak. The pre-peak is not predicted by the standard Debye equation (Table 3.30).

Calculating the interatomic spacings using the Ehrenfest equation predicts values that are not in agreement with any of the atomic radii shown in Table 3.28.

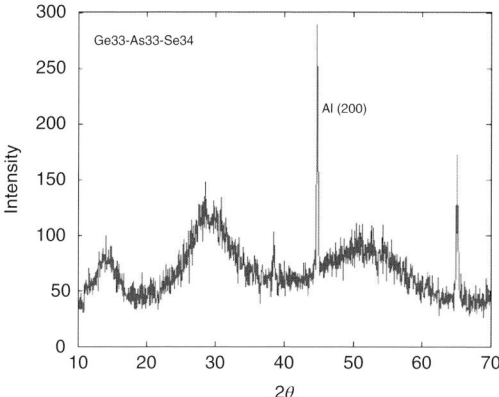

Figure 3.65 Diffractometer X-ray scattering from the AsGeSe glass. Sharp peaks owing to Al sample holder, 200 peak at $2\theta = 44.7°$ with CuKα radiation.

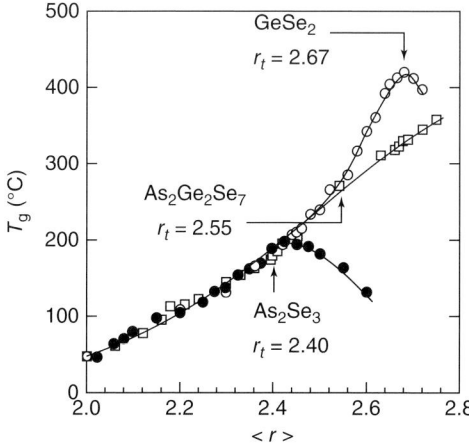

Figure 3.66 Variation of the glass-transition temperature of AsGeSe glasses (2002). (Reprinted with permission from Qu et al. (2003), Copyright (2003) by American Physical Society.)

Table 3.30 Scattering angles of experimental peaks and Ehrenfest analysis (Equation 2.50).

Peak	Measured, 2θ (°)	angle, θ (°)	Ehrenfest spacing L_p (nm)
Pre-peak	14	7	0.777
Main peak	29	14.5	0.378
Second peak	52	26	0.216

From EXAFS, we calculate the radius of the electronic orbit of the corresponding K absorption edge, which is not the same as the distance between the two bonded atoms required in the Debye calculations (Figure 3.66).

3.9.4
Glass-Transition Temperature of AsGeSe Glasses

Glass transition as a function of the average atomic separation in the chalcogenide glasses is shown in Figure 3.66

3.9.5
Models of Atomic Arrangements in AsGeSe Glass

3.9.5.1 IAS Model of AsGeSe Glass

An IAS Round Cell of the AsGeSe glass was created with the atomic radius data in Table 3.28. As the atomic radii for As, Ge and Se are very similar, the resulting atomic arrangements showed coordination distribution similar to that for the monoatomic a-Ar model (Figure 3.15), with predominance of $k = 7$ for direct contact, and $k = 14 - 15$ for the Voronoi method.

In the Debye calculations the average of the atomic scattering factors was taken; a reasonable simplification in view of the similarity of their values (Figure 3.67).

$$f_g = (1/3)(f_{Ge} + f_{As} + f_{Se}). \tag{3.35}$$

The result of these computations is shown in Figure 3.68. Clearly, there is no match with the experimental results shown in Figure 3.65, and therefore, one must conclude that the IAS model is not applicable to the chalcogenide glasses, where the bonding is strongly directional and coordination is limited.

The X-ray scattering from the IAS model of the glass shows main scattering peak at $46°$ (Figure 3.68), in agreement with Ehrenfest prediction with an average radius, $r_{average} = \frac{1}{3}(0.12 \cdot r_{As} + 0.33 \cdot r_{Ge} + 0.55 \cdot r_{Se}) = 0.1209$ nm. This result is in disagreement with the experimentally obtained X-ray scattering pattern as shown in Figure 3.65, in terms of the position of the scattering peaks, the coordination of atoms and the missing pre-peak.

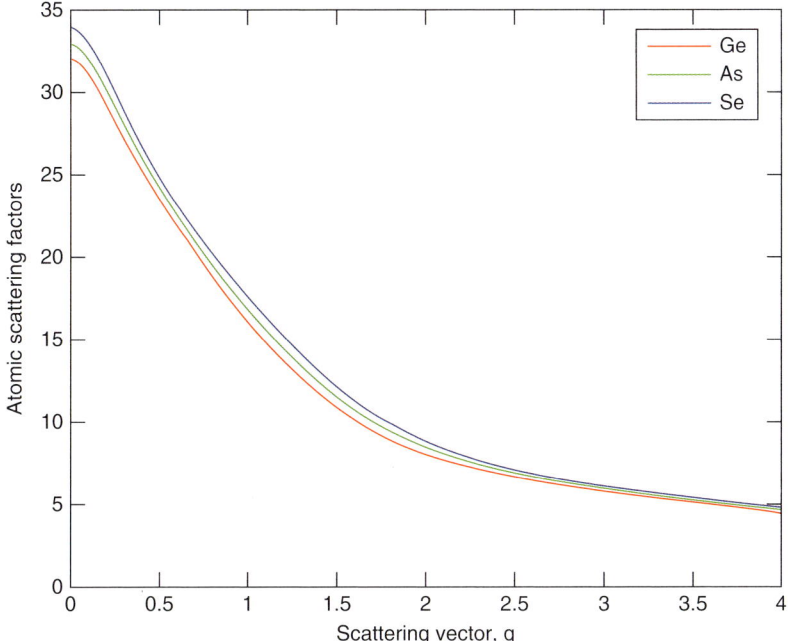

Figure 3.67 The atomic scattering factors of Ge, As and Se.

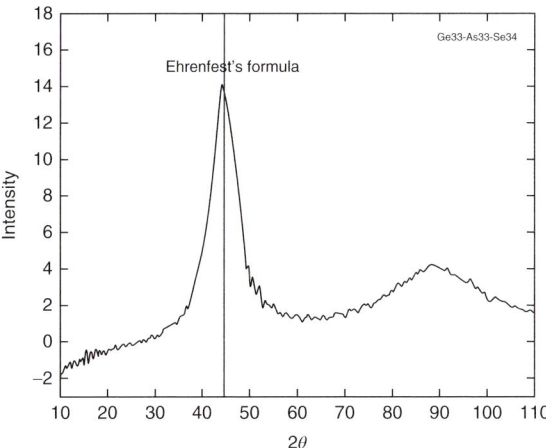

Figure 3.68 Predicted X-ray scattering for the IAS model of AsGeSe glass with CuKα radiation.

Therefore, attempts to predict the correct X-ray scattering peak from the IAS model fail because the nature of this model is to give coordination distribution with an average number between 6 and 7, far from the 2.4 of the glass.

3.9.5.2 Other Models of AsGeSe Glass

An earlier detailed model of the GeAsSe glass has been proposed in which three molecular arrangements were incorporated: (i) tetrahedra based on Ge atom, (ii) As-atomic chains and (iii) Se-atomic chains as shown in Figure 3.69.

To allow for controlled coordination, a modified the IAS model was used to create a two-phase model of the glass. The details of the model were input into the generalized Debye Equation 1.43 with results as shown in Figure 3.70. There is qualitative agreement with experiment.

A model for the FSDP was proposed based on interstitial volume around cation-centred structural units; Ge-centered clusters are the structural units associated with the formation of the FSDP in amorphous $GeSe_2$ glass. The clusters, separated by a distance d, are surrounded by a number of spherical voids at a distance D, resulting in peak position at scattering vector, $Q_1 = 3\pi/2d$.

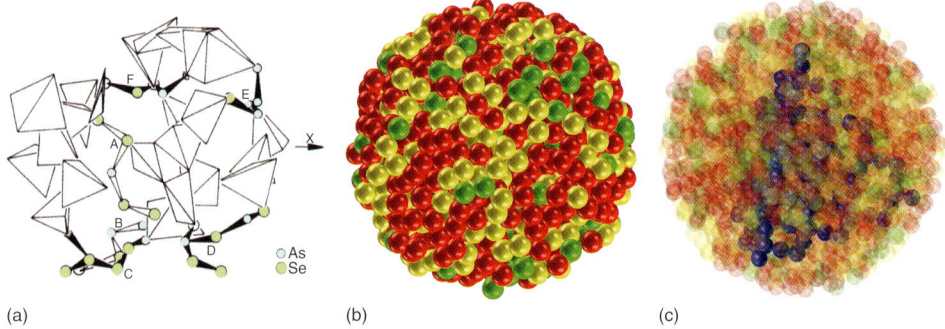

Figure 3.69 (a) A model of atomic arrangement in AsGeSe glass. Chains of As and of Se are dispersed among the $GeSe_2$ tetrahedra. (Reprinted with permission from de la Rosa-Fox et al. (1986), Copyright (1986) by American Physical Society.) (b) Two views of packing of spheres to model the AsGeSe glass: an outside view and a semi-transparent view to reveal a molecular chain formed by As atoms (blue).

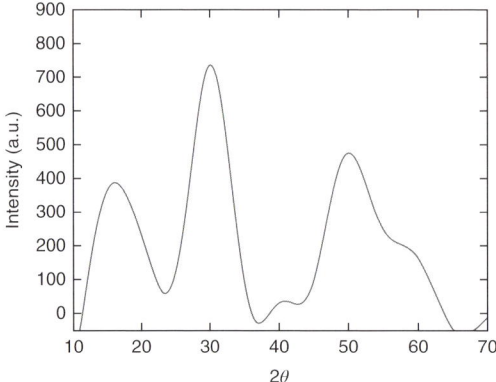

Figure 3.70 Coordination distribution for the packing of spheres in the modified IAS model of AsGeSe.

3.9.6
Summary of a-AsGeSe IAS Structure

Amorphography: tri-atomic IAS Class III
Chemical Variety: $m = 3, R_{As} = 0.1216, R_{Ge} = 0.123, R_{Se} = 0.1183$ nm
Chemical Composition: Arsenic 33%, Germanium 34%, Selenium 33%
Glass-Transition Temperature: $T_g \sim 520$ K
Atomic Positions: SRO and mid-range order evident through clustering of As atoms, Ge and As atoms form heteropolar bonds with Se atoms. Randomly distributed in space in the range [0.5 nm; ∞). The randomness of the atomic positions is evidenced by the X-ray scattering pattern (see Figure 3.63).

3.10
Concluding Remarks

3.10.1
Chapter 3

It is clear from the examples of the IAS models described in this chapter that their applicability is well justified for solids with atoms that can be represented as spheres of appropriate radii, with non-limited coordinations and without strong directional bonding. This applies to all examples from Argon to ZrTiCuNiBe metallic alloy.

It can be argued that amorphous polymethylene and silica belong to Class II of the IAS family, with each CH_2 unit represented as a sphere. This class is distinguished by SRO of units along the molecular chain, but randomly distributed in space.

At this stage of development of the theory of amorphousness, there is no clearly defined ideal model of amorphous AsGeSe chalcogenide glass.

3.10.2
Chapter 2

The IAS model comprising packed spheres, defined as $S_a + X$, is based on the premises that

- all spheres are in fixed positions
- any three adjacent non-touching spheres form an irregular triangle
- scale invariance: if $S_a + X$ is a packing for an IAS, then so is $S_{ca} + cX$ for every positive c.

Then any amorphous solid, $\mathcal{A}_1$, can be shown to be statistically identical to another amorphous solid, $\mathcal{A}_2$.

It is conjectured that the characteristics of the IAS model listed in Chapter 2 are interrelated and that only a limited number of these measures is necessary and sufficient to define the IAS body.

References

Adams, D.J. and Matheson, A.J. (1972) Computation of dense random packings of hard spheres. *Journal of Chemical Physics*, **56** (5), 1989–1994.

Angell, C.A. (1985) *Complex Systems*. National Technical Information Service, US Department of Commerce, Springfield, VA.

Bednarcik, J., et al. (2011) Thermal expansion of a La-based bulk metallic glass: insight from in situ high-energy x-ray diffraction. *J. Phys. Condens. Matter*, **23**, 254204.

Bennett, C.H. (1972) Serially deposited amorphous aggregates of hard spheres. *Journal of Applied Physics*, **43** (6), 2727–2734.

Blessing, R., Guo, D. and Langs, D. (1998) *Intensity Statistics and Normalization*, Kluwer Academic Publishers.

Borodin, V.A. (1999) Local atomic arrangements in polytetrahedral materials. *Philosophical Magazine A*, **79** (8), 1887–1907.

Cahn, R.W. (2001) *The Coming of Materials Science*, Pergamon Press.

Clarke, A.S. and Jónsson, H. (1993) Structural changes accompanying densification of random hard-sphere packings. *Physical Review E*, **47** (6), 3975–3984, doi: 10.1103/PhysRevE.47.3975.

Conway, J.N. and Sloane, N.J.A. (1998) *Sphere Packings, Lattices and Groups*, 3rd edn, Springer-Verlag, New York.

Daley, D.J. and Vere-Jones, D. (2008) *An Introduction to the Theory of Point Process*, Springer, Berlin.

de la Rosa-Fox, N., Esquivias, L., Villares, P. and Jimenez-Garay, R. (1986) Structural models of the amorphous alloy GeAsSe by a random technique. *Physical Review B*, **33**, 4094.

Dobbs, E.R. and Jones, G.O. (1957) Theory and properties of solid argon. *Reports on Progress in Physics*, **20**, 516–566.

Donev, A., Torquato, S. and Stillinger, F.H. (2005) Pair correlation function characteristics of nearly jammed disordered and ordered hard-sphere packings. *Physical Reviews E*, **71**, 0501–0514.

Finney, J.L. (1970) Random packings and the structure of simple liquids. i. the geometry of random close packing. *Proceedings of the Royal Society of London Series A: Mathematical and Physical Sciences*, **319** (1539), 479–493.

Flory, P.J. (1974) Theoretical predictions on the configurations of polymer chains in the amorphous state. *Journal of Macromol. Science, Part B: Physics*, **12**, 1.

Flory, P.J. (1967) *Principles of Polymer Chemistry*, Cornell University Press, Ithaca, NY.

Frank, F.C. and Kasper, J.S. (1958) Complex alloy structures regarded as sphere packings. i. definitions and basic principles. *Acta Crystallographica*, **11** (3), 184–190.

Gan, F.X. (2008) Structure and properties of amorphous thin film for optical data storage. *Journal of Non-Crystalline Solids*, **354**, 1089–1099.

Gibbs and diMarcio (1978) Nature of the Glass Transition and the Glassy State. **28**, 373.

Gleiter, H. (1998) Confined glasses: an equilibrium state of solid matter that melts/solidifies continuously and reversibly?

Hui, X., Fang, H.Z., Chen, G.L., Shang, S.L., Wang, Y., Qin, J.Y. and Liu, Z.K. (2009) Atomic structure of ZrTiCuNiBe bulk metallic alloy. *Acta Materialia*, **57**, 367–391.

Hukins, D.W.L. (1981) *X-ray Diffraction by Disordered and Ordered Systems*, Pergamon Press, Oxford.

Inoue, A. (2000) Stabilization of metallic supercooled liquid and bulk amorphous alloys. *Acta Materialia*, **48**, 279–306.

Jiao, Y., Stillinger, F.H. and Torquato, S. (2008) Modeling heterogeneous materials via two-point correlation functions. ii. algorithmic details and applications. *Physical Review E*, **77** (3), 031135.

Johannessen, B., Kluth, P., Llewellyn, D.J., Foran, G.J., Cookson, D.J. and Ridgway, M.C. (2007) Amorphization of embedded cu nanocrystals by ion irradiation amorphization of embedded cu nanocrystals by ion irradiation amorphization of embedded cu noncrystals by ion irradiation. *Applied Physics Letters*, **90** (7), 072119–073119-3.

Joubert, J.-M., Sundman, B. and Dupin, N. (2004) Assessment of the Ni–Nb system. *Computer coupling of phase diagrams and thermochemistry*, **28**, 299.

Lee, C.Y, Stachurski, Z.H. and Welberry, T.R. (2010) The geometry, topology and structure of amorphous solids. *Acta Materialia*, **58**, 615.

Lu, J., Ravichandran, G. and Johnson, W.L. (2003) Deformation behavior of the zr-ti-cu-ni-be bulk metallic glass over a wide range of strain-rates and temperatures. *Acta Materialia*, **51**, 3429–3443.

Ma, E. (2005) Alloys created between immiscible elements. *Progress in Materials Science*, **50**, 413–509.

MacCrander, A.T. and Crawford, R.K. (1977) Density of solid argon at melting. *Phys. Status Solidi*, **43**, 611–617.

Martin, I., Ohkubo, T., Ohnuma, M., Deconihout, B. and Hono, K. (2004) Nanocrystallisation of zr41.2cu12.5ni10 ti13.8be22.5 metallic glass. *Acta Materialia*, **52**, 4427–4435.

Min, B.G., Stachurski, Z.H. and Hodgkin, J.H. (1993) Cure kinetics of elementary reactions of degeba/dds epoxy resin. *Polymer*, **34** (23), 4908–4912.

Nash, P. and Nash, A. (1986) The Nb–Ni system. *Bulletin of Alloy Phase Diagrams*. **7**, 124.

Nelson, D. (2002) *Defects and Geometry in Condensed Matter Physics*. Cambridge University Press, Cambridge.

Ordway, F. (1964) Condensation model producing crystalline or amorphous tetrahedral networks. *Science*, **143**, 800–801.

Qu, T., Georgiev, D.G., Boolchand, P., Jackson, K., and Micoolaut, M. (2003) The intermediate phase in ternary GeAsSe glasses in Supercooled Liquids, Glass Transition and Bulk Metallic Glasses. (eds T. Egami, A.L. Greer, A. Inoue, S. Ranganthan), *Mater. Res. Soc. Symp. Proc.* 754, CC8.1.1.

Quine, M.P. and Watson, D.F. (1984) Radial generation of n-dimensional poisson processes. *Journal of Applied Probability*, **21** (3), 548–557.

Rahman, A. (1964) Correlations in the motions of atoms in liquid argon. *Physical Review A*, **136** (2), 405–411.

Sen, S. and Aitken, B.G. (2002) Atomic structure and chemical order in Ge–As Selenide and Sulfoselenide glasses. *Physical Review B*, **66**, 1134204.

Speedy, R.J. (1999) Revised Ehrenfest relations for the glass transition. *Journal of Physical Chemistry B*, **103**, 8128–8131.

Speedy, R.J. and Debenedetti, P.G. (1996) The distribution of tertravalent network glasses. *Molecular Physics*, **88**, 1293–1316.

Stachurski, Z.H. (2003) Definition and properties of ideal amorphous solids. *Physical Review Letters*, **90** (15), 5502–5506.

Stoyan, D., Kendall, W.S. and Mecke, J. (1995) *Stochastic Geometry and its Applications*, John Wiley & Sons, Inc., New York, Chichester.

Tanaka, K., Maruyama, E., Shimada, T. and Okamoto, H. (1999) Amorphous Silicon Wiley, New York, Translated by T. Sato.

To, L.Th., Daley, D.J. and Stachurski, Z.H. (2006) On the definition of an ideal amorphous solid. *Solid State Sciences*, **8**, 868–879.

Torquato, S. (2002) *Random Heterogenous Materials*, Springer-Verlag, New York.

Torquato, S., Truskett, T.M. and Debenedetti, P.G. (2000) Is random close packing of spheres well defined? *Physical Review Letters*, **84** (10), 2064–2067.

Trachenko, K. and Brazhkin, V.V. (2009) Understanding the problem of glass transition on the basis of elastic waves in a liquid. *Journal of Physics: Condensed Matter*, **21**, 1.

Wang, G., Jackson, I., FitzGerald, J.D. and Stachurski, Z.H. (2007) Rheology and nanocrystallisation at high hydrostatic pressure of a zr-ti-ni-cu-be bulk metallic glass. *Journal of Non-Crystalline Solids*, **354**, 1575–1581.

Wang, G., Shen, J., Sun, J.F., Zhou, B.D., FitzGerald, J.D., Llewellyn, D. and Stachurski, Z.H. (2005) Isothermal nanocrystallisation behavior of zrticunibe bulk metallic glass in the supercooled region. *Scripta Materialia*, **53**, 641–645.

Wang, G., and Mattern, N., Bednarcik, J., Li, R., and Eckert, J. (2012) Correlation between elastic structural behaviour and yield strength of metallic glasses. *Acta Materialia*, **60**, 3074.

Wang, X.D., Lou, H.B., Wang, G., Xu, J. and Jiang, J.Z. (2011) Atomic packing in Mg61Cu28Gd11 bulk metallic glass. *Applied Physics Letters*, **98**, 031901-1–031901-3.

Warren, B.E. (1934) X-ray determination of the structure of glass. *Journal of American Chemical Society*, **17**, 249–254.

Welberry, T.R. (2004) *Diffuse X-ray Scattering and Models of Disorder*, Oxford Science Publications.

Yavari, A.R. (2007) The changing faces of disorder. *Nature Materials*, **6**, 181–182.

Yavari, A.R., Le Moulenc, A., Inoue, A., Nishiyama, N., Lupu, N., Matsubara, E., Botta, W.J., Voughan, G., Di Michiel, M. and Kvick, A. (2005) Excess free volume in metallic glasses measured by X-ray diffraction. *Acta Materialia*, **53**, 1611–1619.

Yeh, J.-W. (2013) Alloy design strategies and future trends in high-entropy alloys. *Journal of Metals (TMS)*, **65** (12), 1759–1771.

Zachariasen, W.H. (1932) The atomic arrangement in glass. *Journal of the Chemical Society*, **54**, 3841.

Zarzycki, J. (1982) *Les Verres et l'etat vitreux*, Masson, Paris.

Zarzycki, J. (1991) *Glasses and the Vitreous State*, Cambridge University Press, Cambridge.

Zong, C.M. (1999) *Sphere Packings*, Springer-Verlag, New York, ISBN: 978-0-387-22780-1.

4
Mechanical Behaviour

4.1
Introduction

Stone, wood and bone were the first materials at hand accessible to man. With time, a tradition of usage of these materials and knowledge about the strength of the objects that were made had accumulated. The advantage of stronger tools stimulated innate ability to take note of the dependence of strength on the type of stone or wood used. The beginnings of materials science were established.

From the date of its construction in 537 AD, the Hagia Sophia (in present day Istanbul) remained the world's largest cathedral for over 900 years. The creators of the grand design were Isidore of Miletus and Anthenius of Tralles, who were scholars of geometry and physics. During an earthquake in 558, its main dome collapsed completely. The emperor at the time ordered an immediate restoration, appointing Isidorus the Younger (nephew of Isidore of Miletus) to rebuild the dome. Isidorus applied advanced materials science and engineering to its reconstruction. In particular, he used lighter, more porous mortar, and changed the dome design by erecting a ribbed structure with pendentives with diameter of 32 m. The pendentives are triangular segments of a sphere that taper to points at the bottom and spread at the top to establish the continuous circular base needed for the dome. Under Justinian's orders, eight Corinthian columns were disassembled from Baalbek, Lebanon, and shipped to Constantinople around 560. This sixth century reconstruction of the church was completed in 562. It lasted for almost 400 before needing major repairs. Its success has its origins in perfect understanding of geometry and the materials properties used in its construction.

In the first printed book on Mechanics of Materials (1638), Galileo Galilei described mechanical testing of marble beams subjected to bending. The behaviour of these materials is brittle, and at that time, one property of direct interest was the maximum bending load sustained before fracture (catastrophic failure). Galileo correctly established that two beams of identical cross section, as shown in Figure 4.1, would break when their full lengths were equal, even though one is supported in the middle and the other at its ends (Timoshenko, 1983). For a long time, common folklore was contrary to this view. Of course, the columns were used to support compressive loads, to their maximum advantage,

Fundamentals of Amorphous Solids: Structure and Properties, First Edition. Zbigniew H. Stachurski.
© 2015 Wiley-VCH Verlag GmbH & Co. KGaA. Published by Wiley-VCH Verlag GmbH & Co. KGaA.
Boschstr. 12, 69469 Weinheim, Germany, under exclusive license granted by HEP for all media and languages excluding Chinese and throughout the world excluding Mainland China, and with non-exclusive license for electronic versions in Mainland China

192 | 4 *Mechanical Behaviour*

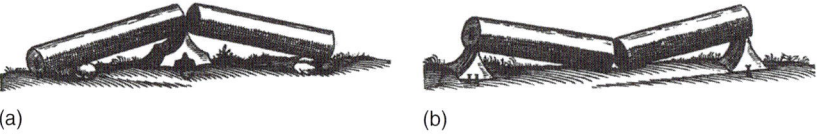

(a) (b)

Figure 4.1 (a) Marble beam supported in the middle. (b) Similar marble beam supported at both ends (see text for explanation).

Figure 4.2 A collection of minerals adopted by Mohs for a scale of hardness, from diamond to talc (Wikipedia).

but storage of columns was usually on supports in that risky horizontal position. Their fracture would occur unexpectedly and seemingly without any reason. This is the emphasis of Section 4.4.

Ductile materials, such as bronze, gold, gut or leather, were used for making tools and objects by craftsmen whose guiding principle was experience of successful designs, rather than engineering understanding of the material properties. However, the relative hardness was important, and to fulfil this need, a scale of hardness was invented by F. Mohs in 1812; the harder mineral will always scratch the one below it (see Figure 4.2) but not the other way round. The method of comparing hardness by scratching minerals against each other is of great antiquity. Later more precise hardness testing machines were invented by Brinell in 1900 in particular for ball bearings and later by Vickers and Rockwell for quality assurance testing. Surprisingly, silica and metallic glasses are some of the hardest materials known.

The twentieth century may be called by future historians the *renaissance period of materials science* because many analytical methods were invented to discover and to ascertain the atomic arrangements in solids (X-ray diffraction being the most direct), thus opening up the new science of structure–property relationship.

The knowledge of the relationship between the nature of the material and its deformation behaviour under load has advanced enormously. The variety of microstructures and resulting atomic rearrangements revealed inherent imperfections that have profound influence on the strength of materials, which is the essence of materials science. This chapter presents a summary of the knowledge on mechanical behaviour with new additions specific to amorphous materials.

4.2 Elasticity

4.2.1 Phenomenology

Elasticity and solidity are natural characteristics of materials. We expect engineering materials to have strength, hardness and stiffness that are constant and unchanging with time, so that the structures we build carry the applied loads indefinitely. In most cases, this is true, but time, temperature and environment have an effect on materials that can change this behaviour, even if the applied loads are well below their elastic design limits.

We think of rocks and stones as materials that can carry load indefinitely. This is true within the time scale of our civilization. The Parthenon temple has been standing supporting its weight for over 3500 years (the deformation (damage) caused by humans only) (Figure 4.3). However, rocks will deform (creep under load) when the time scale is in millions of years. Examples for both cases are shown in the following text, and the reasons for the differences in behaviour relate to the atomic arrangements and interactions.

In metallic alloys, yield strength and strain hardening are measured during mechanical testing at strain rates of the order of 10^{-3} s^{-1}, and typically, at room temperature. This strain rate is achieved when a sample of gauge length 100 mm is stretched in the testing machine by 0.1 mm s^{-1}, or 6 mm min^{-1}; slow enough to avoid dynamic effects, but not too slow to make the test last too long.

The behaviour of a material resisting deformation and returning to original shape after load removal is called *elasticity*. This statement supports the general definition given earlier for a solid as a body with memory of its original configuration. The essential property of elasticity is expressed mathematically by the Hooke's law.

$$P = kx \tag{4.1}$$

(a)

(b)

Figure 4.3 (a, b) The Parthenon temple in Athens is static and immovable, but the rock face, exposed by road cutting for State Circle near Parliament House in Canberra, shows evidence of flow (creep), as well as shear fracture.

where P (for *pondus*) is the force, x the extension and the constant of proportionality, k, is called the *stiffness*. A scientific description of elasticity of materials begins with Robert Hooke's statement in 1678 (initially published as an anagram) of proportionality between extension and force: "Ut tensio, sic vis". The spring, shown in Figure 4.4, besides being a mechanical device, is a mechanical model of elastic, time-independent behaviour. When extension, x, is applied to the spring, it will resist with a force, P, and the force is *always* proportional to the extension.

On atomic scale, this is achieved by ensuring that all atoms remain in their relative positions and preventing all defects in the material from moving (i.e., no dislocation glide, no diffusion of vacancies, no sliding of grain boundaries and no motion of shear bands).

The inherent elastic nature of materials was recognized by Leonhard Euler (1727), Giordano Riccati (1782) and Robert Young (1807), who defined the stiffness, k, from Equation 4.1, in terms of underlying variables, as shown by this inverse Hooke's equation:

$$\Delta l = \left(\frac{l}{AY}\right) P \qquad (4.2)$$

where A is the cross-sectional area, Y is the Young's modulus of elasticity of the material and l is the length of the material (component or sample).

Consider one of the pylons supporting the railway bridge shown in Figure 4.4. It is under a constant load from the weight of the bridge deck and an occasional

Figure 4.4 (a, b) Elastic materials are used to construct a bridge and to make a spring.

additional load when a train is passing over the bridge. Under each load, the pylon is deformed elastically (compressed) by an amount predicted by the Hooke's law (Equation 4.2). The principle of superposition applies to elastic linear behaviour, that is, the total deformation when the train is above the pylon is the sum of the deformation due to the weight of the bridge deck alone plus the deformation caused by the weight of the train. Furthermore, as the train passes on, the deformation of the pylon recovers to its static value caused by the weight of the bridge only.

The quantities in the round brackets are constant for the pylon (its dimensions and the modulus of elasticity); therefore, the deformation is directly proportional to and dependent only on the applied load.

An important point to be made here is that the deformation of the pylon does not depend on how long the bridge has been standing. We refer to this as *time-independent elastic behavior*.

4.2.2
Continuum Mechanics

The deformation of a solid is defined by a *stretch*

$$\frac{l_d}{l_u} = (1 + e) \tag{4.3}$$

where l_d and l_u are the lengths of the deformed and undeformed states, respectively, and $e = \Delta l / l_u$ is called *elongation* (Figure 4.5).

We also define and make use of *quadratic stretch*

$$\lambda = \left(\frac{l_d}{l_u}\right)^2 = (1 + e)^2 \tag{4.4}$$

In any *homogeneous* deformation, points lying on a surface of a sphere in the undeformed state will lie on the surface of an ellipsoid in the deformed state. This ellipsoid is called the *strain ellipsoid*. Conversely, points lying on the surface of a sphere in the deformed state correspond to points lying on the surface of an ellipsoid in the undeformed state.

The strain ellipsoid has three orthogonal planes of symmetry. These planes intersect in three principal diameters of the ellipsoid, and the directions of these diameters are called the *principal directions of strain*, the λ_1 direction, the λ_2 direction and the λ_3 direction.

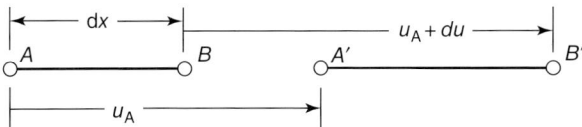

Figure 4.5 Uniaxial deformation involving displacement and elongation.

4 Mechanical Behaviour

For coordinate axes, x, y and z parallel to the principal directions λ_1, λ_2 and λ_3, respectively, each point on the strain ellipsoid has coordinates related to the principal strains by

$$\frac{x^2}{\lambda_1} + \frac{y^2}{\lambda_2} + \frac{z^2}{\lambda_3} = 1 \tag{4.5}$$

In a solid that has been subjected to a homogeneous deformation, the state of strain is specified completely by the orientations and magnitudes of the principal strains.

Equation (4.5) defines general homogeneous deformation that has three components:

- a strain
- a rotation
- a rigid-body translation.

When the strain is the only non-zero component, the deformation is called *pure strain*. The *infinitesimal* pure strain tensor (when $\Delta l/l_u$ is infinitesimally small) is defined by

$$\epsilon_{ij} = \frac{1}{2}\left(\frac{\partial u_i}{\partial x_j} + \frac{\partial u_j}{\partial x_i}\right) \tag{4.6}$$

where the subscripts i, j represent the reference axes x, y, z, and $\partial u_i/\partial x_j$ and $\partial u_j/\partial x_i$ are the displacement gradients (Figure 4.6).

In uniaxial tension (elongation applied in the three-axis direction), the material is subjected to a homogeneous strain field. The components of the strain tensor at each point of the sample are (neglecting sample ends)

$$\epsilon_{ij} = \begin{vmatrix} -\epsilon_{11} & 0 & 0 \\ & -\epsilon_{22} & 0 \\ & & \epsilon_{33} \end{vmatrix} \tag{4.7}$$

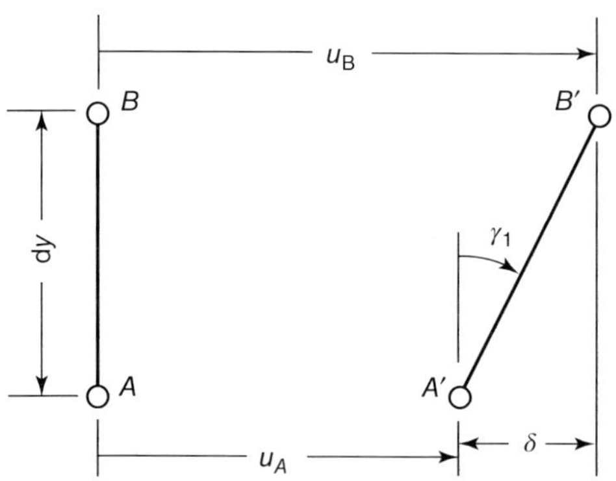

Figure 4.6 Shearing distorsion involving rigid-body translation, rotation and elongation.

where $-\epsilon_{11}/\epsilon_{33} = -\epsilon_{22}/\epsilon_{33} = v$ (Poisson's ratio). The principal strains are equal to the normal strain components, and we derive

$$\frac{\Delta V}{V} = (1 - 2v)\epsilon_{33} > 0, \quad \text{if} \quad \epsilon_{33} > 0 \tag{4.8}$$

Therefore, the relative volume change is proportional to the applied elongation, and a plot of $\Delta V/V$ against ϵ_{33} should be a straight line of constant positive slope if Poisson's ratio is constant. Consequently, volume should increase during the elastic stage of tensile deformation, for which $v < 0.5$. The change of volume with deformation is (as a rule) a fundamental characteristic of elastic behaviour.

The stresses in a solid are defined by the *stress* tensor:

$$t_i = \sigma_{ij} n_j, \quad i, j = 1, 2, 3 \tag{4.9}$$

where t_i denotes surface forces (surface tractions) and n_j denotes the surface normal. In matrix notation, the expression is given as follows:

$$\begin{vmatrix} t_1 \\ t_2 \\ t_3 \end{vmatrix} = \begin{vmatrix} \sigma_{11} & \sigma_{12} & \sigma_{13} \\ \sigma_{21} & \sigma_{22} & \sigma_{23} \\ \sigma_{31} & \sigma_{32} & \sigma_{33} \end{vmatrix} \times \begin{vmatrix} n_1 \\ n_2 \\ n_3 \end{vmatrix} \tag{4.10}$$

of which only six of the stress components are independent because $\sigma_{12} = \sigma_{21}, \sigma_{32} = \sigma_{23}, \sigma_{13} = \sigma_{31}$.

In 1821 Claude-Louis Navier generalized Hookes' law in a mathematically usable form, which is written in modern textbooks as follows:

$$\sigma_{ij} = c_{ijkl} \epsilon_{kl}, \quad i, j, k, l = 1, 2, 3. \tag{4.11}$$

This was followed by rapid developments in the general theory of mathematical elasticity (Jacob Bernoulli, Leonhard Euler, J.F. Lagrange, C.A. Coulomb, Augustin Cauchy, S.D. Poisson, Barrè St-Venant, to mention but a few of that era (Timoshenko, 1983, Hayman, 1998)).

The coefficients, c_{ijkl}, are the elastic constants of a body of general anisotropy. Noting that there are only six independent stress and strain components, the above equation can be written in the so-called short-hand notation $\sigma_n = c_{mn} \epsilon_n$, where $m, n = 1, 2, 6$ with $m = n$ for $i = j \leq 3$, and $m = n = 9 - (i + j)$ for $i, j > 3$.

The transformation rules for elastic constants from one frame of reference to another frame of reference are

$$c_{mn} = a_{mk} a_{ml} a_{nk} a_{nl} c'_{kl} \tag{4.12}$$

$$C_{mn} = a_{mk} a_{ml} a_{nk} a_{nl} C'_{kl} \tag{4.13}$$

where a_{nm} and so on are the direction cosines between the particular z' and z axes, c_{mn} are the elastic stiffnesses and C_{mn} are the elastic compliances.

A single crystal of a specific crystal symmetry has elastic constants reflecting this symmetry. Thus, for a single crystal belonging to face-centred cubic (fcc) crystal system, which possesses three mirror symmetry elements perpendicular to each of the orthogonal axes, the elastic constants become

$$c_1 = c_2 = c_3, \ c_6 = c_5 = c_4 \tag{4.14}$$

4.2.2.1 Calculation of Average Elastic Constants – Aggregate Theory

A bulk polycrystalline material assumes isotropic behaviour because of the fact that the contributions of its numerous individual single crystal grains become spatially averaged. Its mechanical properties are independent of orientation with respect to load application.

In the aggregate theory, (Ward, 1962) it is assumed that a body is made up of individual units, for example, crystalline grains with specific elastic properties of transverse isotropy. We assign a set of local reference axes (x', y', z') to each of the units in such a way that the z'-axis is aligned with the axis of transverse isotropy. The global reference axes are denoted by x, y, z. The angle between z' and z is denoted by θ, and the angle between the axis x' and y or y' and x is denoted by ϕ.

The direction cosines for the x', y', z' axes relative to the x, y, z axes are

$$\begin{vmatrix} -\sin\phi\cos\theta & \cos\phi & \sin\phi\sin\theta \\ -\cos\phi\cos\theta & -\sin\phi & \cos\phi\sin\theta \\ \sin\phi & 0 & \cos\theta \end{vmatrix} \quad (4.15)$$

Next, the body is subjected to uniaxial elongation, and in that process, the units reorient in some as yet undefined way. It is now assumed that on extension along the z-axis, these units reorient according to an *affine* scheme of Kuhn and Grün, which describes the linear transformation of points.

$$x'_i = \alpha_{i0} + (\delta_{ij} + \alpha_{ij})x_j$$
$$x_i = \beta_{i0} + (\delta_{ij} + \beta_{ij})x'_j \quad (4.16)$$

where the coefficients α and β are known and δ_{ij} is the Kronecker delta.

According to Kuhn and Grün, the number of units which lie between θ' and $\theta' + d\theta'$ and ϕ' and $\phi' + d\phi'$ is proportional to the solid angle, $\Omega = \sin\theta' d\theta' d\phi'$, and the angles are related in such a way that $\tan\theta = \lambda^{-3/2} \tan\theta'$ and $\phi = \phi'$.

As there is no preferred orientation perpendicular to the stretching direction, the orientation functions must be of the form

$$I_\theta = \int_0^{\pi/2} f(\theta) \sin\theta' \, d\theta' \quad (4.17)$$

where $f(\theta)$ is a function of θ only. Using the Kuhn and Grün scheme, the individual orientation functions become

$$I_1 = \int_0^{\pi/2} \sin^4\theta \sin\theta' \, d\theta'$$

$$I_2 = \int_0^{\pi/2} \cos^4\theta \sin\theta' \, d\theta'$$

$$I_3 = \int_0^{\pi/2} \sin^2\theta \cos^2\theta \sin\theta' \, d\theta'$$

$$I_4 = \int_0^{\pi/2} \sin^2\theta \sin\theta' \, d\theta'$$

$$I_5 = \int_0^{\pi/2} \cos^2\theta \sin\theta' \, d\theta' \quad (4.18)$$

There are two possibilities to calculate the average of the elastic constants (Ward, 1962):

- by summation of stresses (Voigt average of stiffnesses)

$$c_{33} = I_1 c'_{11} + I_2 c'_{33} + 2I_3(c'_{13} + 2c'_{44})$$

$$c_{44} = \frac{1}{4}(2I_3 + I_4)\, c'_{11} - \frac{1}{4} I_4\, c'_{12} - I_3(c'_{13} - \frac{1}{2}c'_{33}) + \frac{1}{2}(I_1 + I_2 - 2I_3 + I_5)c'_{44}$$

$$c_{11} = \frac{1}{8}(3I_2 + 2I_5 + 3)c'_{11} + \frac{1}{4}(3I_3 + I_4)c'_{13} + \frac{3}{8} I_1\, c'_{33} + \frac{1}{2}(3I_3 + I_4)c'_{44}$$

$$c_{12} = \frac{1}{8}(I_2 - 2I_5 + 1)c'_{11} + I_5\, c'_{12} + \frac{1}{4}(I_3 + 3I_4)c'_{13} + \frac{1}{8} I_1\, c'_{33} + \frac{1}{2}(I_3 - I_4)c'_{44}$$

$$c_{13} = \frac{1}{2} I_3\, c'_{11} + \frac{1}{2} I_4\, c'_{12} + \frac{1}{2}(I_1 + I_2 + I_5)c'_{13} + I_3(\frac{1}{2}c'_{33} - 2c'_{44}) \quad (4.19)$$

- by summation of strains (Reuss average of compliances)

$$C_{33} = I_1 C'_{11} + I_2 C'_{33} + I_3(2C'_{13} + C'_{44})$$

$$C_{44} = (2I_3 + I_4)\, C'_{11} - I_4\, C'_{12} - 2I_3(2C'_{13} - C'_{33}) + \frac{1}{2}(I_1 + I_2 - 2I_3 - I_5)C'_{44}$$

$$C_{11} = \frac{1}{8}(3I_2 + 2I_5 + 3)C'_{11} + \frac{1}{4}(3I_3 + I_4)C'_{13} + \frac{3}{8} I_1\, C'_{33} + \frac{1}{8}(3I_3 + I_4)\, C'_{44}$$

$$C_{12} = \frac{1}{8}(I_2 - 2I_5 + 1)\, C'_{11} + I_5\, C'_{12}$$

$$+ \frac{1}{4}(I_3 + 3I_4)\, C'_{13} + \frac{1}{8} I_1\, C'_{33} + \frac{1}{8}(I_3 - I_4)\, C'_{44}$$

$$C_{13} = \frac{1}{2} I_3\, C'_{11} + \frac{1}{2} I_4\, C'_{12} + \frac{1}{2}(I_1 + I_2 + I_5)\, C'_{13} + \frac{1}{2} I_3(C'_{33} - C'_{44}) \quad (4.20)$$

These averages represent the upper and lower bounds, respectively.

4.2.2.2 Green's Elastic Strain Energy

In equilibrium, the mathematical problem of the theory of elasticity may be stated as follows: Given the externally applied forces or loads and the boundary conditions, determine the stress and strain components at any point of the body and determine the components u_1, u_2 and u_3 of the displacement vector of each particle of the body.

George Green introduced the concept of elastic strain energy in 1837, which allowed him to derive the elastic constants from the principle of virtual work.

$$U_{elastic} = \frac{1}{2} c_{mn} \epsilon_m \epsilon_n \quad (4.21)$$

This fundamental relationship is useful in deriving elastic constants from the knowledge of potential energy and elastic deformation and is used below to estimate the elastic modulus of solid argon in its single crystal and amorphous states.

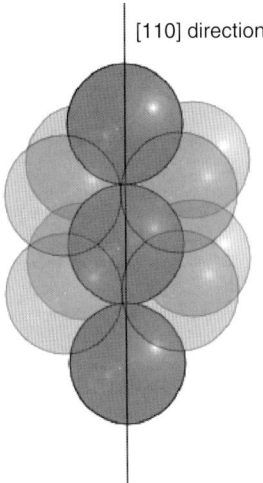

Figure 4.7 A fragment of a unit cell of single crystal of fcc argon, with three atoms along the [110] crystallographic direction.

4.2.3
Atomistic Elasticity

4.2.3.1 Calculation of an Elastic Constant for Single Crystal of Argon

Consider a fragment of the unit cell of solid fcc argon as shown in Figure 4.7. In an fcc cell, atoms touch each other along the [110] direction. We can calculate the elastic constant in that direction from the bond stiffness between the touching atoms.

The stiffness of the bond along the line of three atoms (shown in dark colour) corresponds to the [110] crystallographic direction in the fcc lattice. We use the Lennard-Jones (L-J) interatomic potential to derive bond stiffness. Starting with Equation (1.12) and differentiating it once more with respect to r, we obtain the equation for bond stiffness.

$$\begin{aligned} k_{LJ} &= \frac{df_{LJ}}{dr} \\ &= \frac{4\epsilon_0}{\sigma} \frac{d}{dr}\left[-m\left(\frac{r}{\sigma}\right)^{-m-1} + n\left(\frac{r}{\sigma}\right)^{-n-1}\right] \\ &= \frac{4\epsilon_0}{\sigma^2}\left[m(m+1)\left(\frac{r}{\sigma}\right)^{-m-2} - n(n+1)\left(\frac{r}{\sigma}\right)^{-n-2}\right] \end{aligned} \quad (4.22)$$

where r is the radial distance and σ is a constant.

The bond stiffness is evaluated at the equilibrium separation of the atoms, $r_0 = 2r_{Ar} = 2 \times 0.194 = 0.388$ nm.
The constant, $\sigma = r_0/2^{\frac{1}{6}} = 0.3456$ nm.
The minimum potential, $\epsilon_0 = 1.65 \times 10^{-21}$ J atom pair^{-1} (see Chapter 3).

Substitution of these values into Equation 4.22 gives the bond stiffness:

$$k_{LJ} = \frac{4(1.65 \times 10)^{-21}}{(0.3456 \times 10^{-9})^2}[156(1.1225)^{-14} - 42(1.1225)^{-8}]$$
$$= 0.797 \text{ N m}^{-1} \qquad (4.23)$$

Noting that force = stiffness × displacement, and therefore, force/area = modulus × displacement/length, it follows that modulus = stiffness/length (derived from Equation 4.21). Therefore, as a first approximation to the elastic modulus of argon along the [110] bond axis, we find

$$Y' = 0.797/0.388 \times 10^{-9} = 2.05 \times 10^9 \text{ N m}^{-2} \qquad (4.24)$$

At the second level of approximation, one must consider the interaction with adjacent atoms lying in the (110) plane perpendicular to the [110]-axis. This will have the effect of increasing the stiffness, and consequently increasing the value of the modulus towards that predicted from the experimental data. Simply we consider the additional contribution to be

$$4 \times k_{LJ} \times \sin^2\theta = k_{LJ}$$

With this contribution, the value of the modulus increases to

$$Y'' = 4.10 \times 10^9 \text{ N m}^{-2} \qquad (4.25)$$

The elastic constants of a single crystal of argon have been measured by the acoustic method by Keeler and Batchelder (1970). The velocity of sound in a single crystal of argon was measured for both longitudinal and transverse waves using the pulse echo technique. On analysis of the data, the following elastic constants at 0 K were obtained:

$$c_{11}(= c_{33}) = 4.39 \pm 0.11 \times 10^9 \text{ N m}^{-2} \qquad (4.26)$$
$$c_{12} = 1.83 \pm 0.12 \times 10^9 \text{ N m}^{-2} \qquad (4.27)$$
$$c_{44} = 1.64 \pm 0.04 \times 10^9 \text{ N m}^{-2} \qquad (4.28)$$

To derive the elastic stiffness (modulus) along the [110] direction, the above elastic constants must be transformed by $\theta = 45°$ rotation in the 1-2 or 1-3 plane. In the rotated frame of reference, the elastic constant is given by

$$c'_{11} = (\sin^4\theta + \cos^4\theta)\, c_{11} + 2\sin^2\theta\cos^2\theta\, (c_{12} + 2c_{44}) \qquad (4.29)$$

Substitution of the appropriate values gives

$$c'_{11} = (0.25 + 0.25)\, 4.39 + 2(0.25)(1.83 + 2 \cdot 1.64) = 4.58 \times 10^9 \text{ N m}^{-2} \qquad (4.30)$$

which is an approximation to Young's modulus along the crystallographic [001] direction. The agreement between the calculated value (Equation 4.25) and the experimental value (Equation 4.30) is close and of the right order of magnitude. This agreement provides satisfaction both ways; the method of calculation appears to be correct, but also the value of the L-J constant ϵ_0, obtained by independent means, also seems to be correct.

4.3
Elastic Properties of Amorphous Solids

4.3.1
Elastic Modulus of Amorphous Argon

We may use the aggregate theory to predict the elastic modulus of amorphous solid by assuming the individual units with transverse isotropy are represented by the Ar–Ar atomic bond, which are randomly distributed in space.

We take the elastic stiffness of the bond as

$$c'_{33} = Y'$$

and the other constants, $c'_{33} = c'_{12} = c'_{13} = 0$. Then, according to Equation (4.19), we have

$$c'_{11} = \overline{(\cos^4 \theta)}\, c_{11} = \frac{3\pi}{16} \cdot 4.39 = 2.59 \times 10^9 \text{ N m}^{-2} \tag{4.31}$$

An order of magnitude approach to calculating the elastic modulus of amorphous argon is by utilizing the relationship between elastic constants and density (depending on temperature):

$$Y(T_2) = \frac{Y(T_1)}{(1 + \beta \Delta T)} \tag{4.32}$$

where β is the volumetric coefficient of thermal expansion and $\Delta T = T_2 - T_1$. As the change of volume because of thermal expansion is $\Delta V = V_2 - V_1$ and $\Delta \rho = m/V$, we can derive

$$Y(\rho_2) = \frac{Y(\rho_1)}{(1 + \Delta \rho / \rho_2)} \tag{4.33}$$

From Chapter 3, we find:
$\rho_2 = \rho_{\text{amorphous}} = 1.45 \times 10^3$ kg m^{-3} and
$\rho_1 = \rho_{\text{crystal}} = 1.701 \times 10^3$ kg m^{-3}, therefore
$\Delta \rho = 0.251$ kg m^{-3}.
Calculation by substitution gives

$$Y_{\text{amorphous}} = 3.51 \times 10^9 \text{ N m}^{-2} \tag{4.34}$$

There are no experimental measurements of the elastic constants for amorphous argon, and therefore, the above value must be treated as hypothetical.

Continuum mechanics does not distinguish between isotropic polycrystalline solids and amorphous solids that are isotropic by default.

4.4
Fracture

4.4.1
Phenomenology

The sudden fracture of materials is characterized by being unexpected, unpredictable and catastrophic. Furthermore, fracture strength does not have a unique value but appears to have random variations, when supposedly identical samples are tested. Only a mean value with statistical uncertainty can be given for any particular material.

Components of structures and machinery are frequently subjected to varying loads during service. Stresses, corrosion and accidental overloading can cause surface and internal cracks to appear. Cracks will grow because of the repeating and dynamic nature of the loads, and crack growth will eventually lead to sudden failure unless detected in time to prevent it. The case of railway wheel–axles is a good example of the dynamic and repeating loading (Figure 4.8). An axle is subjected to a four-point bending; the wheels push upwards on the ends, the weight of the wagon pushes down at the bearings. Rotation causes the elements of the axle to be subjected to sinusoidally varying stress.

The crack growth problem is exacerbated by the phenomenon of stress concentration that occurs around holes, at flanges, at sections of change in diameter, at welds and at crack tips (Paterson, 1953). Unbeknown, the periodically recurring concentrated stresses advance the front of the crack through the body until the remaining section is insufficient to carry the design load.

Good engineering strives to minimize the stress concentrations as much as possible or to design the components with sufficient safety factors to account for stress concentrations. Sometimes, a well-designed part may suffer accidental overload. This will cause yielding of the material at the stress concentration points (Figure 4.9). Repeated yielding leads to accumulation of material damage at the microscopic level. With time, this leads to crack initiation and growth. It is the

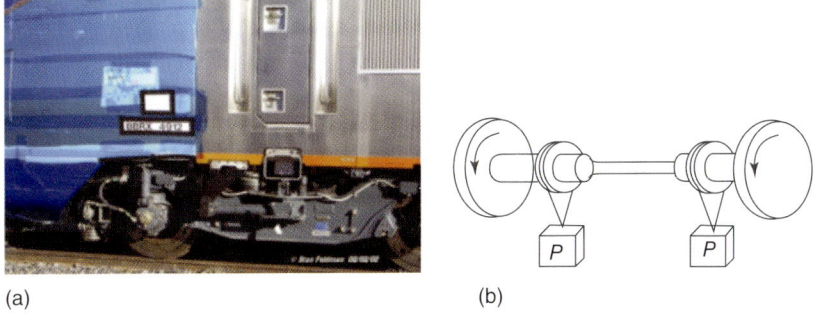

(a) (b)

Figure 4.8 (a) Wheels and axles of railway wagons are subjected to periodically varying loads. (b) An experimental setup for testing fatigue life of axle–wheels assembly.

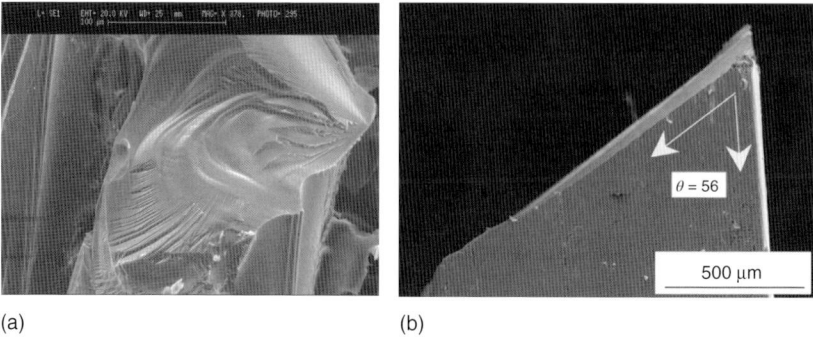

(a)   (b)

Figure 4.9 A typical appearance of fracture surface of a brittle material broken in (a) tension and (b) compression.

aim of fracture mechanics to predict fatigue life of components on the basis of predicted stresses and material properties.

Fracture strength is time and temperature independent to a first approximation.

In 1939, W. Weibull proposed that the probability of survival of a component under stress is described by the expression

$$P_W(\sigma) = \exp\left[-\left(\frac{\sigma}{\sigma_0}\right)^m\right] \qquad (4.35)$$

where σ_0 and m are essentially adjustable parameters. Materials with greater variability of strength have smaller values of m, for example, aluminium alloys have $m = 100$, whereas glasses have $m = 10$.

This relationship can be described by expressing the Weibull probability in terms of a volume $V = nV_0$:

$$P_W = [P_W(V_0)]^n = [P_W(V_0)]^{V/V_0} \qquad (4.36)$$

so that

$$P_W(\sigma) = \exp\left[-\frac{V}{V_0}\left(\frac{\sigma}{\sigma_0}\right)^m\right] \qquad (4.37)$$

hence, the probability of fracture increases exponentially with the volume of the material component (Roylance, 1996). The probability of surface cracks diminishes as the dimension of the object decrease. Testing of many fibres of different diameters will results in a remarkable relationship as shown in Figure 4.10. Evidently, the fibres become stronger the thinner they are, corroborating Weibull's theory.

4.4.2
Continuum Mechanics

4.4.2.1 Definition of Fracture Mechanics: Fracture Toughness

Fracture toughness is the property of solid materials that defines their resistance to crack propagation (and, therefore, their resistance to sudden failure). It is one

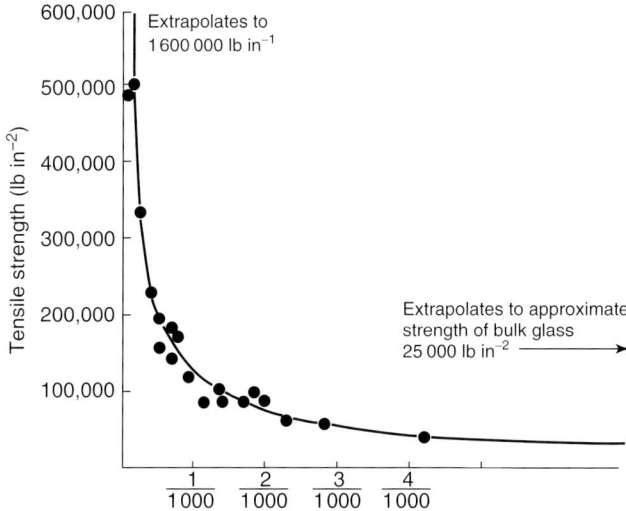

Figure 4.10 Tensile strength of glass fibres measured as a function of decreasing fibre diameters. Fibre diameter is in thousands of an inch. (*Source:* Adapted from the book The New Science of Strong Materials by J.E. Gordon .)

of the three fundamental mechanical properties of materials, which are listed in the Table below

Property	Symbol	Units
Modulus of elasticity	E	N m^{-2}
Tensile strength	σ_t	N m^{-2}
Fracture toughness	G_{Ic} (or R)	J m^{-2}

Fracture toughness of materials is measured by mechanical testing of samples containing cracks (Atkins and Mai, 1985). This is illustrated by the simple case of a rectangular plate of uniform thickness, t, which has a crack of length, $2a$, in the middle of it, as shown in Figure 4.11. If the crack propagates to the edges of the plate, the plate will fail.

The plate is loaded in uniaxial tension with force, P. Just above the crack two force vectors are shown in opposition, R and G. They act on the tip of the crack (indicated as a point) but have been moved up so as not to obscure the crack. A similar situation exists at the other tip of the crack (by symmetry around the vertical axis). The force, G, is called a *crack extension force*, and we assume that it is caused by the applied force P. The other force, R, is called *crack resistance force*, and we say that it results from the material's resistance to crack propagation (its fracture toughness).

Notice that $G = 0$ if $P = 0$, and G increases if P increases. Fracture toughness, R, is a material property and, therefore, of constant value (at constant temperature).

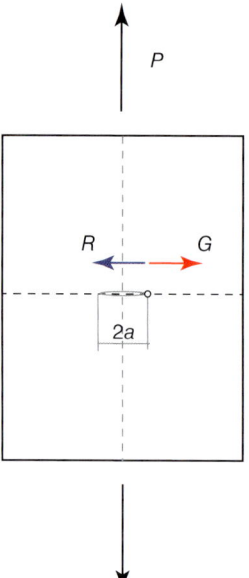

Figure 4.11 Balance of forces on a plate with a crack.

We have the following two conditions:

- the crack will grow if $G > R$
- the crack will not grow if $G < R$.

The problem The problem is to derive an expression relating G to P, so that the conditions for fracture can be predicted from the applied loading. In 1921, Griffith published a theory for fracture of brittle materials that provided a solution to this problem. The Griffith's theory is based on an energy criterion involving three concepts:

1) an elastic material subjected to load stores elastic strain energy, U_{el},
2) the elastic strain energy is released (reduced) if the crack increases in size,
3) the total surface energy of the material component increases in proportion to the size of the crack.

Then the Griffith's criterion for sudden (uncontrolled) fracture is expressed as

$$\frac{d}{da}(U_{el} - U_s) < 0 \tag{4.38}$$

In the above equation, U_{el} represents the elastic strain energy stored in the plate component and U_s is the surface energy associated with the cracked surfaces.

The graph in Figure 4.12 shows how the stored elastic energy changes with increasing crack size and fracture toughness. The interpretation of the differential equation is that with a change in crack length, da, if the change of the difference between the elastic and surface energy is negative, then the crack will become

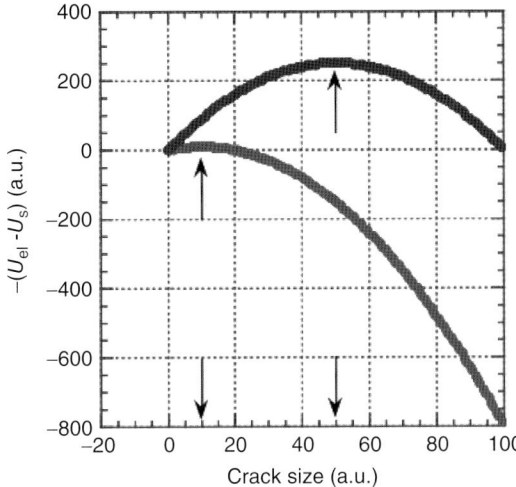

Figure 4.12 The variation of elastic strain energy stored in the plate as a function of crack length.

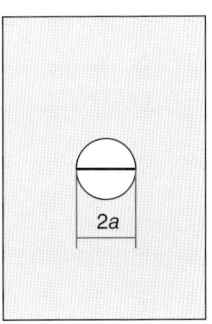

Figure 4.13 Elastic plate under tensile stress with a central crack. It is assumed that internal stress around the crack relaxes to zero as shown by the white area.

unstable and will grow in a catastrophic manner, resulting in a sudden brittle failure.

4.4.2.2 Elastic Strain Energy Release

The elastic strain energy in a plate of uniform dimensions, and loaded by uniaxial force P, is given by

$$U_{el} = \frac{1}{2} P \cdot \Delta l = \frac{1}{2} \frac{P}{A} \frac{\Delta l}{l} V_{plate} = \frac{1}{2} \sigma \epsilon V_{plate} = \frac{1}{2} \frac{\sigma^2}{E} V_{plate} \qquad (4.39)$$

where $V_{plate} = A \times l = W \times t \times l$. We refer to the change of the elastic strain energy with a change in crack size as the strain energy release rate:

$$G = \frac{d(U_{el})}{da} \qquad (4.40)$$

As the crack is a material discontinuity, it cannot carry stresses across its faces. Then a certain amount of the stresses around the crack must relax. The elastic strain energy stored in the plate with a crack is less than that in a plate without a crack. We assume that the length of the plate does not change with the introduction of the crack ($\Delta l = $ const).

As a first approximation, we calculate the volume over which the stress relaxes by assuming that it is equal to the circular disk as shown in the diagram in Figure 4.13. In this volume, the tensile stress is assumed to relax completely to zero (white area). Everywhere else (grey area) the stress is $\sigma = P/A$, where $A = (W - 2a)\,t$. Therefore, the amount of elastic strain energy released is given by

$$U_{el} \text{ (released)} = \frac{1}{2}\frac{\sigma^2}{E} V \text{ (around crack)} = \frac{1}{2}\frac{\sigma^2}{E} \pi a^2\, t \qquad (4.41)$$

therefore,

$$G = \frac{dU_{el} \text{ (released)}}{da} = \frac{\sigma^2}{E} \pi a\, t \qquad (4.42)$$

We make an assumption that the volume of the plate is much larger than the stress free volume around the crack (i.e., $2a \ll W$). The exact calculation of the amount of released elastic strain energy around the crack requires

- an analytical solution for the stresses around the crack
- integration of the stress over total volume of the plate.

4.4.2.3 Solid Surface Energy

Solids, similar to liquids, have surface tension associated with their surfaces. Surface tension causes local deformation over a short distance (atomic size) just below the surface and, therefore, has surface energy associated with it. The surface energy of a solid is denoted by γ, and it is expressed in units of joules per metre square. As a crack has surface area, it has surface energy associated with it. The surface energy of a crack is given by

$$U_s = 2\,(2a\,t)\,\gamma = 4\,a\,t\gamma \qquad (4.43)$$

As the crack increases in size, new surface area is created, requiring additional amount of surface energy (note that the outside surface area of the plate does not change, and therefore, only the additional surface energy owing to creation of the crack need be considered.)

Substitution of the terms for released elastic strain energy and surface energy into the fracture Griffith's criterion (Equation 4.38) gives the following:

$$\frac{d}{da}\left(\frac{1}{2}\frac{\sigma^2}{E}\pi\,a^2 t - 4\,a\,t\,\gamma\right) < 0 \qquad (4.44)$$

The above equation can be solved for the critical crack size, a_c, below which the crack will not grow:

$$a_c \leq \frac{4}{\pi}\frac{E\,\gamma}{\sigma^2} \qquad (4.45)$$

As the applied load, P, is directly related to the stress and plate cross-sectional area, we can write the final relationship as

$$a_c \leq \frac{4}{\pi} \frac{E\gamma}{(P/A)^2} \qquad (4.46)$$

This is the solution to the problem.

The fracture toughness, R, should now be identified. We observe that when $G = R$, then the elastic strain energy release rate becomes zero, $dU_{el}/da = 0$. Its value becomes critical: $R = G_c$.

Example

A plate of glass is subjected to tensile stress of 40 MN m^{-2}. The specific surface energy for this glass is 300 mJ m^{-2}. Young's modulus is 69 GN m^{-2}.
Calculation of the minimum critical crack size gives

$$a_c = \frac{4 \times 69 \times 10^9 \times 0.3}{\pi (40 \times 10^6)^2} = 16.4 \times 10^{-6}, \quad \text{that is } 16.4 \text{ μm} \qquad (4.47)$$

This result shows that even microscopic scratches on the surface of the glass can act as critical cracks that lead to brittle fracture if the stress exceeds 40 MN m^{-2}. The lack of plasticity in the glass makes it a brittle material.

4.4.2.4 Griffith's Fracture Stress

The Griffith's equation for critical crack size can be solved for stress required to fracture the material containing a crack of length, a. The solution is

$$\sigma_f = 1.13 \sqrt{\frac{E\gamma}{a}} \qquad (4.48)$$

The above equation shows that the strength of a body containing a crack decreases with increasing crack size. The stress at fracture, σ_f, is proportional to the load at fracture, P_f. The load at fracture varies with crack length as a^{-2}. The above relationship is true for a plate in which the crack size is very small compared to its width. For finite width plates, the constant in front of the square root becomes a function, $f(a/W)$, where W is the width of the plate.

Further Developments in Linear Elastic Fracture Mechanics During the 1950s, NASA began the development of space flight program. In the UK, the first commercial airliner with jet engines, the Comet, suffered major losses because of fracture in flight. Several industrial countries committed to the development of nuclear power stations requiring safe steel containment vessels. These and other engineering developments provided a powerful impetus to the study and application of the new science of fracture mechanics.

In the 1950s, Irwin proposed that the Griffith fracture criterion should be extended to ductile materials, for example, steel, by the inclusion of work of plastic deformation during crack propagation. He proposed that

$$R = 2(\gamma_s + \gamma_{pl}) \qquad (4.49)$$

Ductile materials resist crack propagation not only by the increase in surface energy but also by the work of plastic deformation that occurs around the propagating crack. According to this theory, the fracture stress is now given by

$$\sigma_f = f\left(\frac{a}{W}\right) \sqrt{\frac{ER}{\pi a}} \qquad (4.50)$$

Exact solution gives $f(a/W) = [(W/\pi a) \tan(\pi a/W)]$ Approximate solution is $f(a/W) = 1/[1-(2a/W)^2]$ for $a/W < 0.5$. The variation of the function with crack length is shown in Figure 4.14(d). Notice that the approximation is reasonably accurate up to $a/W = 1.005$.

Typical values of fracture toughness for selected materials.

R	Alloy steel	Aluminium alloy	Mild steel	Rubber	Wood	Glass
(kJ m^{-2})	100	20	15	15	0.1	0.01

The surface energy of solids varies from a minimum to a maximum value by a factor of approximately 10. However, the fracture toughness varies by a factor of approximately 10^4, as shown in the table. This large increase is due to the plastic work around the crack tip during crack propagation.

For ductile (elastoplastic) materials, the strain energy release rate is now given by

$$G = \frac{d(U_{el} - W_{pl})}{da} \qquad (4.51)$$

We note that the crack will grow and propagate only if,

$$G > R \qquad (4.52)$$

As R is so much larger for ductile materials, they are much more resistant to crack propagation. Figure 4.14 show a fracture surface of a shaft that started growing a crack at the keyway. Figure 4.14b is magnified views of the shaft surface. The step-like growth pattern of the crack is clearly visible. Figure 4.14c is a highly magnified view of the ductile fracture surface, with dimples and voids.

4.4.2.5 The Role of Defects

The strength of real materials is *lowered by orders of magnitude* by the presence of defects (imperfections), such as microcracks, dislocations, grain boundaries and other (Atkin and Mai (1985), Flynn (1972), Kelly and Knowles (2012) Nelson, 2002). Also, the presence of defects such as dislocations allows for plasticity in some materials, particularly metals (Cottrell (1965), Hull (1975)).

Surprisingly, we purposefully modify the arrangement and structure (i.e., introduce more) of the defects to increase the strength of metallic alloys. The mechanisms of strengthening are described later.

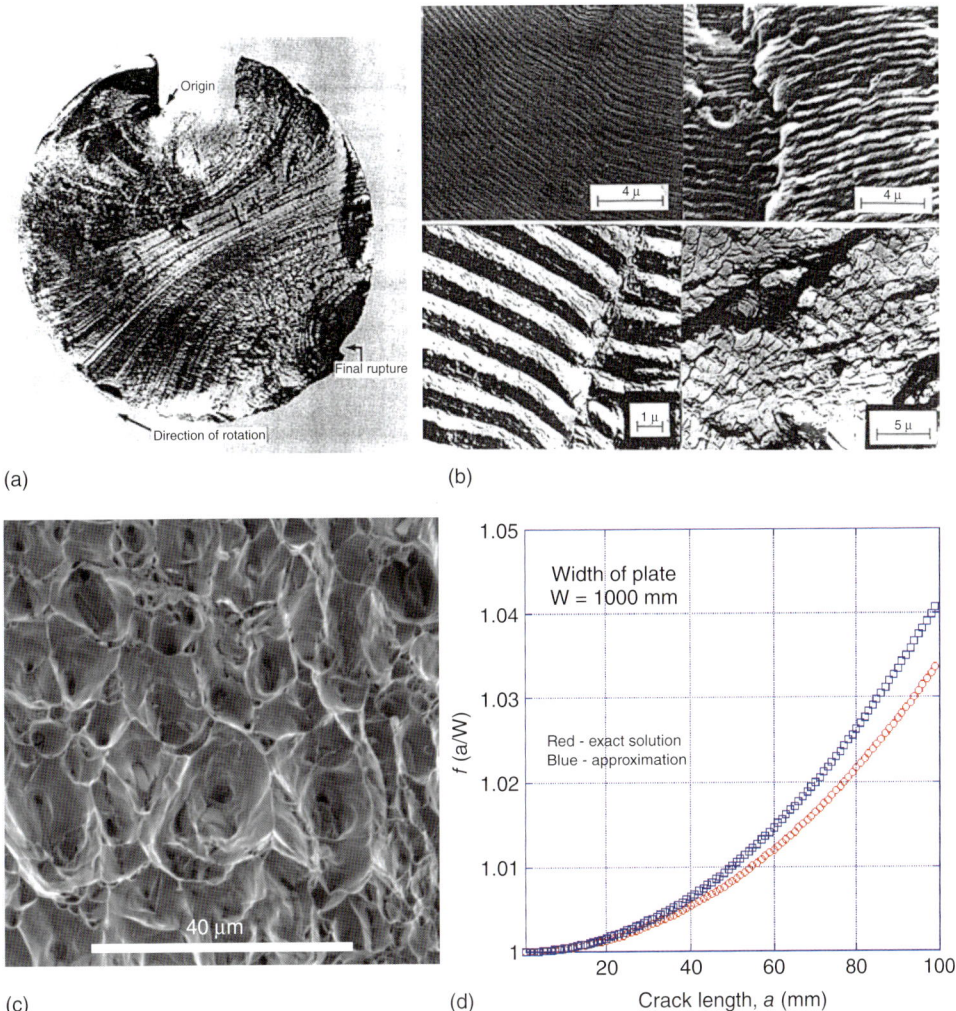

Figure 4.14 Surface morphology after fatigue fracture: (a) surface of a broken shaft, (b) magnied views of surface striations, (c) dimpled surface of ductile fracture. (d) A graph of the function given in Equation (4.50) for exact and approximate solutions.

Microcracks The fracture strength of a material containing a crack and subjected to a tensile force acting perpendicular to the crack as shown in Figure 4.15 is given by the Griffith theory:

$$\sigma_f = \beta \sqrt{\frac{E\gamma}{a}} \qquad (4.53)$$

where β is the geometrical constant, E is the (Young's) modulus of elasticity of the material, γ is the surface tension (energy) of the material and a is the length of the existing crack.

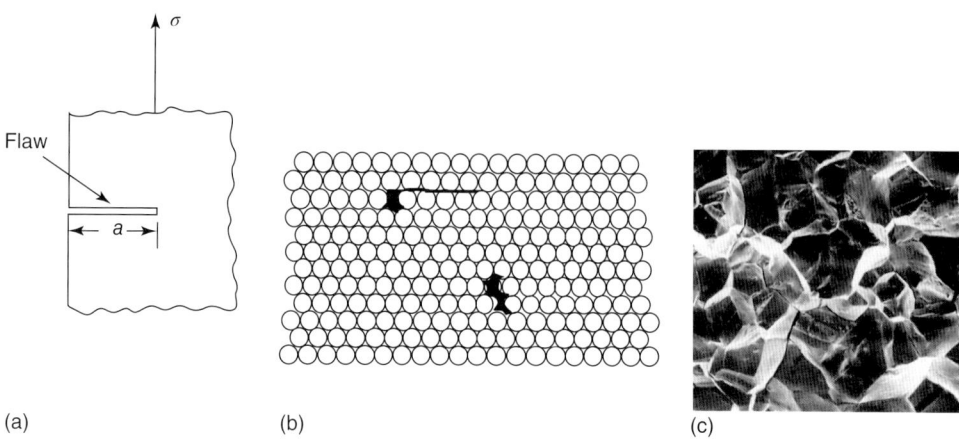

Figure 4.15 Imperfections (defects) in crystalline materials: (a) surface microcrack, (b) internal dislocations and (c) internal grain boundaries.

Consider metallic glass ZrTiCuNiBe of composition as described in Chapter 3. It has modulus of elasticity $E = 120$ GPa, specific surface energy $\gamma = 2$ J m^{-2} and a measured fracture strength $\sigma_f = 2$ GN m^{-2}. Using again Equation 4.53, the minimum critical crack length for fracture at this stress will be

$$a_c = \frac{\sigma_f^2}{(\beta^2 \gamma E)} = \frac{(2 \times 10^9)^2}{(4 \times 2 \times 1.2 \times 10^{11})} = 0.043 \times 10^{-6} \text{ m} \tag{4.54}$$

Surface cracks that small (0.043 μm) are not visible to the naked eye. Fracture will occur without any obvious defect, and moreover, the microcracks can vary in size and, therefore, result in variations of fracture strength.

4.4.3
Atomistic Fracture Mechanics of Solids

4.4.3.1 Theoretical Cleavage Strength

A fragment of a perfectly crystalline solid is represented by the atoms (spheres) arranged in close packing. Under an applied tensile force, the two halves of the solid separate as shown in Figure 4.16. The separation occurs along a perfect crystallographic plane, called the *cleavage plane*. The tensile force has to increase until a maximum is reached beyond which the solid will fracture.

Note: The separation will occur in all parallel planes simultaneously, not only in the one plane as shown. At the critical point when the maximum force is reached, the work done by the external force should be sufficient to practically vaporize the solid.

The theoretical cleavage strength was estimated by Orowan in 1949 and given as

$$\sigma_{\text{cleavage}} = \sqrt{\frac{E\gamma}{a_0}}, \tag{4.55}$$

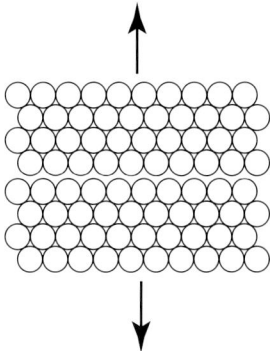

Figure 4.16 An ideal crystalline solid is being pulled apart. Bonds between atoms are stretched until broken.

Table 4.1 Calculated theoretical cleavage strength of selected materials.

Material	Modulus (GN m^{-2})	Surface energy (mJ m^{-2})	Atomic distance (nm)	σ_{cleave} (GN m^{-2})
Silk fibre	240	24	0.154	6.12
Copper	120	45	0.125	6.57
Vitreloy BMG	120	45	0.145	6.10
Diamond [100]	1075	6000	0.154	204

where E is the modulus of elasticity, γ is the surface energy of the solid and a_0 is the distance between layers of atoms. Typical values calculated for several materials are given in Table 4.1.

4.4.3.2 Theoretical Shear Strength

An applied shearing force causes two halves of an ideal crystalline solid to slide past each other as shown in Figure 4.17. To cause increasing displacement, the shear force must increase until a maximum is reached at or before $\frac{1}{2}b$ distance. The separation is in a perfect crystallographic plane, called the *slip plane*. Theoretical shear strength of crystalline materials was estimated by Frenkel in 1926 to

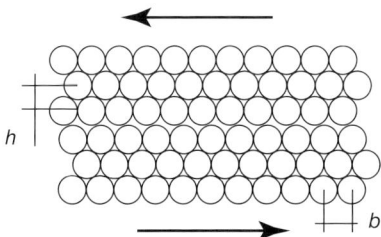

Figure 4.17 An ideal crystalline solid is sheared. The bonds between atoms are broken and remade for every increment of distance b.

4 Mechanical Behaviour

Table 4.2 Calculated theoretical shear strength of selected materials.

Material	Shear modulus (GN m^{-2})	τ_{max}/G	τ_{max} (GN m^{-2})
Copper	33	0.039	12.9
Silicon	57	0.24	13.7
Iron	60	0.11	66.0
Diamond	505	0.24	115
Vitreloy BMG	40	0.25	10

be (Table 4.2)

$$\tau_{max} = \frac{Gb}{2\pi h} \tag{4.56}$$

Note: The incremental sliding, repeated for every step of magnitude b, results in decreasing force since fewer and fewer atoms are connected in the plane subjected to shear. Eventually, the solid will fracture (separate). Also, the sliding must occur for all the (parallel) planes simultaneously, not just the one plane as shown. Then the work done by the force is equal to the work necessary to liquify the solid.

Clearly, diamond is exceptional. Cutting diamond single crystal is to a large degree a shearing process. Crystallographic planes with the lowest density of atoms require least work. As a rule, the smaller the number of atoms per unit area the lower the shear strength. The crystal unit cell of diamond is shown in Figure 4.18, and the corresponding area and number of atoms in (100), (110) and (111) planes (per unit cell) are shown in the Table 4.3.

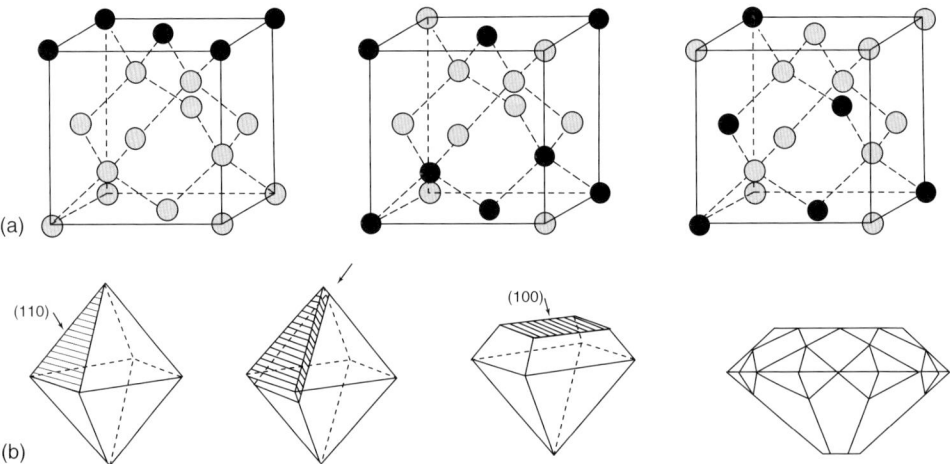

Figure 4.18 (a) (001), (110) and (111) crystallographic planes in diamond, with emphasized atoms lying in the plane. (b) The relationship between crystallographic and cutting planes for diamond single crystal.

Table 4.3 Properties of crystallographic planes in diamond.

Plane	Area, a^2	Number of atoms	Atoms/a^2
(100)	1	2	2
(110)	$\sqrt{2}$	4	2.83
(111)	$\sqrt{3}$	2	1.15

There are other ways of estimating theoretical strength of solid materials. One has to find the strength of chemical bonds between atoms and then multiply by the number of atomic pairs acting across unit area. Methods to measure strength of bonds are by means of heat of combustion or vaporization, spectroscopy of atomic vibrations, atomic force microscopy and others. The strength of interatomic and intermolecular bonds can be found in scientific literature. According to the data, a solid made from nitrogen atoms ought to have (theoretically) the highest strength.

We conclude that as a first approximation, the mechanisms of fracture at the atomic level are similar in both crystalline and amorphous solids.

4.5 Plasticity

4.5.1 Phenomenology

The fundamental characteristics of plastic deformation is that it is non-recoverable and much larger (typically 20% for ductile metals) than the elastic deformation (0.01%). Plasticity belongs to a regime of conditions involving temperatures not too high (not too close to the melting point or the glass-transition temperature), and stresses above the elastic limit, that is, above the yield stress.

During a tensile or compression test, the deformation applied on the specimen and the corresponding load are measured. The deformation perpendicular to the tensile axis can also measured, as shown in Figure 4.19. From these data, the nominal stress and nominal strain are calculated as

$$\sigma_{nom} = \frac{P}{A_0}, \quad \epsilon_{nom} = \frac{\Delta \ell}{\ell_0} \tag{4.57}$$

A plot of σ_{nom} versus ϵ_{nom} is called the *nominal* stress–strain curve. The nominal stress–strain graph shows the so-called *ultimate* tensile strength at the maximum stress point.

From the graphs in Figure 4.20, the ultimate strength of the normalized 4135A steel is approximately 1140 MN m^{-2} and that of which is one of the strongest steels in quenched state (4135B) is 2100 MN m^{-2}. The last value is only three times less than the theoretical cleavage (tensile) strength.

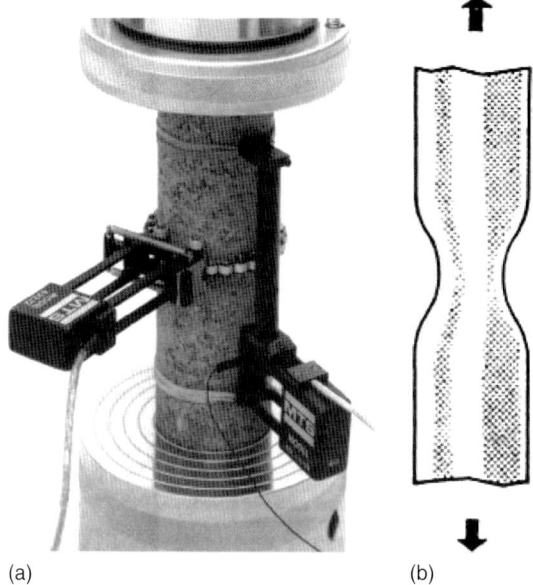

(a) (b)

Figure 4.19 (a) Compression test with extensometers attached to the specimen. (b) Non-homogeneous plastic deformation resulting in necking.

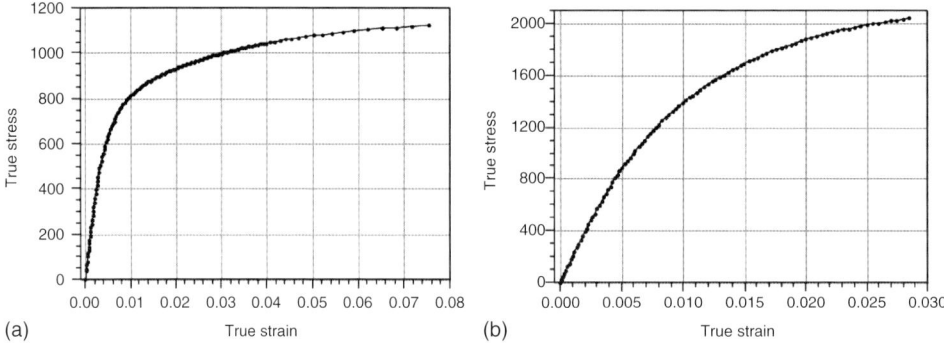

(a)   (b)

Figure 4.20 Experimentally derived true stress - true strain curves for engineering steels. (a) Annealed (normalized) steel 4135A (2). (b) After quenching (quenched) steel 4135B. Note the differences in scale and compare with theoretical strengths from Table 4.1.

The specimen extends during the test and, beyond the elastic limit, plastic deformation occurs at constant volume. Then the true stress and true strain are related as shown below:

$$\sigma_t = \frac{\sigma_{nom}}{(1 + \epsilon_{nom})}, \quad \epsilon_t = \ln(\frac{\ell}{\ell_0}) \tag{4.58}$$

A plot of σ_t versus ϵ_t is called the *true stress–true strain* graph. The true stress–true strain curve for ductile metals and alloys is usually described above the yield point by the following expression.

$$\sigma_t = \sigma_y + K \cdot \epsilon_t^n \tag{4.59}$$

where σ_y is the yield point (nominal and true values are the same), K is the strengthening coefficient and n is the strengthening exponent. Both are determined from experimental testing by plotting Equation 4.59 on logarithmic scale (Ruiz Ocejo and Gutirrez-Solana 1998).

4.5.2 Continumm Mechanics

The constitutive equations describing the plastic behaviour in an ideal plastic body are

- machine applied elongation:

$$\sigma_{appl} = \sigma_y = \text{Constant} \tag{4.60}$$

- machine applied load:

$$\epsilon_{ij}^p = s_{ij}\, d\lambda \tag{4.61}$$

where ϵ_{ij}^p are the plastic strain increments, s_{ij} are the deviatoric stress components and $d\lambda$ is a proportionality factor that may change during loading. Equation 4.61 is known as the *Prandtl-Reuss* relations.

Yield strength is usually measured during machine-imposed deformation (strain control).

4.5.2.1 Tresca Yield Criterion

The appearance of slip steps on the surface of deformed bodies that coincide approximately with the direction of maximum shear stress has lead in 1865 French engineer Henri Tresca to propose a *maximum shear stress* theory as the criterion for plastic failure of materials: The material will yield if

$$\tau_{appl} > \tau_{max}, \text{ where } \tau_{max} = \frac{1}{2}(\text{max - min}) \text{ Principal stresses} \tag{4.62}$$

4.5.2.2 Huber–von Mises Criterion

In 1904, M.T. Huber proposed that general yielding should occur when the *distortional elastic strain energy*, U_{elas}^d reaches a certain critical value. He argued that the strain energy associated with change of volume (under hydrostatic pressure) should not contribute to yielding in accordance with experimental observations. This yield criterion is expressed as

$$U_{elas}^d > U_{crit} \tag{4.63}$$

Huber derived the distortional elastic strain energy in terms of principal stresses and strains as

$$U_{elas}^d = \frac{1}{6G}(\sigma_1^2 + \sigma_2^2 + \sigma_3^2 + \sigma_1\sigma_2 + \sigma_2\sigma_3 + \sigma_3\sigma_1) \tag{4.64}$$

For the case of uniaxial tension, for a material with yield stress, σ_y, the above criterion becomes

$$(\sigma_1 - \sigma_2)^2 + (\sigma_2 - \sigma_3)^2 + (\sigma_3 - \sigma_1)^2 \geq 2\sigma_y^2 \tag{4.65}$$

The above equation has been published also by von Mises in 1913, and it has become to be known as the *von Mises yield criterion*.

4.5.3
Atomistic Mechanics of Crystalline Solids

The minimum stress needed to move a dislocation in a pure crystal is called the *Peierls–Nabarrow stress* (or strength). The shear strength of a single crystal with a dislocation is given by Cottrell (1965)

$$\tau_{min} = \frac{2G}{1-v} \cdot \exp\left[\frac{-2\pi d}{(1-v)b}\right] \tag{4.66}$$

where v is the Poisson ratio of the material and b and d are constants.

The stress required to glide dislocations defines the yield strength of metals. Therefore, any changes of the metal's crystal structure that impede the motion of dislocations will result in an increased yield strength. Such modifications, called the *mechanisms of strengthening*, are usually grouped under four mechanisms:

1. Strain hardening (= work hardening = cold work hardening) 2. Grain size strengthening (= grain boundary strengthening) 3. Solid solution hardening (= alloying strengthening) 4. Precipitation hardening (= dispersion hardening).

4.5.3.1 Strain Hardening

Enforcing plastic deformation in a metal will result in its hardening (increased yield strength). The greater the amount of plastic deformation, the greater the effect.

Figure 4.21 shows several edge dislocations in a single crystal on parallel crystallographic planes. The dislocations are gliding under the action of an applied shear stress, τ. A dislocation is characterized by its Burgers vector, $\vec{b}$. The amount of plastic shear deformation caused by one dislocation moving n times through the

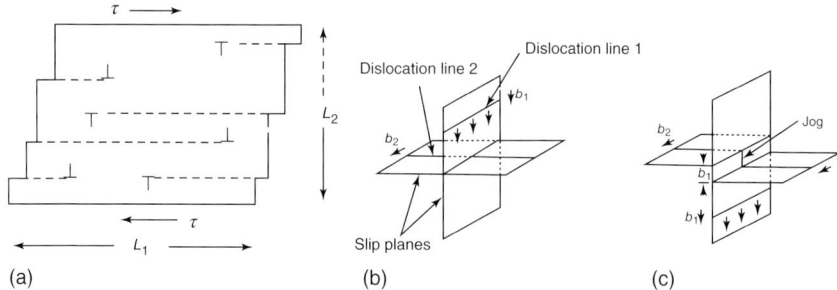

Figure 4.21 (a–c) Dislocations glide in a crystal.

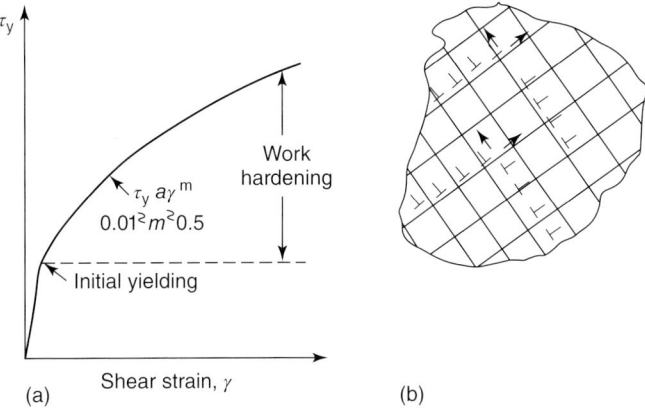

(a) (b)

Figure 4.22 (a, b) Stress–strain curve with hardening.

crystal (or n dislocations moving once) is (Cottrell, 1965, Hull, 1975)

$$\epsilon_{plastic} = \frac{n|\vec{b}|}{L_2} \tag{4.67}$$

where $|\vec{b}|$ is the size of the step produced when the dislocation exits the crystal.

In fcc structure, there are four equivalent non-parallel (111) planes on which the dislocations can glide simultaneously in any of the three direction $<110>$. In body-centred cubic (bcc) structure, there are six slip planes of type (110), each with two $<111>$ directions (12 systems). Two dislocations on such planes are shown schematically in Figure 4.21. The dislocations will cross at some stage of their glide. Crossing of the dislocations causes the so-called jogs and kinks to appear along the dislocation line. The movement of a dislocation with a jog or a kink is hindered as compared to that of a straight dislocation. It will require a higher stress to move it.

This hindrance of the dislocation's movement results in hardening of the metal (higher resistance to plastic deformation). A multiplying effect comes from the fact that in a polycrystalline metal, new dislocations are continuously generated at grain boundaries during plastic deformation. Consequently, many dislocations will cross, thus obstructing each other's ease of glide. This is shown schematically in Figure 4.22 together with the corresponding stress–strain diagram. It shows that the stress beyond the initial yield point increases continuously because of strain hardening.

The increasing dislocation density with plastic deformation, and the increasing resistance to dislocation motion because of jogs, is described to a first approximation by the Taylor equation (Gottstein, 2004):

$$\tau_p = \tau_0 + \alpha G|\vec{b}|\sqrt{\rho} \tag{4.68}$$

where τ is the stress needed to plastically deform the metal, τ_0 is the Peierls–Nabarro stress, α is the geometrical factor taking into account

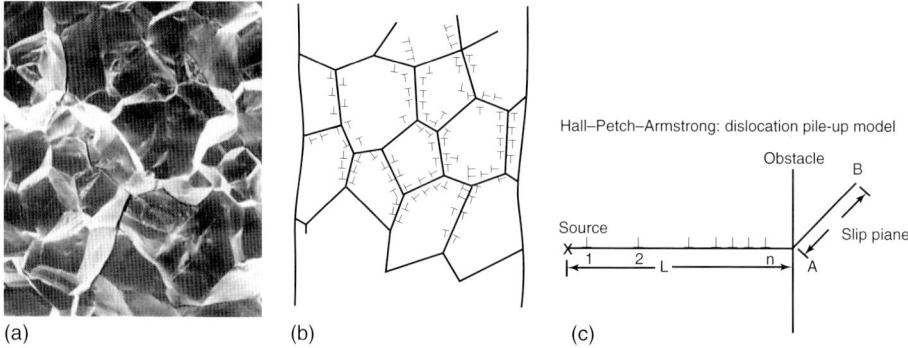

Figure 4.23 (a–c) Dislocations motion in a polycrystalline metal. Grain boundary pile-up.

the specifics of crystal structures and relative grain orientations, G is the shear modulus of the metal and ρ is the dislocation line density. Clearly, ρ is a function of the extent of plastic deformation, although too involved to derive the relationship here.

In a pure single crystal carefully annealed, $\rho \approx 0$, and then $\tau_\rho = \tau_0$.

In a heavily deformed metal dislocation, density can be as high as 10^{18} lines cm^{-2}.

This is the most well known and practiced strengthening of metals, and it has been known for ages. Hardening of cutting blades to keep them sharp is achieved by hammering the blade edge. It has been practiced by farmers on their scythes. It has been practices by smiths hammering horse shoes and swords. It is done presently to rails and bars by cold rolling and to wire by cold drawing and so on.

4.5.3.2 Grain Boundary Strengthening

Grain boundaries are the main source and sink of dislocations. Every time a dislocation comes to the boundary and leaves the grain, it causes step-like displacements on its surface, as shown in Figure 4.23. The dislocation's glide causes the grain to change shape. The change of shape exerts pressure on neighbouring grains. This has two results. First, a dislocation coming against grain boundary experiences a barrier to exit the grain because the grain deformation is resisted by its neighbour. The magnitude of the barrier depends on the relative crystallographic orientation of the neighbouring grains. Second, the pressure on its neighbouring grain causes a new dislocation to appear in that grain. So, this is the sink and source mechanism for dislocations, as shown in Figure 4.23.

The barrier to exit the grain causes the so-called pile-up at the grain boundary, as shown in Figure 4.23. Dislocations gliding towards the grain boundary on the same crystallographic plane accumulate and cause a dislocation pile-up. The resistance to move the dislocations is directly related to the ratio of the size of the deformation step produced by the dislocation to the diameter of the grain (see Figure 4.24). This mechanism was analysed by Petch and Hall who proposed an equation, known as the Petch–Hall formula, describing this effect (McClintock

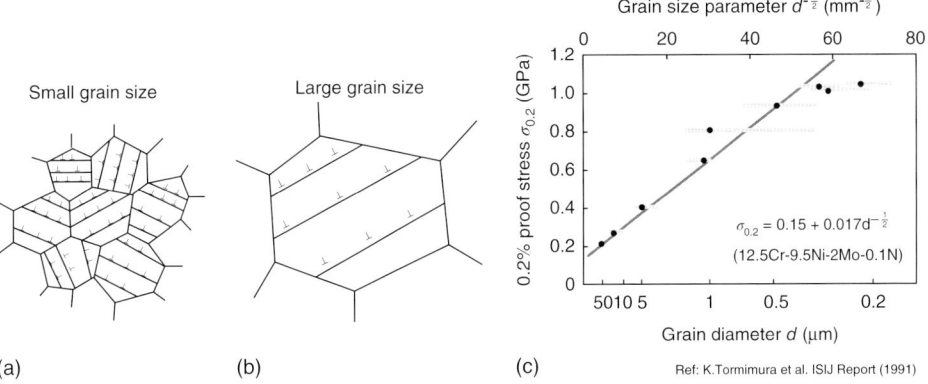

Figure 4.24 (a–c) Dislocations motion in a polycrystalline metal.

and Argon, 1966):

$$\tau(d) = \tau_0 + \frac{k}{\sqrt{d}} \qquad (4.69)$$

where k (in Nm$^{3/2}$) is a constant characteristic of a particular alloy and d (in m) is the average grain diameter. Therefore, the strengthening is proportional to $1/\sqrt{d}$.

A plot of the Petch–Hall relationship for an alloy steel is shown, confirming the inverse square root relationship.

4.5.3.3 Solid Solution Hardening

All pure metals with fcc crystal structure (Al, Cu, Ag, Au, Pt, Pb, Ni) have the lowest yield strengths of all metals. Pure gold and silver are commonly known to us as very soft and ductile metals. A jeweller will usually recommend 18 or 22 carat gold ring rather than 24 karat (pure gold) because pure gold ring or necklace chain will scratch, deform and break easily. All pure metals with bcc crystal structure come next (Fe, Cr, W) in terms of their yield strengths.

A way to strengthen metals (i.e., increasing yield strength) is to hinder dislocation motion by introducing alloying elements that form solid solutions, for example, C in Fe, or Ni in Cu, or Ag in Au, or Cu in Al. Solid solution means that the atoms of the alloying elements either substitute for the matrix atoms in the lattice or occupy interstitial spaces between the atoms (Figure 4.25).

As the alloying element has atoms of different size, the atoms will produce a misfit, called a *point defect*. A dislocation gliding on a plane containing such a defect will experience additional resistance to glide.

Both the dislocation and the point defect mismatch produce stress fields around them. Therefore, to a first approximation, the resistance force is proportional to the mismatch and to the distance between them. The amount of resistance will also be related to the degree of misfit and, therefore, to the degree of disruption of the crystalline order. All together this can be described by the following expression:

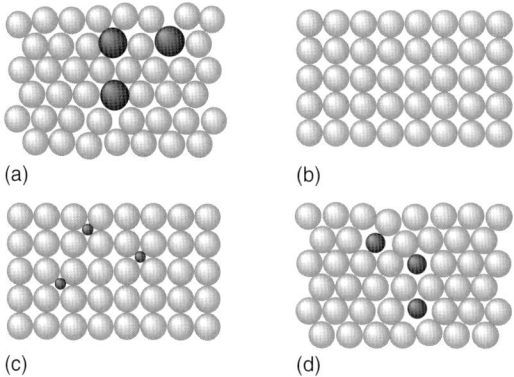

Figure 4.25 (a–d) Superstitials and interstitials.

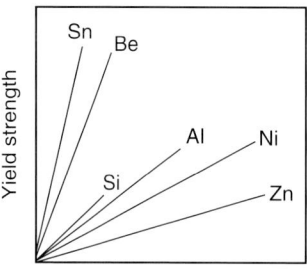

Figure 4.26 Schematic plot of yield strength of copper with additions of alloying elements.

$$\tau(\text{SolS}) = \tau_0 + Gb\frac{n}{l_d}\frac{\Delta\epsilon^m}{\Lambda} = \tau_0 + Gb\Delta\epsilon^m \sqrt{c} \quad (4.70)$$

where n is the number of solute atoms along a dislocation line of length l_d, Λ is the average distance between solute atoms, $\Delta\epsilon^m$ is the strain misfit caused by the presence of the point defect and c is the concentration of the solute in the alloy. The exponent, $m \approx 1.5$, takes into account the nonlinearity of this effect.

Figure 4.26 shows the increase in yield strength of copper with the addition of alloying elements. The plot shows increasing yield strength with increasing concentration of the solute atoms. Also, it shows increasing yield strength with increasing misfit between the matrix and solute atoms (see Table 4.4).

Table 4.4 Atomic radii for a number of metallic elements.

Atom type	Cu	Zn	Ni	Al	Si	Be	Sn
Atom radius (nm)	0.128	0.133	0.125	0.143	0.117	0.114	0.158
% Change in radius		3.90	2.34	11.7	8.59	10.9	23.4

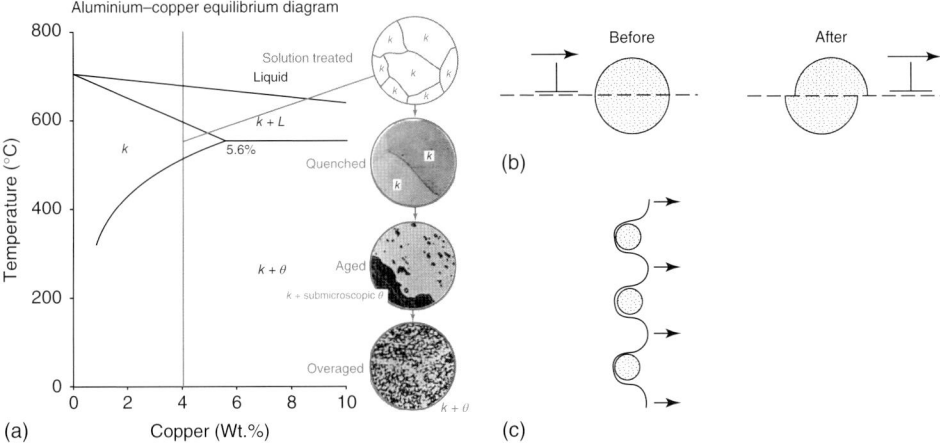

Figure 4.27 (a–c) Strengthening in Al–Cu alloys by precipitation of Al_2Cu particles.

4.5.3.4 Precipitation Hardening

The presence of a precipitated particle (second phase) in the matrix metal presents a barrier to a dislocation gliding on a plane containing the particle. Particles precipitate in alloys of specific compositions (see Figure 4.27). The dislocation can either bow around the particle or can cut through the particle (see Figure 4.28). In both the cases, the hindrance to dislocation glide is increased and, therefore, the strength of the metal alloy is increased.

If a dislocation cuts through a particle, it displaces one half of the particle relative to the other half. Such a displacement requires extra work of deformation and, consequently, a higher stress to move the dislocation. This is described by the following equation:

$$\tau_{dc} = \tau_0 + \frac{\pi \gamma r_p}{bL} \sqrt{v} \tag{4.71}$$

where γ is the interfacial surface energy between the particle and the matrix, r_p is the radius of the particle and v is the volume fraction of the particles in the alloy.

In the second case, the dislocation line bows around the particle (as shown in Figure 4.28). The stress to bend and bow the dislocation line around the particles is less than the stress to cut through the particles. The equation describing this mechanism is

$$\tau_{db} = \tau_0 + \beta \frac{Gb}{\lambda} \tag{4.72}$$

where β is a geometrical factor characteristic of a given alloy and λ is an average distance between the particles in the matrix.

Precipitation of particles is the most effective way to increase yield strength of metals (high-strength steels, high-strength Al alloys, etc.)

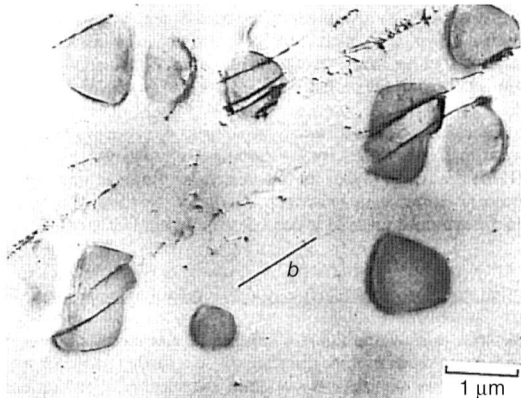

Figure 4.28 Example of dislocations cutting through particles.

4.5.3.5 Mechanisms of Plastic Flow in Crystalline Materials

As deformation occurs by the movement of dislocations, the critical stress is determined by the stress required to move dislocations through obstacles in the material.

$$\tau = \tau_G + \tau_S \tag{4.73}$$

The process of plastic flow deformation by dislocation glide is self-sustaining because dislocations are able to multiply under the application of the flow stress.

Obstacle-limited dislocation glide kinetics is described by

$$\dot{\epsilon} = b\, \rho_m\, \bar{v} \tag{4.74}$$

where b is the Burgers vector length, ρ_m is the density of dislocations and $\bar{v}$ is the average dislocation velocity.

4.5.3.6 Displacement of Atoms Around Dislocations

Consider a model of an edge dislocation as shown in Figure 4.29.

Place the origin of x–y-axes on the core atom. Denote the horizontal displacement of atoms in the upper crystal (in the joining plane) by $u(x)$. The centres of the crystal has moved, therefore, at $x = 0, u(x) = 0$. However, at $x = \pm\infty$ (i.e., the edges of the crystal), the atoms joined to form perfect lattice, so $u(x) = \pm b/4$.

The displacement function is

$$u(x) = \frac{b}{2\pi} \tan^{-1}\left[\frac{2(1-v)x}{a}\right] \tag{4.75}$$

where

v is the Poisson's ratio,

b and a are the lattice constants (see drawing), when [ ] becomes large, $\tan^{-1}$ [ ] tends to $\pi/2$, and $u(x \rightarrow \text{large}) \rightarrow b/4$.

The magnitude of the mismatch in the plane is given by the relative displacement, $b/2 - 2u(x)$. Note that the mismatch diminishes quickly with distance away from

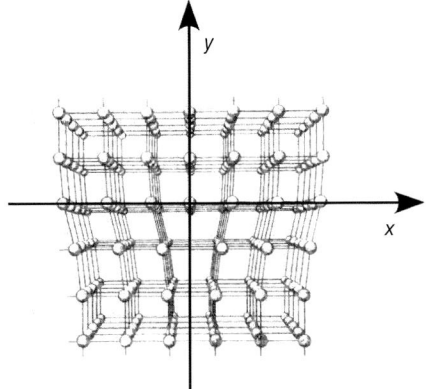

Figure 4.29 A view of edge dislocation in a crystal.

the core of the dislocation. Atoms in layers below and above the joining plane also suffer relative displacements that decay rapidly in a similar manner.

The gradient of the lattice distortion determines the width of a dislocation. Foreman, Janson and Wood give the following definition:

$$w = \frac{Gb}{2\pi(1-\nu)\tau_{FTS}} \tag{4.76}$$

where G is the elastic shear modulus, τ_{FTS} is the critical shear stress required to move the dislocation, from one position to next.

Note that

- elastic energy of dislocation line/unit length is:

$$U_{el} \approx Gb^2 \ln\left(\frac{b}{w}\right) \tag{4.77}$$

for $G = 40$ GPa and $w = 1$ nm, the value of this energy is of the order of 10^{-8} Nm m^{-1}, or 10^{-18} J atom^{-1}.

- thermal energy per atom is

$$U_{therm} \approx \frac{3kT}{2} \tag{4.78}$$

at 300 K this is of the order of 10^{-20} J atom^{-1}.

- as $U_{therm} \ll U_{el}$, therefore, dislocation motion is not affected by temperature variations (to a first approximation),
- from the principles of crystallographic repetition and structure invariance, it can be inferred that dislocation glide occurs at constant volume (no volume dilatation during glide):

$$\delta V = 0 \tag{4.79}$$

- as dislocation glide occurs under zero volume dilatation, Therefore, it must be independent of hydrostatic pressure (to a first approximation).

4 Mechanical Behaviour

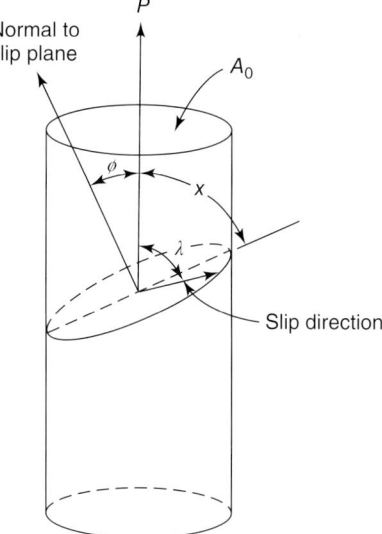

Figure 4.30 Schmidt factor.

4.5.3.7 Critical Shear Stress to Move Dislocation

The critical value of shear stress (in the glide plane) required to move a dislocation is

$$\tau_{PN} = \frac{2G}{(1-v)} \exp\left[-\frac{2\pi w}{b(1-v)}\right] \qquad (4.80)$$

for $a = b$ and $v = 0.35$ calculation gives $\tau_{PN} = 2 \times 10^{-4} G$

for $a = b\sqrt{2}$ and $v = 0.35$ calculation gives $\tau_{PN} = 2 \times 10^{-6} G$.

Note that τ_{PN} decreases rapidly with increasing dislocation width, therefore, observed yield strength of soft metals is $\approx 10^{-5} G$. Cottrell observed in 1950s that the critical stress needed to move a dislocation through a lattice is considerably smaller than the observed yield strengths. The fact that the observed value is sensitive to the purity of the metal (and density of dislocations) shows that this limiting value has not yet been reached experimentally and that other effects present stronger obstacles to the slip process than the force caused by the periodic lattice field.

In a tensile specimen, the maximum resolved shear stress is always at 45° to the tensile axis; its magnitude is equal to half of the applied tensile stress. This means that it occurs on conical surfaces as shown in Figure 4.30. However, a given dislocation is most unlikely to have its slip plane and direction coinciding with this surface.

The tensile stress required to move the dislocation is related to the critical shear stress through appropriate cosine factors, called *Schmidt factor* (Figure 4.30):

$$\sigma = \tau_{PN} \cos \alpha \sin \beta \qquad (4.81)$$

4.6
Plasticity in: Amorphous Solids

The most important statement to be made at the start of this section is that amorphous materials do not have dislocations or grain boundaries. This eliminates most of the plastic deformation mechanisms and explanations for the weakening and strengthening methods described above for crystalline solids.

Nevertheless, plasticity is observed in amorphous metallic glasses as shown in Figures 4.31 and 4.32. In all the cases shown, the testing was in compression, showing at first elastic behaviour up to its limiting value, followed by plastic deformation up to the point of fracture. The figures show dependence of plastic behaviour

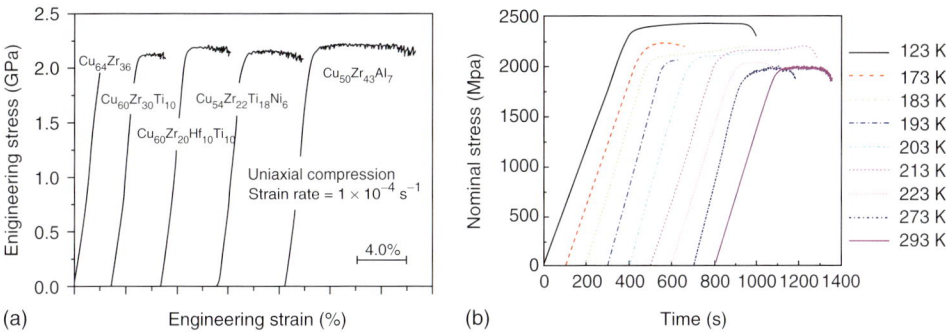

Figure 4.31 (a, b) Plasticity in bulk metallic glasses as a function of composition and temperature of testing. (Part (a) reprinted with permission from Mattern et al. (2009), Copyright (2009) by American Physical Society and part (b) reprinted with permission from Liu et al. (2013), Copyright (2013) by American Physical Society.)

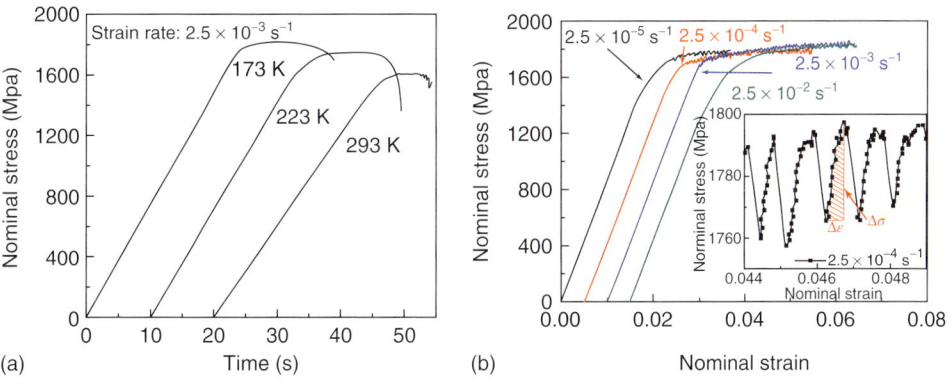

Figure 4.32 (a, b) Variation of plasticity in bulk metallic glass as a function of applied strain rate and details of serrations in magnified portion of the curve. (Part (a) reprinted with permission from Mattern et al. (2009), Copyright (2009) by American Physical Society and part (b) reprinted with permission from Liu et al. (2013), Copyright (2013) by American Physical Society.)

on variations in the composition of the alloys, temperature of testing as well as applied strain rate.

4.6.1
Plastic Deformation by Shear Band Propagation

It has been established that the plastic behaviour in bulk metallic glasses is a result of shear band propagation in the material. Shear band propagation is a mechanism of non-elastic permanent deformation found in practically all types of materials as shown by the examples in Figure 4.33.

The shear band shown in Figure 4.34 was formed during compressive deformation of the bulk metallic glass. The plane of the shear band runs along an axis

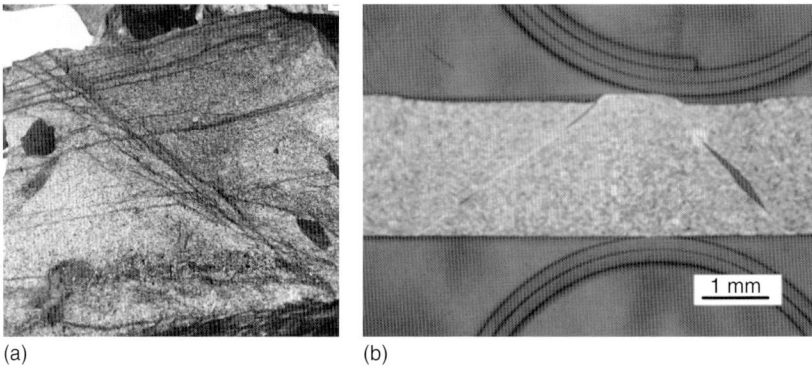

Figure 4.33 Shear bands found in (a) graphite and (b) steel.

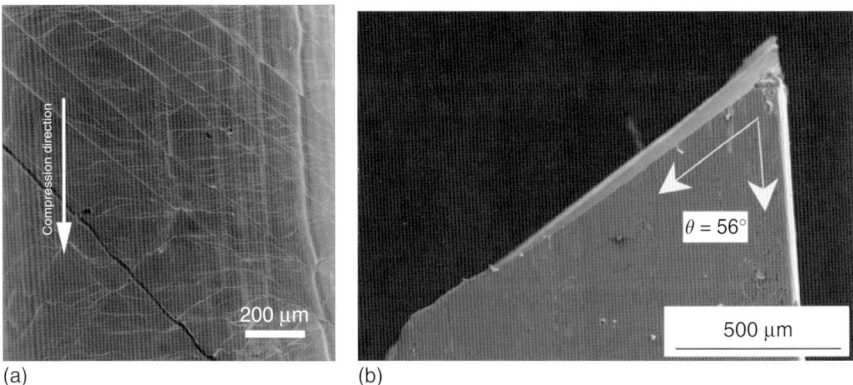

Figure 4.34 Inhomogeneous deformation along a shear band in (a) compression with shear band angle close to 44° and (b) tension with shear band angle close to 56°. (Part (a) reprinted with permission from Wang et al. (2005a), Copyright (2005) by American Physical Society and part (b) reprinted with permission from Wang et al. (2005b), Copyright (2005) by American Physical Society.)

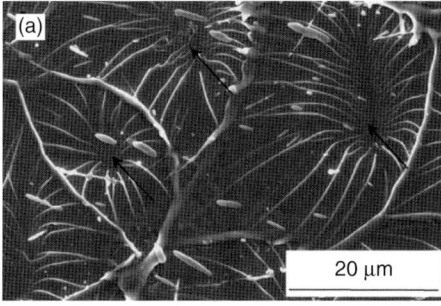 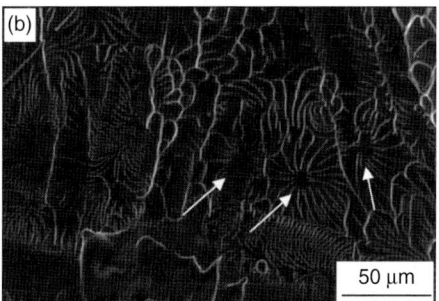

Figure 4.35 Morphology of shear band fracture surfaces in (a) tension and (b) compression. Note the similarity of patterns. (Part (a) reprinted with permission from Wang et al. (2005a), Copyright (2005) by American Physical Society and part (b) reprinted with permission from Wang et al. (2005b), Copyright (2005) by American Physical Society.)

inclined to the axis of deformation that follows the Tresca criterion. However, no general plasticity has been observed in bulk metallic glasses and, therefore, no conclusive analysis can be carried out.

Under uniaxial loading, maximum shear stress develops on planes at an angle of 45° to the load axis, although the angle of the shear band can deviate from that value because of specific materials properties and loading conditions (Wang et al., 2005a,b). To a first approximation, the phenomenology of failure is described by the Coulomb–Mohr equation:

$$\tau = k_0 - \sigma_n \tan \phi \qquad (4.82)$$

where ϕ is the internal friction angle, as shown in Figure 4.34.

Special alloying and careful heat treatment can result in very high compressive strengths, reaching above 5 GPa (Inoue et al. (2003)). Of course, shear bands can be also induced under simple shear loading. It should also be noted that shear bands occur in most materials, rocks, metals, polymers, that is, crystalline or not.

In amorphous materials, fracture surfaces of the shear bands show characteristic vein-like features with round cores, and subsidiary shear band slides, as seen in Figure 4.35. It has been noted that the vein pattern is similar to that obtained by pulling apart two hard surfaces with a viscous layer between them. Shear band melting has been correlated with adiabatic temperature increase at the interface and found that for thin shear band of 50 nm, the temperature can reach as high as 1300 K (Georgarakis et al., 2008).

The presence of imperfections in the amorphous structure, as described in Chapter 1, gives rise to a localized intense deformation gradient that develops into a shear band. Shear bands are thin layers containing material that is inherently different from the glassy matrix. It requires thermal activation process. In the local region, the free volume content increases within the shear band.

The modelling of shear bands has been done both theoretically and by computer simulations as shown in Figure 4.36a and b.

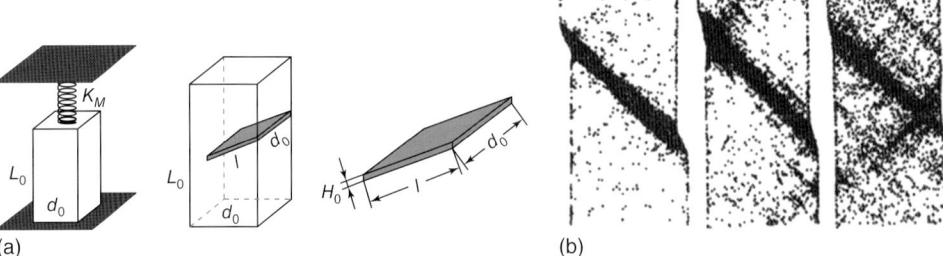

Figure 4.36 (a) Model of a shear band formation. (Reprinted with permission from Liu et al. (2011), Copyright (2011) by American Physical Society.) (b) Computer simulation of a shear band. (Reprinted with permission from Cheng et al. (2011), Copyright (2011) by Elsevier Ltd.)

The bands, which are subjected to a high shear stress with heating under a high shear strain rate, are expected to have a lower density and lower elastic modulus and assume a higher energy state compared to the surrounding glassy matrix. Under an applied stress, the elastic energy stored in the sample and testing machine system will increase with increasing stress until it approaches a critical point, that is, shear band formation.

To model shear band nucleation and propagation, we consider a system comprising a sample in the form of a rectangular prism under uniaxial compression and an elastic spring representing the compliance of the testing machine as shown in Figure 4.36a. Under load, the system will contain stored elastic and free energy. A change in the potential energy of the system can be described by

$$\Delta G_{system} = \Delta G_{elast} + \Delta G_{vol} + \Delta G_{inter} \tag{4.83}$$

where the individual terms correspond to elastic energy released because of load relaxation when the shear band moves, volume free energy change because of difference between the shear band and surrounding glassy matrix and interfacial free energy change because of shear band propagation causing increase in the interfacial area, respectively.

The driving force for the shear band propagation is the elastic stored energy in the combined sample and machine system, whilst the resisting force consists of the increase in the volume free energy and the newly generated interfacial free energy. Equation 4.83 is quadratic in the length l of the shear band, showing characteristic resistance to propagation until a critical size is reached, similar to that of critical crack size in Equation 4.54.

The activated (moving) shear band can propagate and stop within the material, or in extreme case, it will run through the material and cause fracture, depending on the inherent properties of the metallic glass. Consider the former case, with a shear band nucleating and propagating over a relatively short distance. For every propagation, the movement of the shear band will result in a sudden release of the load, which will manifest itself as a stress drop as can be seen in Figure 4.32.

Such a drop in load (stress) will occur with every nucleation and propagation of individual shear bands, resulting in the so-called serrated stress–strain curve. The main research efforts are directed to the understanding of the effect of atomic arrangements on the capacity of the metallic glass to nucleate as high a density of shear bands as possible with limited propagation distances, resulting in increased plasticity of the material.

4.7 Superplasticity

4.7.1 Phenomenology

Superplasticity is the property of a material to exhibit large plastic deformations (of the order of 1000%), examples of which are shown in Figure 4.37. This phenomenon is both of academic and practical interest, because it provides the capability for forming complex parts from sheet materials.

In polycrystalline metals, superplasticity refers to the phenomenon of extraordinary ductility exhibited by some alloys with extremely fine grain size, when deformed at elevated temperatures and in certain ranges of strain rate. Microstructural observations reveal the characteristics that are required for superplastic behaviour. Amongst the macroscopic characteristics, there is a need

(a) (b)

Figure 4.37 (a, b) Superplastic deformation of polyethylene film, and the ubiquitous aluminium can manufactured by the so-called deep drawing relying on superplastic behaviour of the aluminium.

for a proper equation for the strain rate as a function of stress, grain size and temperature. Any satisfactory theory must also arrive at the dependence of the superplastic behaviour on the various microstructural characteristics. Theories presented so far for microstructural characteristics may be divided into two classes:

- those that attempt to describe the macroscopic behaviour
- those that give atomic mechanisms for the processes leading to observable parameters.

The former sometimes incorporate micromechanisms. The latter are broadly divided into those making use of dislocation creep, diffusional flow, grain boundary deformation and multimechanisms. The theories agree on the correct values of several parameters, but in matters that are of vital importance such as interphase grain boundary sliding or dislocation activity, there is continuing disagreement.

However, superplasticity is also exhibited by amorphous metallic alloys, as well as amorphous polymeric materials. Examples of superplastic behaviour in bulk metallic glass are shown in Figures 4.38 and (4.39). The graphs show stress–strain curves recorded during extensional tests as a function of strain rate or by varying the temperature of the sample. The as-cast samples are 2.5 cm long before extension. In all cases, the stress increases linearly in the elastic region until a critical stress is reached. Above the critical stress, the sample begins to deform homogeneously, but being constrained by the grips at the ends, its shape develops as shown in Figure 4.39.

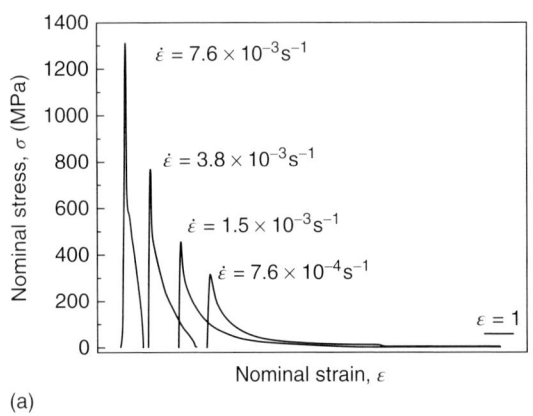
(a)

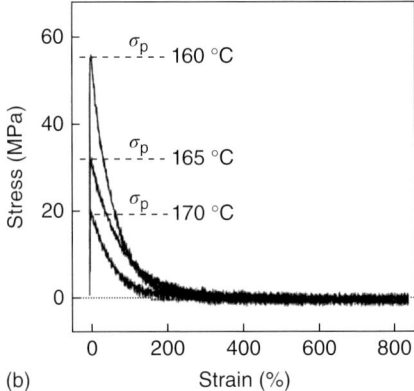
(b)

Figure 4.38 (a, b) Superplastic behaviour of bulk metallic glasses: Zr-based Vitreloy and Mg-based amorphous alloy. Effect of strain rate and temperature. Note the very large plastic strains. (Part (a) reprinted with permission from Shen et al. (2005), Copyright (2005) by American Physical Society and part (b) reprinted with permission from Gun et al. (2006), Copyright (2006) by American Physical Society.)

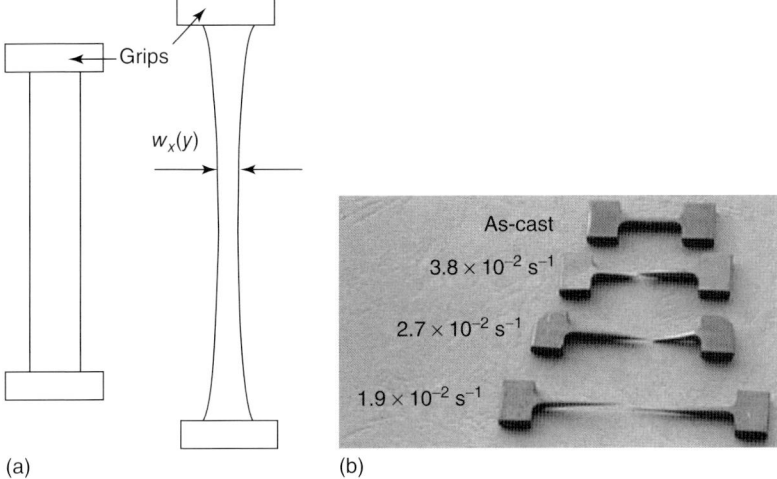

Figure 4.39 (a, b) Change of shape of elongated samples under superplastic deformation.

At the point mechanical instability, described by Considere's construction, flow of the material leads to reduction of cross section and, therefore, reduction of applied stress as seen in the graphs.

Since amorphous materials have no dislocations and no grain boundaries, the mechanism of homogenous deformation can only be by vacancy–atom exchange or by a cooperative atomic motion not requiring vacancies. Superplasticity in bulk metallic glasses can be achieved, even at room temperature (Liu *et al.* (2007)). This can lead to metallic glasses of high ductility and toughness (Hoffmann *et al.* (2008), Greer *et al.* (2011)).

4.7.2
Continuum Mechanics

The rheological state of superplasticity is equivalent to solid state flow. The results shown in Figure 4.38 were obtained with a fixed extension rate applied to a sample and the sample was homogeneously extending with time. Following an initial response, the rate of deformation of the sample will reach a steady state (i.e., the sample will elongate at a constant rate). The apparent extensional viscosity of the sample is calculated according to this equation (Wang *et al.* 2008):

$$\eta_{elong} = \frac{\sigma_{applied}}{\dot{\epsilon}} \tag{4.84}$$

where

$$\dot{\epsilon} = \left(\frac{1}{l}\right)\left(\frac{dl}{dt}\right) \tag{4.85}$$

4 Mechanical Behaviour

This is a constant volume deformation, $\Delta V/V = 0$, unlike elastic deformation described by equation (4.8). Newtonian viscosity is equal to one third of the extensional viscosity.

The equation of flow in an incompressible continuum

Associated with every material continuum, there is the property of conservation of mass:

$$m = \int_V \rho(x,t)\, dV \qquad (4.86)$$

in which $\rho(x, t)$ is called the *mass density*. The law of conservation of mass requires that the mass of the continuum remains constant with time and, hence, that the material derivative of Equation 4.86 be zero.

$$\frac{dm}{dt} = \int_V \left[\frac{d\rho}{dt} + \rho \frac{\partial v_k}{\partial x_k}\right] dV = 0 \qquad (4.87)$$

It follows that

$$\frac{d\rho}{dt} + (\rho v_k)_{,k} = 0 \qquad (4.88)$$

For an incompressible continuum, $d\rho/dt = 0$ and, therefore,

$$\mathrm{div}(\mathbf{v}) = 0 \qquad (4.89)$$

where $\mathbf{v}$ is the velocity vector of the continuum.

4.7.3 Superplasticity in Bulk Metallic Glasses

4.7.3.1 Calculation of Strain Rate for Superplasticity

A model of homogeneous deformation in amorphous materials based on vacancy/atom exchange has been proposed by Spaepen (1977), as shown in Figure 4.40. The exchange of atom/vacancy pair can be considered as flow of atoms and vacancies in opposite directions. Vacancies (on average) have spherically symmetrical tensile stress field around them and, therefore, will be attracted towards hydrostatic compression and repelled by hydrostatic tension. For this mechanism to produce elongational strain in a sample subjected to uniaxial tension, the atoms must flow towards the top and bottom surfaces of the sample and vacancies towards the side surfaces. The diffusion of one vacancy to side surface (and one conjugate atom to the top surface) will produce volumetric displacement of the order of d^3 and contribute to longitudinal strain of magnitude:

$$\epsilon_{\mathrm{pair}} = \frac{1}{3} g \frac{d^3}{A\ell} = \frac{1}{3} g \frac{d^3}{V}, \qquad (4.90)$$

where d is the atomic/vacancy diameter, A is the cross-sectional area of the sample, ℓ is its length and g is a geometrical factor, related to atomic packing factor.

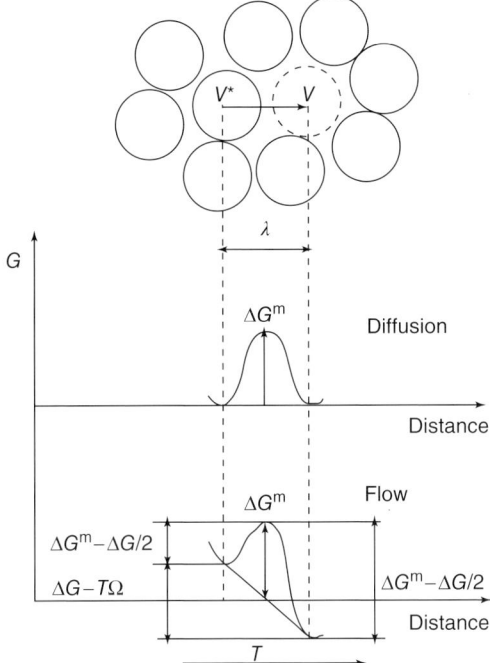

Figure 4.40 Atom/vacancy exchange by diffusion activated by applied stress. (Reprinted with permission from Spaepen et al. (1977), Copyright (1977) by American Physical Society.)

Then the predicted strain rate in the sample caused by the diffusion of vacancies and atoms will be

$$\dot{\varepsilon}_{sample} = \frac{1}{3} c \frac{\bar{v}}{\bar{\lambda}} = c \frac{3D}{\ell d} = 3 \cdot 10^{-9} \, s^{-1}, \tag{4.91}$$

where $c = n\, d^3/V = v_f$ is the concentration of vacancies, equivalent to free volume, $\bar{v}$ is the average velocity of diffusion of the atom/vacancy pair and $\bar{\lambda}$ is the average distance of diffusion. The average velocity is derived from diffusion equation: $\bar{v} = 3D/d$, and the average distance for diffusion is taken as approximately one-third of the sample diameter, $\bar{\lambda} = \ell/3$. The diffusivity constant for atoms in an amorphous metallic alloy, at 300 K, is taken as $D = 10^{-19}$ m^2 s. For the other quantities, we take $\ell = 10$ mm, $d = 0.2 \times 10^{-9}$ m and $c = 0.02$. Clearly, the strain rate as shown by calculation in Equation 4.91 is far too low to be observed.

The original description by Spaepen includes stress and temperature activated diffusion, so that the diffusivity, D, in Equation 4.91 should be changed to $D_T \exp[(Q - \sigma\Omega)/RT]$, with Q being the activation energy for self-diffusion, σ the applied stress and Ω the activation volume (Figure 4.41).

There are at least two fundamental problems with this mechanism of creep applied to amorphous metals. First, as in a single crystal, vacancies in a glass

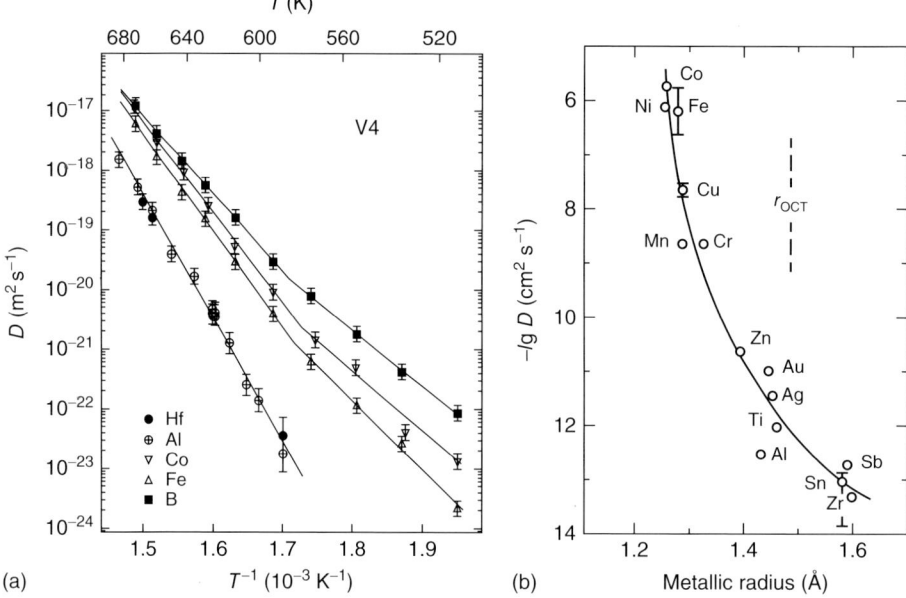

Figure 4.41 (a, b) Diffusivity as a function of temperature and atomic species. (Part (a) reprinted with permission from Zumkley et al. (2002), Copyright (2002) by American Physical Society and part (b) reprinted with permission from Hood et al. (1978), Copyright (1978) by The Institute of Physics.)

cannot be regenerated because there are no grain boundaries, so the strain rate should diminish with time as the content of vacancies in the sample is exhausted. Second, the free volume is dispersed throughout the body rather than concentrated in vacancies (Yavari et al. 2005). Consequently, the effective value of c is much less than the apparent free volume. Even with stress activation included, the experimentally observed strain rates cannot be accounted for by this mechanism (Wang et al. 2005). Therefore, other mechanisms of faster flow must be considered to explain the behaviour observed in Figure 4.39. Two approaches have emerged. Argon has anticipated the dispersed free volume and proposed a diffused shear zone mechanism (Argon, 1979), and Ekambaram et al. (2008) developed a model for free volume evolution during deformation of glasses. Ekambaram et al. assumed that metallic glasses have an amorphous structure without dislocations and point defects, but that generation of free volume is accomplished by plastic deformation and application of hydrostatic pressure. Furthermore, atomic rearrangement causes concentration of free volume, thus leading to creation of vacancies.

The new 'concordant' deformation mechanisms for superplastic flow, requiring no vacancies (no free volume), can account for the observed rates of deformation. Comparison of the strain rates for the two models shows clearly its advantage, as will be shown later.

4.7.4
Concordant Deformation Mechanism

Consider a cluster of atoms, taken from the ideal amorphous solid (IAS) model, as shown in Figure 4.42.

The irregular primary cluster, representative of such clusters as building blocks of the IAS, has the inner sphere marked as A. The outline of the whole primary cluster is shown by the larger broken line circle. Three of the outer spheres of this cluster are shown in colour, the remainder are schematically indicated by empty circles. There is an atom marked B, belonging to an adjacent primary cluster, but touching the three outer spheres of the first cluster. All five spheres in blue form a five-atom subcluster, outlines by the smaller broken line circle. The whole random packing can be thus subdivided into these five-atom subclusters. The five-atom subcluster is envisaged as the minimum and essential grouping of atoms capable of permanent plastic deformation, activated by strain energy.

Figure 4.42 also shows the five-atom subcluster transformed into its α formation by elongation along the vertical z-axis, and another transformed into its β formation by compression along the same axis. It is further envisaged that an amorphous solid has a high density of such effective subclusters and that the process of α–β transformation is regenerative, that is, it can repeat itself continuously, thus yielding unlimited plastic deformation.

We generalize the process by representing the five-atom subclusters by circles, as shown in Figure 4.43, transformed into α and β formations when subjected to

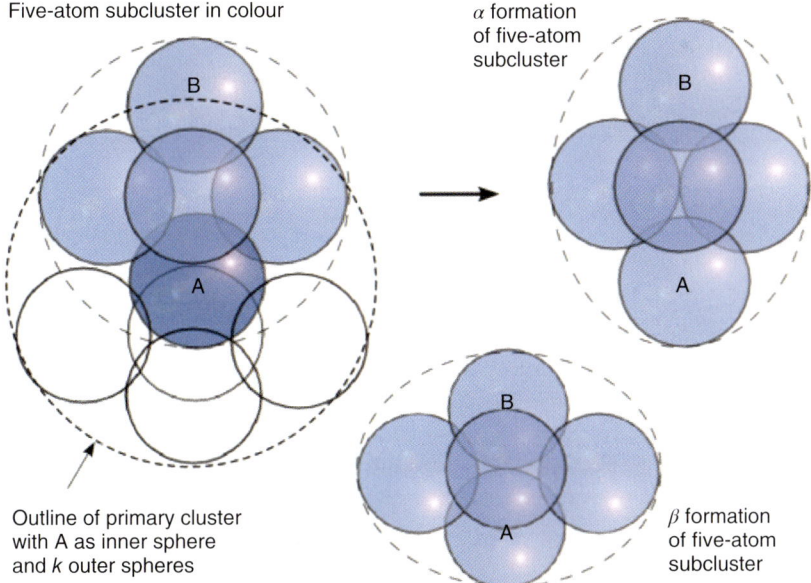

Figure 4.42 A five-atom subcluster of a primary irregular cluster from IAS packing. Transformation of the cluster into the α and β forms.

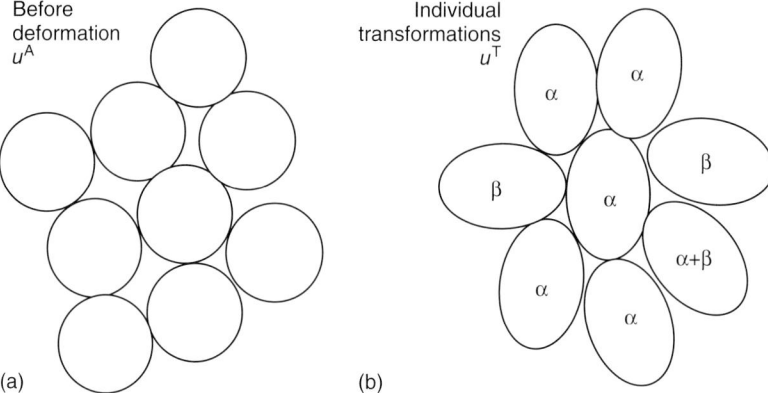

Figure 4.43 Schematic representation of the five-atom subclusters into α and β formations. (a) The five-atom subclusters represented as circles before elongation. (b) Transformations activated by the applied elongation.

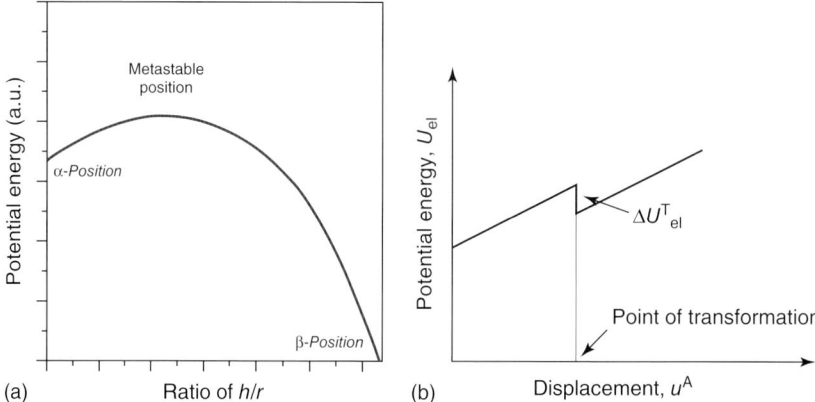

Figure 4.44 (a, b) Potential energy variation of a five-atom subcluster on deformation from α to β formation, and the corresponding change in potential energy of the system of matrix and inclusion.

uniaxial elongation. We use the term *concordant* for this mutually accommodating process of deformation.

It is assumed that the transformation from one to the other formation involves an energy barrier as shown in Figure 4.44. It can be observed in Figure 4.42 that in the α formation, the subcluster has nine direct contacts, and one separation, whereas in the β formation, it has seven direct contacts and three separations. Consequently, the potential energy of the two should differ, with an activation energy barrier between them, as shown schematically in Figure 4.44.

When such an amorphous solid is subjected to deformation, the five-atom subclusters transform suddenly at a critical level of strain, specific to the individual local arrangement of the atoms. The variation of the potential energy with the level of deformation is shown schematically in Figure 4.44. This process is similar to that of a dislocation or of a point defect (such as vacancy/atom pair) being pushed over an energy barrier until it reaches its point of instability and then moves suddenly to its next stable position. In addition to the activation of the transformation by applied strain, the clusters themselves create local strain gradients that may facilitate such transitions.

4.7.4.1 Density Variation in Amorphous Solids

There is sufficient evidence to show that there are significant local density variations in what may appear on a surface as a homogeneous glassy solid. This evidence comes mainly from modelling of glasses at the atomic level, although corroborating evidence also comes form light scattering in optical glass fibres.

Figure 4.45 shows the distribution of local densities in Zr-based metallic glasses measured by Sun *et al.*, who studied computer-generated models of these materials. They divided the whole simulation cell into subcells of varying sizes. The graph shows the distribution of the local normalized density as a function of the different subcell edge lengths, denoted by λ. It turns out that the distribution of local number density can be described by the normal distribution function, with standard deviation found to increase with decreasing size of subcell length, suggesting that the atomic distribution is more heterogeneous on small scale and that structural variation is obscured by averaging over a larger system (smoothing process). The size of the dense packing region is larger than that of an individual Kasper polyhedral cluster; hence, the density distribution is evaluated under the condition that the cell edge length is six subdivided. This information is already anticipated in the description of the random packing of spheres presented in Chapter 2, Figures (2.6) and (2.8). Therefore, it is important to bear in mind that atoms in an amorphous solid reside in a continuously varying field of density variations and density gradients with wavelength of the order of several atomic radii.

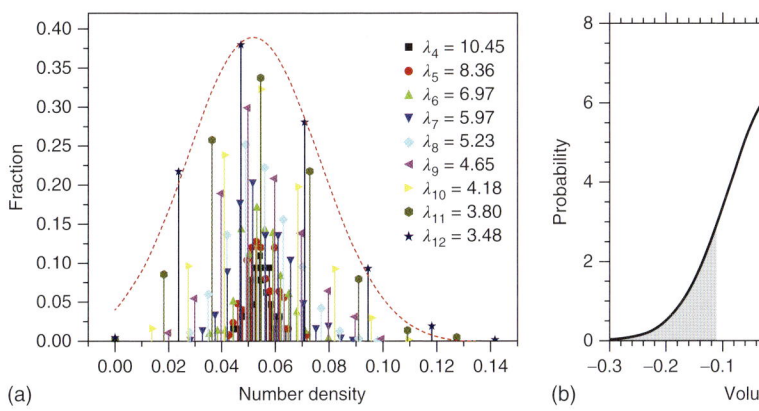

Figure 4.45 (a, b) Variation of density in amorphous metallic glasses. (Part (a) reprinted with permission from Sun *et al.* (2010), Copyright (2010) by American Physical Society and part (b) reprinted with permission from Egami *et al.* (2007), Copyright (2007) by American Physical Society.)

Figure 4.45 also includes a schematic graph showing a distribution of inverse densities, analogous to the graph in the previous figure, with indication of the limits on the stability of local clusters. This graph, adapted from Suzuki 1987, was drawn to support the topological fluctuation theory for the behaviour of glasses. Here it is shown to emphasize that if the local volumetric strain is more than approximately 10%, the strength of the gradient is sufficient to rearrange the clusters spontaneously.

The concordant deformation process as described above is the main atomistic model of superplastic deformation in amorphous metals. It operates when the material is subjected to deformation and is capable of producing large elongations (see Figure 4.39) at high strain rates ($\sim 10^{-1}$ s^{-1}) in the temperature region above the glass-transition temperature up to its melting point (supercooled liquid region).

With this background, answers to the following questions are sought:

- How does external stress bias a local molecular rearrangement?
- How does local molecular rearrangement give rise to macroscopic deformation?
- Why is this deformation irrecoverable?

4.7.4.2 The 'Inclusion' Problem

A small region of a body, representing an amorphous metallic glass, comprises an *inclusion* within a *matrix*. The inclusion has the same elastic constants as the matrix. The matrix and the inclusion form an elastic *system*.

The inclusion is considered to act as having two equilibrium formations, separated (in the deformation space) by an energy barrier. The volume of the inclusion before and after the transformation is the same, that is, change of shape but no change of volume. The two positions are stress free, but the potential energy of each may differ. It is assumed that an applied elongation of the system causes the transformation of the inclusion from one to the other position.

The aim is to show that when the system is subjected to elongation, the potential energy of the system drops when the inclusion is transformed.

4.7.4.3 The System without Transformation

Initially, the inclusion is in the shape of a sphere. Let S represent the surface between the inclusion and the matrix. The system is subjected to an extension, u_i^A, defined by the strains,

$$\epsilon_{ij}^A = \begin{vmatrix} -v\epsilon_{33}^A & 0 & 0 \\ 0 & -v\epsilon_{33}^A & 0 \\ -0 & 0 & \epsilon_{33}^A \end{vmatrix} \quad (4.92)$$

where v is the Poisson's ratio.

Consequently, the strains in the inclusion are those given by Equation 4.92, and with the deformation the inclusion will gradually change shape to a prolate ellipsoid, accordingly.

The shapes of a sphere and an ellipsoid are described by

$$x^2 + y^2 + z^2 = \rho^2 \tag{4.93}$$

$$\frac{x^2}{a^2} + \frac{y^2}{b^2} + \frac{z^2}{c^2} = 1, \tag{4.94}$$

where ρ is the radius of the sphere and $a = b < c$ are the axes of the ellipsoid.

The potential energy of the system is given by

$$V_{pot} = V_{int} + U_{el} \tag{4.95}$$

where V_{int} is the enthalpy of the system, and U_{el} is the Green's strain energy. The potential energy of the system increases linearly with strain.

4.7.4.4 The System with Transformation

1) Following Robinson/Eshelby method, cut the inclusion out of the matrix when it is spherical in shape (Robinson, 1951, Eshelby, 1957).
2) Apply surface tractions to the inclusion to change its shape from a sphere to an ellipsoid. Traction acting at a point on the surface, S with normal $\vec{n}$, is given by

$$t_i = \sigma_{ij} n_j$$

The principal strains of the transformation are

$$\frac{x^2}{\lambda_1} + \frac{y^2}{\lambda_1} + \frac{z^2}{\lambda_3} = 1 \tag{4.96}$$

As the transformation is at a constant volume, the axes of the ellipsoid relate to the principle strains as

$$c = \rho\sqrt{\lambda_3}, \quad a = b = \frac{\rho}{\lambda_1}$$

where $\lambda_i = (1 + \epsilon_i)$. The deformation of the inclusion because of the transformation, u_i^T, is described by the strain components:

$$\epsilon_{ij}^T = \begin{vmatrix} -\epsilon_{33}^T/2 & 0 & 0 \\ 0 & -\epsilon_{33}^T/2 & 0 \\ -0 & 0 & \epsilon_{33}^T \end{vmatrix} \tag{4.97}$$

This transformation is stress free because the inclusion has transformed in the absence of the matrix.

3) Insert the inclusion into the matrix and make perfect adhesion on the common surface. Relax the system. As a result, the matrix will apply surface tractions onto the inclusion, and the inclusion will resist until mechanical equilibrium is reached. It can be asserted that in equilibrium,

$$t_i^I = -t_i^M \quad \text{M is for matrix}$$

$$u_i^I = u_i^M \quad \text{I is for inclusion}$$

$$t_i^M \sim \frac{1}{r^2}, \text{ for } |r| \text{ outside } S$$

4) Next, the system is subjected to an extension, u_i^A, defined by the strains (Equation 4.92).
5) The strain energy within the inclusion is

$$U_{el}^I = \frac{1}{2}\int_S (u_i^I - \epsilon_{ij}^I r_j)\sigma_{ij} n_j \, dS, \qquad (4.98)$$

where $\int_S$ is surface integral, and $\epsilon_{ij}^M = \epsilon_{ij}^A$, and the strain energy in the matrix is

$$U_{el}^M = \frac{1}{2}\int_S (u_i^M - \epsilon_i^M r_j)\sigma_{ij} n_j \, dS \qquad (4.99)$$

As $u_i^M = u_i^I$, the total elastic strain energy of the system is

$$U_{el} = U_{el}^M + U_{el}^I = \frac{1}{2}(\epsilon_{ij}^M + \epsilon_{ij}^I)\int_S r_j \sigma_{ij} n_j \, dS. \qquad (4.100)$$

Using Green's theorem, the surface integral is transformed into volume integral, giving

$$U_{el} = \frac{1}{2}(\epsilon_{ij}^M + \epsilon_{ij}^I)\int_V (\sigma_1^P + \sigma_2^P + \sigma_3^P) \, dV = \frac{1}{2}V(\epsilon_{ij}^M + \epsilon_{ij}^I)\sum_{i=1}^{3}\sigma_i^P \qquad (4.101)$$

where σ_i^P's represent principal stresses.

6) At the instant of transformation, the strain in the inclusion suddenly decreases to

$$\epsilon_{ij}^I = \epsilon_{ij}^A - \epsilon_{ij}^T \qquad (4.102)$$

Consequently, at the point of transformation, the potential energy in the system drops to a lower value as indicated in Figure 4.44.

The main points of this deformation mechanism have been proposed previously by Suzuki et al. (1987) and Wang et al. (2011).

4.7.4.5 Conclusions
Within the elastic matrix of the surrounding atoms, the five-atom subcluster will change its shape by the interaction of interatomic forces with an applied field, assisted by the density differences between adjacent main clusters. When the hard sphere regime is relaxed by introduction of central atomic force fields, the configuration of the five-atom subclusters becomes unstable except at the α and β positions.

It is conjectured that the bistable action of the five-atom subcluster is the main mechanism for atomic rearrangements in amorphous bulk metallic glasses. Furthermore, it can be noted that this adjustment of the relative atomic sites occurs without any long-distance diffusion. It should occur most vigorously for clusters

Table 4.5 Strain rates for vacancy and concordant deformation mechanisms.

Quantity	Vacancy/atom model	Concordant model
Concentration, c (%)	2	30
Diffusivity, D (m² s)	10^{-19}	10^{-19}
Diffusion distance, λ (m)	10^{-3}	$0.1 \cdot 10^{-9}$
Predicted creep rate, $\dot{\varepsilon}$ (s^{-1})	$3 \cdot 10^{-9}$	$1 \cdot 10^{-5}$

of lowest coordination number, surrounded by clusters of high coordination numbers (high density). It is presumed that this occurs as a cooperative movement with the surrounding atoms.

Without stress, the glassy phase is believed to have an isotropic structure. However, applying a stress induces local reorientations and anisotropy. The structural evolution in the first nearest-neighbour shell is well documented.

The atomic reorientation in the first nearest-neighbour shell can relax the external stress and reduce local energy concentration. In order to counterbalance this drop in local stress, other atoms in a surrounding region must bear more stress. The incremental stress imposed on the surrounding atoms will concordantly move the atoms. This movement corresponds to the relaxation process that is necessary to compensate the local stress increment because of the atomic reorientation in the first shell.

Summarizing the properties of the proposed concordant deformation mechanism, we note that

- it requires no vacancies (no free volume),
- every atom/cluster is a potential source of concordant deformation.

The strain rate Equation 4.91 (at $T \ll T_g$) can now be re-evaluated with new parameters for the concordant mechanism. In particular, the concentration of atoms taking part in simultaneous displacement can be as high as 10–100%, and the diffusion distance need be only of the order of $0.1 \cdot 10^{-9}$ nm. Substitution of these new values in Equation 4.91 gives predicted strain rate as

$$\dot{\varepsilon}_{sample} = \frac{1}{3} c \frac{\bar{v}}{\lambda} = c \frac{3D}{\ell d} = 1 \cdot 10^{-5} \text{ s}^{-1} \qquad (4.103)$$

A comparison of the strain rates calculated for the two mechanisms is shown in Table 4.5.

4.8 Viscoelasticity

Viscoelastic behaviour manifests itself mainly through two phenomena:

- *creep*, which is the flow of materials at imperceptibly slow rates and at stress levels well below that of yield strength;

(a)   (b)

Figure 4.46 Flow of condensed matter: (a) a liquid (courtesy of Peter Muesburger) and (b) a solid.

- *stress relaxation*, which is the gradual loss of elastic resistance of a material with time.

It is a combination of both elastic and viscous flow behaviour within a solid material (Figure 4.46).

4.8.1
Phenomenology

Viscoelastic behaviour is characteristic of materials that show both elastic and viscous behaviour simultaneously. Of all materials, the greatest propensity to viscoelastic behaviour at normal atmospheric temperatures is shown by polymeric materials, and the least for ceramic materials. This can be generalized in reference to the ratio of ambient temperature of the material to its melting point. For polymers, the melting points are around 470 K (200°C) so the ratio is $290/470 \approx 0.6$. For steel, the melting point is 1600 K, so that the ratio is $290/1600 \approx 0.2$. For ceramics with melting points above 3000 K, the ratio is $290/3000 \approx 0.1$. As a rule, if the ratio is more than 0.5, the material will behave in a viscoelastic manner.

The plot of strain for the creep of the Velveeta cheese shows a continuing increase, which may be understood as an unlimited flow. By contrast, the creep behaviour of concrete appears to be asymptotic towards a constant value at sufficiently long time (Figure 4.47). Similarly, the stress relaxation of the stainless steel levels off to a constant value.

A measure of creep is the *creep strength*, defined by

$$\frac{\Delta l_{creep}}{l_u} \tag{4.104}$$

where Δl_{creep} is the steady-state value of the elongation (or contraction) by creep and l_u is the undeformed (initial) length of the sample.

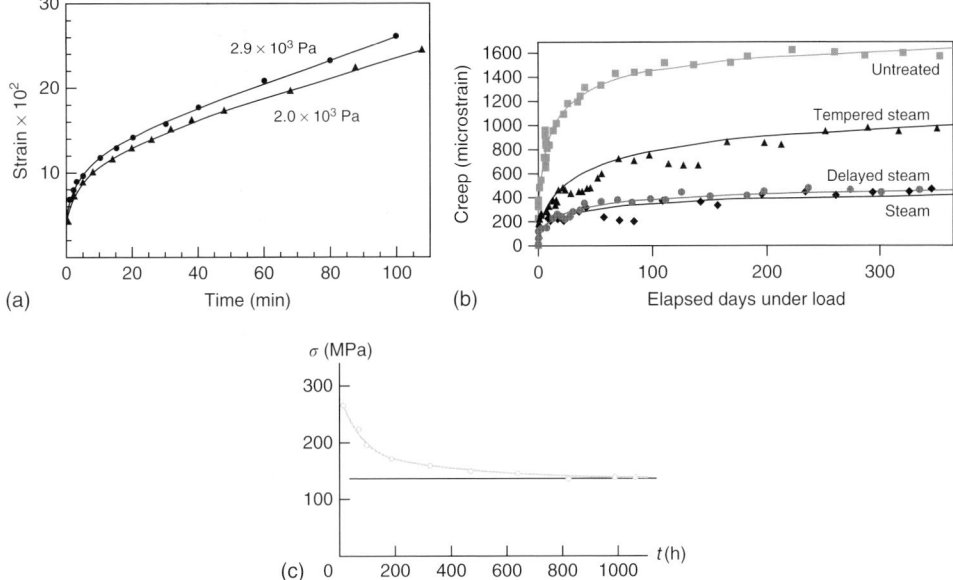

Figure 4.47 (a, b) Creep behaviour of Velveeta cheese at 295 K (melting point ≈ 350 K), subjected to stresses as shown, and creep behaviour of concrete subjected to a constant stress and different treatments during curing. (c) Stress relaxation in samples of stainless steel at a temperature of 920 K (melting point ≈ 1800 K).

And the *retardation time*

$$\tau_{ret} = \frac{\eta}{E} \quad (4.105)$$

where η is the elongational viscosity measured during creep and E is the modulus of elasticity of the material measured without the creep effect. For creep experiment that shows a limiting asymptotic value, the retardation time is equal to the time at which the creep strain reaches a value of $(\epsilon_{asym} - \epsilon_0)$ divided by 2.72 (base number for natural logarithm).

By analogy, a measure of stress relaxation is the *relaxation strength*, defined by

$$\frac{\Delta\sigma_{relax}}{\sigma_u} \quad (4.106)$$

where $\Delta\sigma_{relax}$ is the amount of the loss of elastic stress and σ_u is the unrelaxed (initial) value of stress in the sample at the beginning.

And the *relaxation time*

$$\tau_{relax} = \frac{\eta_{fict}}{E} \quad (4.107)$$

where η_{fict} is a fictitious viscosity during stress relaxation (there is no external flow) and E is the modulus of elasticity of the material measured before the relaxation effect. In the theory of viscoelasticity, $\tau_{ret} \neq \tau_{relax}$.

4.8.2
Time- and Temperature-Dependent Behaviour

Flow is a natural characteristics of fluids. Fluidity is the physical property of a substance that enables it to flow. A liquid is a continuous, amorphous substance whose molecules move relative to and past one another. In a liquid, its atoms or molecules (whichever may apply) are close enough to each other to give rise to a bulk modulus of the order of 1 GPa or higher, and to be called colloquially 'incompressible'. At the same time, the atoms/molecules are mobile, and exchange places with each other continuously. The flow of a liquid is through the exchange of atomic positions, defining molecular mobility, and at the macroscopic level, it gives rise to a property called *viscosity*.

By definition, *viscosity* is the reciprocal of *fluidity*.

4.8.2.1 Definitions of Viscosity

1) *Newton's Definition.* A liquid of viscosity, η, subjected to shear flow at a constant shear rate, $d\gamma/dt$, require shear stress of the magnitude given below to maintain that flow:

$$\tau = \frac{\eta \, d\gamma}{dt} \tag{4.108}$$

Inverting the equation gives the amount of flow, which depends on the applied stress and on time (during which the viscosity is assumed to remain constant):

$$\gamma(t) = \frac{1}{\eta} \int_0^t \tau \, dt \tag{4.109}$$

2) *Stokes' Definition.* Drag force on a ball of diameter D, falling through a liquid under the influence of gravity:

$$F_d = 6\pi\eta v D \tag{4.110}$$

3) *Poiseuille's Definition.* Pressure required to pump liquid through a pipe of fixed diameter:

$$P = \frac{8\eta q L}{\pi R^4} \tag{4.111}$$

where F_d is drag force experienced by a ball of diameter D flowing with velocity v through a liquid of viscosity η and P is the pressure required to maintain flow rate, q through a pipe of length L and radius R. In all cases, laminar flow is assumed.

Liquids have an enormous range of viscosities, depending on chemistry, pressure and temperature, ranging from, for example, 10^{-3} Pa s for water at 20° C, to 10^6 Pa s for pitch, to 10^{12} Pa s for an organic or inorganic glass (at its glass-transition temperature), to values much higher, as 10^{21} Pa s, for solids such as rocks.

4.8.2.2 Order of Magnitude Calculations

In a typical laboratory experiment, the employed strain rates are of the order of 10^{-3} s^{-1}. At room temperature, the magnitude of stress required to maintain extensional flow of these substances will be ($\sigma_\eta = 3 \times \eta \times \tau$):

for water: $\sigma_\eta = 3 \times 10^{-3}$(Pa s) $\times 10^{-3}$[s^{-1}] = 3 (μN m^{-2})
for pitch: $\sigma_\eta = 3 \times 10^{3}$(Pa s) $\times 10^{-3}$[s^{-1}] = 3 (Nm^{-2})
for glass: $\sigma_\eta = 3 \times 10^{12}$(Pa s) $\times 10^{-3}$[s^{-1}] = 3 (GN m^{-2})
for rock: $\sigma_\eta = 3 \times 10^{21}$(Pa s) $\times 10^{-3}$[s^{-1}] = 3×10^9 (GN m^{-2})

In inorganic glass, such high stresses are never achieved (see Section 4.4.2.5); instead glass will fracture well before this level of stress is reached. However, the glass can flow if it is tested at much slower strain rates. Rocks will flow under a stress of 100 (MN m^{-2}), for which the corresponding flow rate is 10^{-13} s^{-1}. For a slab of rock, 1 m thick to flow 1 m in shear at this strain rate will require 3 000 000 years (geological time).

The viscosity of liquids relates to fundamental interactions between the atoms and molecules. During flow (at shear rates which are not extreme), all of the external work involved in making the liquid flow is dissipated in the liquid, mainly as heat.

4.8.3
Temperature Effect on Viscoelastic Behaviour

For the standard linear viscoelastic solid (SLVS) model, it is assumed that a change of temperature will affect the viscosity component of the behaviour but will have no effect on the elastic component.

4.8.3.1 Arrhenius Behaviour
The variation of viscosity of liquids as a function of temperature is described by the Arrhenius relationship:

$$\eta(T) = \eta(T_R) \exp\left(\frac{Q}{RT}\right) \qquad (4.112)$$

where T_R is a reference temperature, Q is the activation energy for viscous flow, R is the universal gas constant and T is absolute temperature.

The graph shown in Figure (4.48) shows the viscosity changes with a change in reciprocal temperature.

4.8.3.2 Vogel-Fulcher–Tammann Behaviour
For glasses, the temperature effect on their viscoelastic behaviour is described by the Vogel-Fulcher-Tammann (VFT) equation:

$$\eta(T) = B \exp\left(\frac{A}{T - T_0}\right) \qquad (4.113)$$

where A, B and T_0 are material constants. The variation of viscosity with reciprocal temperature as described by Equation 4.113 is shown in Figure 4.48. Each flow

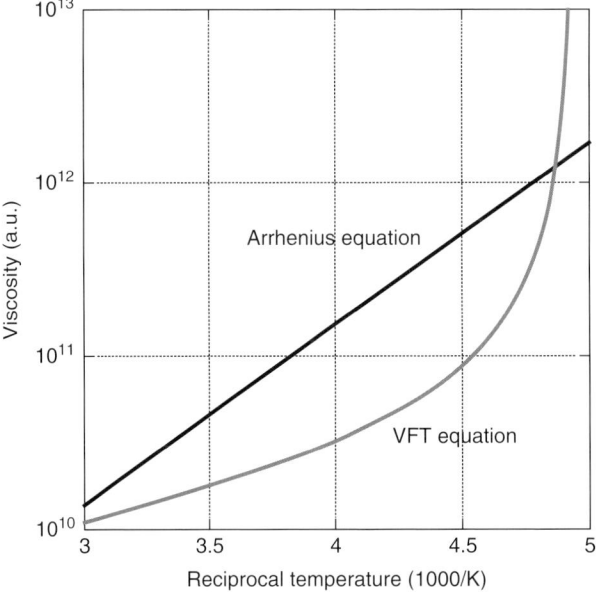

Figure 4.48 The graph shows variation of viscosity versus inverse temperature as predicted by Arrhenius equation (4.112) and by VFT equation (4.113).

event in the supercooled is a jump of an atom from its cage to another cage, accompanied by a rearrangement of the surrounding atoms, as shown in Figure (3.2). A local relaxation event (LRE) involves restructuring of the cage that involves atomic motions of approximately 0.1–0.2 nm. We can call this process an LRE, following a recent theory by Trachenko and Brazhkin (2009). The duration of an LRE lasts is of the order of the Debye vibration period $\tau_0 \sim 0.1$ ps.

Elastic interaction between LREs is the physical origin of *co-operativity* of relaxation in a liquid and glass. The range of co-operativity is given by

$$d_{el} = c\tau \tag{4.114}$$

where d_{el} is the distance over which a high-frequency wave propagates and c is the speed of sound in the matrix. It follows the above relationship that the range of co-operativity increases with decreasing temperature because the relaxation time increases with lowering the temperature.

For each event, the volume of the surrounding atomic cage must increase in order to allow for the escape of the central atom. On the very short timescale of a LRE ($\tau_0 < \tau$), the surrounding supercooled liquid behaves like an elastic medium. The work done against the elastic matrix is equal to the activation barrier for an LRE, U_{LRE}. This barrier is surmounted by temperature fluctuations, so that the relaxation time of the event is

$$\tau = \tau_0 \exp\left(\frac{U_{LRE}}{k_B T}\right) \tag{4.115}$$

Therefore, the atomic motion from one LRE deforms the surrounding atoms, inducing elastic waves. Because their wavelengths are of the order of interatomic distances, the frequency of these waves, ω, is of the order of corresponding the Debye frequency. The propagating high-frequency waves from an LRE have an effect on other LREs. This affects the relaxation times of the other flow events because the jump of an inner atom depends on its surrounding outer atoms. Therefore, LREs interact via the elastic waves they create.

The number of compressive wavefronts that pass through a potential LRE during time τ increases the activation energy:

$$U_{LRE} = U_0 + \sum \Delta v_a p_i(r) \tag{4.116}$$

where U_0 is the high temperature activation barrier owing to thermal interactions and p_i is the compressive pressure stress-wave caused by an adjacent LRE. The changing range of co-operativity affects the relaxation time and changes the Arrhenius law into the VFT law.

References

Argon, A.S. (1979) Plastic deformation in metallic glasses. *Acta Materialia*, **27**, 47–58.

Atkin, A.G. and Mai, Y.W. (1985) *Elastic and Plastic Fracture: Metals, Polymers, Ceramics, Composites, Biological Materials*, John Wiley and Sons, New York.

Cheng, Y.Q. and Ma, E., (2011) Atomic-level structure and structure–property relationship in metallic glasses. *Progress in Materials Science*, **56**, 379–473.

Cottrell, A. (1965) *Dislocations and Plastic Flow in Crystals*, Clarendon Press, Oxford, New York.

Egami, T., Poon, S.J., Zhang, Z. and Keppens, V. (2007) Glass transition in metallic glasses: a microscopic model of topological fluctuations in the bonding network. *Physical Review B*, **76**, 024203-1-024203-6.

Ekambaram, R., Thamburaja, P., and Nikabdullah, N. (2008) On the evolution of free volume during the deformation of metallic glasses at high homologous temperatures. *Mechanics of Materials*, **40**, 487–506.

Eshelby, J.D., (1957) The determination of the elastic field of an ellipsoidal inclusion, and related problems..

Frenkel, J. (1926) Zur Theorie der Elastizitatsgrenze und der Festigkeit kristallinischer Korper. *Z. Phys.*, **37**, 752.

Flynn, C.P. (1972) *Point Defects and Diffusion*, Clarendon Press, Oxford.

Georgarakis, K., Aljerf, M., Li, Y., LeMoulec, A., Chartol, F., Yavari, A.R., Chornohkvostenko, K., Tabachnikova, E., Evangelakis, G.A., Miracle, D.B., Greer, A.L., and Zhang, T., (2008) Shear band melting and serrated flow in metallic glasses. *Applied Physics Letters*, **93**, 31907.

Greer, A.L. (2011) Damage tolerance at a price. *Nature Materials*, **10**, 8889.

Gottstein, G. (2004) *Physical Foundations of Materials Science*, Springer-Verlag.

Gun, B., Laws, K.J., and Ferry, M. (2006) Superplastic flow of a Mg-based bulk metallic glass in the supercooled liquid region. *J. Non-cryst. Solids*.

Hood, G.M. (1978) An atom size effect in tracer diffusion. *Journal of Physics, F: Metal Physics*, **8**, 1677.

Hoffmann, D.C., Suh, J.-Y., Wiest, A., Duan. G., Lind, M.-L, Demetriou, M.D. and Johnson, W.L. (2008) Designing metallic glass matrix composites with high toughness and tensile ductility. *Nature*, **451**, 1085–1089.

Hull, D. (1975) *Introduction to Dislocations*, Pergamon Press, Oxford.

Inoue, A., Shen, B., Koshiba, H., Kato, H., and Yavari, A.R. (2003) Cobalt-based bulk glassy alloy with ultrahigh strength and soft magnetic properties. *Nature Materials*, **2**, 662–663.

Kanninen, M.F. and Popelar, C.H. (1985) *Advanced Fracture Mechanics*, Oxford University Press.

Keeler, G.J. and Batchelder, D.N. (1970) Measurement of the elastic constants of argon from 3 to 77° K. *Journal of Physics C: Solid State Physics*, **3**, 510.

Kelly, A. and Knowles, K.M. (2012) *Crystallography and Crystal Defects*, 2nd edn., John Wiley & Sons,, New York.

Lawn, B.R. (1992) *Fracture of Brittle Solids*, Cambridge University Press.

Liu, Z.Q., Li, R., WAng, G., Wu, S.J., Lu, X.Y., and Zhang, T. (2011) Quasi phase transition model of shear bands in metallic glasses. *Acta Materialia*, **59**, 7416.

Liu, Z.Y., Wang, G., Chan, K.C., Ren, J.L., Huang, Y.J., Bian, X.L., Xu, X.H., Zhang, D.S., Gao, Y.L., and Zhai, Q.J., (2013) Temperature dependent dynamics transition of intermittent plastic flow in a metallic glass. Experimental investigations. *J. Appl. Phys.*, **114**, 033520–8.

Liu, Y.H., Wang, G., Wang, R.J., Zhao, D.Q., Pan, M.X., and Wang, W.H. (2007) Superplastic bulk metallic glasses at room temperature. *Science*, **315**, 1385.

Mattern, N., Bednarcik, J., Pauly, S., Wang, G., Das, J. and Eckert, J. (2009) Structural evolution of Cu–Zr metallic glasses under tension. *Acta Materialia*, **57**, 4133.

McClintock, F.A. and Argon, A.S. (1966) *Mechanical Behaviour of Materials*, Addison-Wesley Publishing.

Nelson, D.R. (2002) *Defects and Geometry in Condensed Matter Physics*, Cambridge University Press, Cambridge.

Patterson, R.E. (1953) *Stress Concentration Design Factors*, John Wiley & Sons, Inc., New York.

Robinson, K. (1951) Elastic energy of an ellipsoidal inclusions in an infinite solid. *Journal of Applied Physics*, **25**, 1045–1054. *Proceedings of the Royal Society A*, 376–396.

Roylance, D. (1996) *Mechanics of Materials*, John Wiley & Sons, Inc., New York.

Ruiz Ocejo, J. and Gutirrez-Solana, F. (1998) On the Strain Hardening Exponent. Report/SINTAP/UC/07, Universidad de Cantabria.

Sun, Y.-L., Qu, D.-D., Sun, Y.-J., Liss, K.-D. and She, J. (2010) Inhomogeneous structure and glass-forming ability in zr-based bulk metallic glasses. *Journal of Non-Crystalline Solids*, **356** (1), 39–45, doi: 10.1016/j.jnoncrysol.2009.09.021.

Spaepen, F. (1977) A microscopic mechanism for steady state inhomogeneous flow in metallic glasses. *Acta Metallurgica*, **25**, 407–415.

Suzuki, Y., Haimovich, J., and Egami, T., (1987) Bond-orientational anisotropy in metallic glasses observed by X-ray diffraction. *Physical Review B*, **53**, 2162–2168.

Trachenko, K., and Brazhkin, V.V. (2009) Understanding the problem of glass transition on the basis of elastic waves in a liquid. *Proceedings of the Royal Society A*, 376–396.

The New Science of Strong Materials. 2nd edn, Pittman Pub. Ltd., Colchester and London.

Timoshenko, S. (1983) *History of Strength of Materials*, McGraw-Hill Book Company Inc., New York.

Wang, G., Shen, J., Sun, J.F., Zhou, B.D., FitzGerald, J.D., Llewellyn, D. and Stachurski, Z.H. (2005) Isothermal nanocrystallisation behavior of zrticunibe bulk metallic glass in the supercooled region. *Scripta Materialia*, **53**, 641–645.

Ward, I.M. (1962) Optical and mechanical anisotropy in crystalline polymers. *Proceedings of the Physical Society*, **80**, 1176.

Wang, G., Shen, J., Sun, J.F., Lu, Z.P., Stachurski, Z.H., and Zhou, B.D. (2005a) Tensile fracture characteristics and deformation behavior of Zr-based bulk metallic glass at high temperature. *Intermetallics*, **13**, 642–648.

Wang, G., Shen, J., Sun, J.F., Lu, Z.P., Stachurski, Z.H., Zhou, B.D., and Liu, C.T. (2005b) Compression fracture characteristics of Zr-based bulk metallic glass at high test temperatures. *Materials Science and Engineering: A*, **398**, 82–87.

Wang, G., Mattern, N., Bednarcik, J., Li, R., Zhang, B., and Eckert, J. (2011) Correlation between elastic structural behaviour and yield strength of metallic glasses. *Acta Materiala*, **60**, 3074–3083.

Wang, G., Jackson, I., FitzGerald, J.D., Shen, J., and Stachurski, Z.H. (2008) Rheology and nano-crystallization of a ZrTiNiCuBe bulk metallic glass. *Journal of Non-Crystalline Solids*, **354**, 1575–1582.

Yavari, A.R., Le Moulenc, A., Inoue, A., Nishiyama, N., Lupu, N., Matsubara, E., Botta, W.J., Voughan, G., Di Michiel, M. and Kvick, A. (2005) Excess free volume in metallic glasses measured by X-ray diffraction. *Acta Materialia*, **53**, 1611–1619.

Zumkley, T., Naundorf, V., Macht, M.P., Fielitz, P., and Frohberg, G. (2002) Relation between time and temperature dependence of diffusion and the structural state in ZrTiCuNiBe bulk glasses. *Materials Transactions*, **43** (8), 1921–1930.

Index

a

aggregate of spheres   8, 34, 187
aggregate theory   198, 202
Agricola, G.   5
amorphous glasses   107
amorphous solids   1, 8, 43, 75, 81, 92, 100, 123, 141, 151, 202, 215, 227, 236
– atomic arrangements   3
amorphous solids model
– disordered clusters   121
– geometrical model   121
– irregular clusters   123
– molecular dynamics   123
– Monte Carlo method   124
– packing of spheres   119
– regular, incongruent clusters   121
– X-ray diffraction patterns   120
amorphous thin films   119
amorphousness   1, 59, 63, 113, 120, 187
Anglo-Saxon glass   107
Angstrom, A. J.   7
angular separation   10, 15, 19, 55
Archimedes density   92
argon glass
– atomic radius   126
– crystalline and amorphous cluster   132
– density   126
– physical parameters   126
– round cell simulation
– – configuration distribution functions   129
– – coordination distribution function   129
– – vs. MD simulation   129
– – radial distribution function   129
– – Voronoi volume   129
– volume–temperature graph   126
– X-ray scattering   131
argon, amorphous   15, 125, 131, 199–201
AsGeSe glass

– atomic arrangements
– – FSDP   186
– – IAS model   184
– – modified IAS model   186
– coordination distribution   182
– enthalpy of bond formation   181
– glass transition temperature   184
– interatomic distances   180
– X-ray scattering pattern   183
atomic scattering factor   99
atomistic, simulation   3, 200, 212, 218, 239
Avogadro, A.   7
axiom   59, 60

b

Bernal, J.D.   3, 6, 22, 46
Biruni, Al   5
blocking function   20, 56
blocking model   19, 20, 23, 86
Born–Oppenheimer approximation   5
Bragg's law   96
Bravais, A.   1, 31, 37, 45, 52, 68
Brownian motion   38
Buckingham potential   14

c

calcite   6
calibration   104, 149
calorimetry, for glass transition measurement   103
chalcogenide glass   116, 180, *see also* AsGeSe glass
– IR optoelectronics and photonics   180
– rewriteable (DVD+RW) optical disk   180
chaos   38
chord, distance   10
circumradius   76

close packing 4, 19, 23, 30, 35, 64, 80, 83, 87, 88, 120, 125, 215, 245
closed packing 40, 88, 95, 120, 132
closing vector 26, 30, 32
cluster(s)
– configuration
– – irregular crystals 26
– – radial vector construction 27, *see also* random walks
– – regular crystals 25
– – spherical harmonics 33
– coordination number
– – blocking model 19
– – definition 18
– – Fürth model 22
– irregular 17, 26, 40, 55, 123, 148, 236
– – configuration 26
random 18, 22, 110, 124
– regular
– – configuration 25
– – cubic geometry 16
– – definition 15
– – tetrahedral geometry 15
– of spheres 8, 11, 15, 25, 36, 52, 75, 88, 132, 143, 236, 243
coercivity 33
combinatorics 22
concept, of atoms 4
concordant 237, 242
condensed matter 9, 50, 244
configuration 17, 25, 31, 41, 50, 75, 79, 81, 115, 129, 143, 171, 175, 193, 242
configuration distribution functions 82
– argon glass 129
– MgCuGd alloy 145
conical angle 10
contact configuration function 50
contact, point 10, 65, 68, 71, 85
convex polyhedron 78
convex, body 9
cooling, rate 4, 111, 125
coordination distribution function 75
– argon glass 129
– AsGeSe glass 182
– MgCuGd alloy 143
– NiNb alloy 133
– ZrTiCuNiBe alloy 159
coordination, number 75, 81, 85, 114, 123, 129, 132, 181, 242
coordination, shell 10, 41, 44, 48, 80, 121, 181
correspondence, principle 63
crystal, lattice 2, 37
crystallography 66, 72, 81, 95, 111
– geometrical 1

cubic, packing 5, 16, 21, 80, 126, 133, 180, 197, 219
cumulative distribution function 20, 22, 25

d

datum 1
Debye equation 23, 33, 52, 98, 100, 120, 137, 138, 151, 165, 174, 177, 183, 184, 186
Delaunay simplexes 52
Delaunay tessellation 76
density 5, 17, 24, 35, 47, 67, 80, 103, 108, 112, 126, 132, 138, 141, 155, 167, 174, 202, 219, 233
– MgCuGd alloy
– – amorphous alloy 156
– – crystalline alloy 156
– NiNb alloy
– – amorphous alloy 139
– – crystalline alloy 139
– of amorphous solid 92
– of composite 93
– of crystalline solid 92
– of single phase 92
– ZrTiCuNiBe alloy
– – amorphous alloy 168
– – crystalline alloy 168
– – Vitreloy alloy 169
density functional theory (DFT) 15
differential scanning calorimeter (DSC)
– for crystalline sapphire and diamond 104
– for metallic glass and semi-crystalline polymers 104
– parts of 103
– types of 103
– for Zr-based amorphous metallic glass 104
diffraction 7, 96, 102, 116, 122, 149, 159, 183
diffusion 67, 110, 194, 232, 234, 243
dislocation 194, 209, 218, 225
disordered crystalline solids 2
disordered materials 2
disordered sphere packing 37
DSC, *see* differential scanning calorimeter (DSC)
ductile materials 192

e

Ehrenfest formula 100, 101, 131, 183
elasticity
– atomistic elasticity
– – acoustic method 201
– – bond stiffness 200
– – solid fcc Argon 200
– continuum mechanics
– – aggregate theory 198

– – elastic constants   199
– – elastic strain energy   199
– – homogeneous deformation   196
– – strain ellipsoid   195
– – transverse isotropy   198
– – undeformed state   195
– – uniaxial elongation   198
– – uniaxial tension   196
– phenomenology   193
electron, density   5, 98
electronic, orbitals   4
end-to-end distance   28, 31, 82
entropy   50
equilateral, triangle   15, 76, 80, 121
equilibrium, separation   13, 39, 92, 109, 170, 199, 240
Eshelby inclusion   241
Euclidean distance, space   1, 9, 68, 77
Euler polyhedra theorem   78
exclusion, condition, zone   10, 19, 39

f

Fürth model   22
first sharp diffraction peak (FSDP)   183, 186
fixed sphere packing   30, 36, 40, 41, 67
flat plate camera, side-on geometry of   98
flaw, imperfection   66, 168, 215
fluctuations, density   24, 38, 88, 103, 108, 143, 158, 178, 245
Fourier transform   96, 99
fracture mechanics   204
– defects, role of   210
– elastic strain energy   207
– fracture toughness   204, 209
– Griffith's fracture stress   209
– phenomenology   203
– solid surface energy   208
– theoretical cleavage strength   212
– theoretical shear strength   213
free volume   51, 68, 86, 113, 132, 229, 234, 243
Frenkel theoretical strength   213
FSDP, *see* first sharp diffraction peak (FSDP)
Furth, R.   22, 40

g

Galileo Galilei   191
Gaussian function   42, 63, 110, 154
Gaussian random walk (GRW)   28, 29, 32, 82, 175
glass   47, 57, 68, 77
inorganic   64
metallic   64, 71, 104
transition   68, 103, 156
glassmaking   107

graphical user interface (GUI)   55, 142, 159
Green's strain energy   240
Green's theorem   242
Griffith's theory   206

h

Hamiltonian   51, 108
hard sphere(s)   12, 13
– condition   62, 70, 83, 91, 120, 140
– random sequential addition   38
Hauy, R.J.   1, 5, 6
heat flux calorimeter   103
heavy metal fluoride glass   116
hexagonal, arrangement   4, 123
hexagonal, packing   4, 35, 80, 180
high entropy alloys (HEA)   115
homogeneity of point field   35, 61
Hooke's law   193
Hooke, R.   4, 193
Huygens sum   52

i

Ideal amorphous solids (IAS)   52, 56, 60, 65, 69, 100, 143, 148, 188
– class I   64
– class II   64
– class III   66
– imperfections
– – categories   66
– – density   69
– – geometrical flaws   67
– – statistical flaws   68
– statistical flaws
– – kappa function   68
– – zeta function   69
Ideal amorphous solids (IAS) model
– argon
– – atomic radius   126
– – crystalline and amorphous cluster   132
– – density   126
– – physical parameters   126
– – round cell simulation   127
– – volume–temperature graph   126
– – X-ray scattering   131
– AsGeSe glass   184
– construction
– – IAS parameters   58
– – IAS software   53
– – outer spheres   58
– – primary cluster creation   54
– – Round Cell   59
– coordination numbers   52
– geometrical construction   52
– MgCuGd alloy

Ideal amorphous solids (IAS) model (*contd.*)
– – physical properties   140
– – round cell simulation   141
– – X-ray scattering   149
– NiNb alloy
– – physical properties   133
– – primary uses   133
– – round cell simulation   133
– – X-ray scattering   136
– packed spheres   187
– polyethylene
– – crystal structure   170
– – radial distribution function   171
– – specific volume   171
– – uses   170
– – X-ray scattering   173
– random atomic arrangement   51
– silica
– – formation   176
– – modified glasses   176
– – molecular parameters   176
– – round cell simulation   177
– – short range order   179
– – united atom model   177
– – X-ray scattering   177
– statistical functions   52
– X-ray scattering   52
– ZrTiCuNiBe alloy
– – mechanical properties   157
– – physical properties   157
– – round cell simulation   159
– – transmission electron microscopy   157
– – X-ray scattering   165
ideal crystal   72, 83, 111
ideal Greek solids   1, 53
imperfection, defect   66, 102, 192, 210, 229
incongruent   121
indivisible, object   4
inherently non-crystallizable (equable) glasses   114
inhomogeneity   51
inner sphere   10, 21, 36, 45, 55, 75, 83, 134, 159, 236
inorganic glasses   116
inorganic silica-based glasses   107
intermolecular, forces   14, 112, 215
irregular clusters   17, 40, 45, 54, 79, 236
– blocking function   21
– configuration   26
irregular, triangle   56
isosceles triangle   40, 56, 80, 121
isotropy, point field   42, 61, 198, 202

j
jammed, configuration   68, 77

k
Kepler, J.   4

l
lattice   7, 31, 45, 68, 69, 95, 112, 120, 200, 221
Lennard–Jones potential   12, 14, 15, 200
local relaxation event (LRE)   248
London dispersion forces   14
loose sphere packing   19, 36, 51, 67, 83
Lorentz function   153
Lucretius   4

m
Markov chains   43
mechanical behavior
– elastic modulus of amorphous solids   202
– elasticity   193
– fracture   203
– viscoelasticity   243
metallic glass   49, 64, 71, 104, 114, 157, 166, 192, 239
metastable   114, 238
metastable (crystallizable) glasses   114
Mg65-Cu25-Gd10 alloy   149, 155
MgCuGd alloy
– density
– – amorphous alloy   156
– – crystalline alloy   156
– round cell simulation
– – atomic parameters   141
– – cluster composition   148
– – configuration distribution functions   145
– – coordination distribution functions   143
– – interatomic distances   146
– – *vs.* MD simulation   143
– – probability of contacts   147
– – radial distribution functions   143
– X-ray scattering   149
Miller indices   2, 6
Miller, W.H.   6
mineralogy   5
Minkowski distance   78
molecular dynamics   3, 13, 34, 46, 120, 123, 163
Monte Carlo method   124
motion invariance   42

n
neighbour, touching   8, 15, 29, 36, 64, 69, 75, 143, 160, 187, 200, 236

Index | 257

neighbouring, sphere   2, 9, 19, 43, 53, 79, 124, 220
network   64, 76, 114, 120, 177
Ni62-Nb38 alloy   138
NiNb alloy
– density
– – amorphous alloy   139
– – crystalline alloy   139
– round cell simulation
– – atomic parameters   133
– – coordination distribution functions   133
– – pair distribution function   135
– – probability of contacts   135
– – Voronoi volume   135
– X-ray scattering   136
2-norm distance   77

o
ordered cluster   17, 45, 70, 80, 121
ordered packing   6, 36, 45, 76, 87
ordered sphere packing   36
organic glasses   117
outer sphere   10, 18, 27, 36, 55, 75, 81, 134, 236

p
packing, dense   49, 91, 119, 239
packing, disordered   37
packing, fraction   8, 35, 54, 68, 80, 85, 86, 129, 139, 157, 169
packing, geometric   13
packing, of spheres   3, 18, 34, 40, 41, 61, 69, 71, 82, 95, 119, 143, 171, 185
packing, random   18, 34, 44, 60, 76, 84, 95, 113, 124, 164, 175
pair distribution function (PDF)   49
– NiNb alloy   135
pattern, repeating   1
Peierls–Nabarrow stress   218
percolation   47
plasticity   215, 227
– amorphous solids   227
– continumm mechanics   217
– crystalline solids, atomistic mechanics
– – atoms displacement   224
– – critical shear stress   226
– – grain boundary strengthening   220
– – plastic flow mechanism   224
– – precipitation hardening   223
– – solid solution hardening   221
– – strain hardening   218
– phenomenology   215
– superplasticity
– – bulk metallic glasses   234

– – concordant deformation mechanism   236
– – continuum mechanics   195
– – phenomenology   193
Platonic solids   53
Pliny the Elder   5
polarity, cluster   33
polyethylene
– crystal structure   170
– radial distribution function   171
– specific volume   171
– uses   170
– X-ray scattering   173
polygon   15, 43, 53, 76
polyhedron   44, 53, 78
– convex   78
– topology of   79
polytope   76
potential   12, 95, 113, 129, 146, 161, 173, 199, 237
power compensation calorimeter   103
primary cluster   9, 18, 41, 54, 129, 148
principle   2, 15, 53, 115, 120, 199, 225, 233
probability   17, 22, 31, 43, 72, 81, 98, 124, 135, 146, 160, 204, 239
– density   17, 38
probability of contacts   50
– MgCuGd alloy   147
– NiNb alloy   135
– ZrTiCuNiBe alloy   162
properties, of matter   4
pseudo-random   38
– packing   41
pseudo-randomness   38
pulse echo technique   201
Pythagoras   2

r
radial distribution function
– argon glass   129
– MgCuGd alloy   143
– polyethylene   171
– silica glass   116
– ZrTiCuNiBe alloy   161
radial pair distribution function   50
random closed packing   8, 24, 34, 38, 40, 46, 64, 76, 85
random sequential addition (RSA)   38
random sequential packing   39
random sphere packing   38
random walk, Gaussian (GRW)   28
random walk, self avoiding, space limited (SASLRW)   28
random walk, self-avoiding (SARW)   28

random walks   28, 43, 83
– Markov chains   43
random, addition   16
random, atomic arrangement   3, 21, 58, 69, 92, 103
random, configuration   17, 61
random, patterns   2, 66, 82
randomness   21, 38
rational ratios   2
regular clusters   15, 21, 25, 32, 56, 122, 133
– blocking function   21
– configuration   25
– cubic geometry   16
– definition   15
– polygon   15
– shape   15
– tetrahedral geometry   15
representative volume element (RVE)   53, 71, 91
rigidity   12, 52, 94
rotational symmetry   16, 25, 71, 122, 132
round cell   8, 54, 126, 134, 161, 176, 184
round cell simulation
– argon glass
– – configuration distribution functions   129
– – coordination distribution function   129
– – vs. MD simulation   129
– – radial distribution function   129
– – Voronoi volume   129
– MgCuGd alloy
– – atomic parameters   141
– – cluster composition   148
– – configuration distribution functions   145
– – coordination distribution functions   143
– – interatomic distances   146
– – vs. MD simulation   143
– – probability of contacts   147
– – radial distribution functions   143
– NiNb alloy
– – atomic parameters   133
– – coordination distribution functions   133
– – pair distribution function   135
– – probability of contacts   135
– – Voronoi volume   135
– silica glass   177
– ZrTiCuNiBe alloy
– – atomic parameters   159
– – coordination distribution function   159
– – graphical user interface   159
– – radial distribution functions   161
– – Voronoi volume distribution   161

s

saturation density   39
scale invariance   35, 61
self-avoiding random walks (SARWs)   29, 31, 64
self avoiding space limited random walk (SASLRW)   29
silica glass
– formation   176
– modified glasses   176
– molecular parameters   176
– radial distribution function   116
– round cell simulation   177
– short range order   179
– united atom model   177
– X-ray scattering   177
simplex, simplical   52, 76
soft spheres   12
solid, angle   19, 198
solidification   108, 111
solidity   36, 51, 62, 94, 193
solidity of packing   94
space-filling model   7
sphere packing characteristics   75
– geometrical properties   75
– – configuration distribution function   82
– – coordination distribution function   75
– – density   92
– – packing fraction   85
– – representative volume element   91
– – solidity of packing   94
– – tetrahedricity   76
– – topology of clusters   79
– – volume fraction   83
– – Voronoi polyhedra notation   78
– glass transition   103
– X-ray scattering   96
sphere packings   35, 48, 53, 69, 87, 156
– disordered packing   37
– fixed   31, 36, 40, 67
– loose   11, 19, 36, 39, 51, 67
– ordered packing   36
– random closed packing   40
– random packing   38
spheres
– definition   9
– force
– – electrical and magnetic forces   14
– – gravitational forces   14
– – interatomic forces   14
– – work function   13
– hard   4, 12, 13, 35
– neighbours
– – categories   9

– – coordination shell   47
– – touching neighbours   10
– – Voronoi tessellation   43
– random assemblage   18
– random closed packing   40
– soft   12
– unequal sizes   11
spherical cap   10, 19
spherical harmonics   27, 33
statistical mechanics   47, 158
statistics, mathematical   3, 71, 82
Steno, N.   6
strain ellipsoid   195
superplasticity   231
– bulk metallic glasses   234
– concordant deformation mechanism
– – density variation, amorphous solids   238
– – transformation   241
– continuum mechanics   195
– phenomenology   193
superplasticity: concordant deformation mechanism: 5-atom subcluster   236
symmetry
– rotational   1, 16, 25, 71, 122
– translational   1, 52, 71

t

tessellation   10, 43, 52, 76, 133
tetrahedral   15, 44
tetrahedricity   76
tetrahedron   53
tetrahedron, irregularity of
– Gamma parameter   77
– Minkowski measure   76
theory of amorphousness   1
– axioms   60
– conjecture   61
– rules   62
– statistical correspondence   63
topology   78
topology of clusters
– classification of   80
– crystalline clusters   81
– irregular clusters   81
– ordered clusters   80
touching spheres   10
translational symmetry   1, 52, 71
triangulation   44

u

united atom model   177
unpredictable   3

v

van der Waals force   14
vector, polygon   10, 26, 54, 59, 72
viscoelasticity
– phenomenology   243
– temperature effect, viscoelastic behaviour   247
– time and temperature dependent behaviour   245
viscosity   15, 111, 176, 233, 244
Vitreloy   157
volume fraction   35, 83, 111, 139, 155, 167, 223
– irregular polyhedra   84
– regular polyhedra   84
– Voronoi polyhedron   84
volume, occupied   35, 92
volume, unoccupied   35
Voronoi polyhedra notation   78
Voronoi tessellation   35, 43
Voronoi volume distribution
– argon glass   129
– NiNb alloy   135
– ZrTiCuNiBe alloy   161
Voronoi, H.   46

w

Warren method   179
Weibull probability   204
Wigner-Seitz cells   45

x

X-ray diffractometer
– for amorphous samples   96
X-ray scattering   7, 52, 70, 96, 113, 131, 137, 150, 165, 175, 183
– argon glass   131
– diffraction and scattering, geometry of   96
– factors affecting integrated scattered intensity   102
– MgCuGd alloy   149
– NiNb alloy   136
– polyethylene   173
– scattered wave, intensity of   97
– silica glass   177
– ZrTiCuNiBe alloy   165

y

Young's modulus   201

z

Zarzycki, J.   188
ZrTiCuNiBe alloy
– atomic probe   162

ZrTiCuNiBe alloy (*contd.*)
– clusters 165
– density
– – amorphous alloy 168
– – crystalline alloy 168
– – Vitreloy alloy 169
– mechanical properties 157
– physical properties 157
– probability of contacts 162
– round cell simulation
– – atomic parameters 159
– – coordination distribution function 159
– – graphical user interface 159
– – radial distribution functions 161
– – Voronoi volume distribution 161
– transmission electron microscopy 157
– X-ray scattering 165